T0202674

Lecture Notes in Computer Science　　12754

More information about this subseries at http://www.springer.com/series/7411

Karima Echihabi · Roland Meyer (Eds.)

Networked Systems

9th International Conference, NETYS 2021
Virtual Event, May 19–21, 2021
Proceedings

 Springer

Editors
Karima Echihabi (iD)
Mohammed VI Polytechnic University
Ben Guerir, Morocco

Roland Meyer (iD)
Technische Universität Braunschweig
Braunschweig, Niedersachsen, Germany

ISSN 0302-9743   ISSN 1611-3349 (electronic)
Lecture Notes in Computer Science
ISBN 978-3-030-91013-6   ISBN 978-3-030-91014-3 (eBook)
https://doi.org/10.1007/978-3-030-91014-3

LNCS Sublibrary: SL5 – Computer Communication Networks and Telecommunications

This Springer imprint is published by the registered company Springer Nature Switzerland AG
The registered company address is: Gewerbestrasse 11, 6330 Cham, Switzerland

Preface

In May 2021, the 9th Edition of the International Conference on Networked Systems NETYS 2021, was held as a virtual event. While the scope of NETYS is distributed and networked systems, the conference is known for its broad perspective on the topic. NETYS solicits submissions from practice and theory and aims to foster the exchange of ideas between these directions. In recent years, data management and the associated security aspects have increasingly attracted attention in distributed systems. With the 9th edition, NETYS decided to adjust its orientation in that direction.

The present book contains revised versions of selected contributions to NETYS 2021. The conference received 30 high-quality submissions from all around the globe that were reviewed by a Program Committee consisting of 33 international experts covering all branches in the spectrum of distributed and networked systems. Each paper received at least three reviews and underwent a critical discussion of its merits. Based on this, 15 submissions were accepted as full papers (a 50% acceptance rate) and another two as short papers. The Program Committee also selected a Best Paper and two Best Student Paper awards. The Best Paper was awarded to Jean-Philippe Abegg, Quentin Bramas, and Thomas Noel for their paper Blockchain using Proof-of-Interaction. The two Best Student Paper awards were conferred to Joseph Oglio, Kendric Hood, Gokarna Sharma, and Mikhail Nesterenko for their paper Byzantine Geoconsensus and to Elena Yanakieva, Michael Youssef, Ahmad Hussein Rezae, and Annette Bieniusa for their paper On the Impossibility of Confidentiality, Integrity and Accessibility in Highly-available File System. Joseph Oglio, Elena Yanakieva, Michael Youssef, and Ahmad Hussein Rezae were all full-time students at the time of submission.

The COVID-19 pandemic had a serious impact on the organization of NETYS. It was decided not to organize the METIS Spring School that otherwise accompanies NETYS and serves as a social platform for PhD students. The conference was implemented as an asynchronous event and the videos of all presentations are available at the following link: https://www.youtube.com/channel/UCuut9KIWEy6nvywXd 9HuOnA/featured.

As the program chairs of NETYS 2021, we thank the authors for their high-quality submissions, the external reviewers for their support, and the Program Committee for their careful evaluations and the lively discussions. All this occurred under difficult circumstances, and we are grateful to all of you for making NETYS 2021 happen. Special thanks go to the NETYS General Chair Mohammed Erradi (ENSIAS, Morocco) and the General Co-chairs Ahmed Bouajjani (Université de Paris, France) and Rachid Guerraoui (EPFL, Switzerland) for numerous helpful suggestions on the organization. The Organization Committee brought the online event to life, and we thank you for your support. Finally, we thank Springer for their help in assembling the proceedings.

September 2021   Karima Echihabi
                                                                       Roland Meyer

Organization

General Chair

Mohammed Erradi   ENSIAS, Morocco

General Co-chairs

Ahmed Bouajjani   Université de Paris, France
Rachid Guerraoui   EPFL, Switzerland

Program Committee Chairs

Karima Echihabi   Mohammed VI Polytechnic University, Morocco
Roland Meyer   TU Braunschweig, Germany

Program Committee

Parosh Aziz Abdulla	Uppsala University, Sweden
Mohamed Faouzi Atig	Uppsala University, Sweden
Slimane Bah	Mohammed V University, Morocco
Salima Benbernou	Université de Paris, France
Yahya Benkaouz	Mohammed V University, Morocco
Ismail Berrada	Mohammed VI Polytechnic University, Morocco
Erik-Oliver Blass	Airbus Group Innovations, Germany
Ahmed Bouajjani	Université de Paris, France
Emanuele D'Osualdo	Max Planck Institute for Software Systems, Germany
Carole Delporte-Gallet	Université de Paris, France
Amr El Abbadi	University of California, Santa Barbara, USA
Bernd Freisleben	TU Darmstadt, Germany
Pierre Ganty	IMDEA Software Institute, Spain
Matthew Hague	Royal Holloway, University of London, UK
Maurice Herlihy	Brown University, USA
Mohamed Jmaiel	University of Sfax, Tunisia
Ahmed Khaled	Northeastern Illinois University, USA
Mohammed-Amine Koulali	Mohammed I University, Morocco
Michele Linardi	Université de Paris, France
Aziz Mohaisen	University of Central Florida, USA
K. Narayan Kumar	Chennai Mathematical Institute, India
Guevara Noubir	Northeastern University, USA
Ismail Oukid	Snowflake Computing, Germany
Mourad Ouzzani	Qatar Computing Research Institute, HBKU, Qatar
Andreas Podelski	University of Freiburg, Germany
Sayan Ranu	IIT Delhi, India
Michel Raynal	IRISA, Université de Rennes, France

François Taïani   IRISA/Inria, Université de Rennes, France
Klaus v. Gleissenthall   Vrije Universiteit Amsterdam, Netherlands
Josef Widder   Informal Systems, Austria
Kostas Zoumpatianos   Snowflake Computing, Germany

Organizing Committee

Khadija Bakkouch   IRFC, Morocco
Yahya Benkaouz   Mohammed V University, Morocco
Abdellah Boulouz   Ibn Zohr University, Morocco
Rachid El Guerdaoui   Mohammed VI Polytechnic University, Morocco
Mustapha Hedabou   Mohammed VI Polytechnic University, Morocco
Zahi Jarir   Cadi Ayyad University, Morocco
Mohammed Ouzzif   Hassan II University, Morocco

Sponsors

Fondation Hassan II

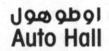

King Abdullah University of
Science and Technology

Université Mohammed V de Rabat

Contents

Distributed Systems

On the Impossibility of Confidentiality, Integrity and Accessibility
in Highly-Available File Systems 3
    Elena Yanakieva, Michael Youssef, Ahmad Hussein Rezae,
    and Annette Bieniusa

Byzantine Geoconsensus 19
    Joseph Oglio, Kendric Hood, Gokarna Sharma, and Mikhail Nesterenko

Loosely-self-stabilizing Byzantine-Tolerant Binary Consensus
for Signature-Free Message-Passing Systems 36
    Chryssis Georgiou, Ioannis Marcoullis, Michel Raynal,
    and Elad M. Schiller

Leader Election in Arbitrarily Connected Networks with Process Crashes
and Weak Channel Reliability. 54
    Carlos López, Sergio Rajsbaum, Michel Raynal, and Karla Vargas

Fault-Tolerant Termination Detection with Safra's Algorithm 71
    Georgios Karlos, Wan Fokkink, and Per Fuchs

Weak Amnesiac Flooding of Multiple Messages 88
    Zahra Bayramzadeh, Ajay D. Kshemkalyani, Anisur Rahaman Molla,
    and Gokarna Sharma

Optimal Exclusive Perpetual Grid Exploration by Luminous Myopic
Robots Without Common Chirality 95
    Arthur Rauch, Quentin Bramas, Stéphane Devismes, Pascal Lafourcade,
    and Anissa Lamani

AUCCCR: Agent Utility Centered Clustering for Cooperation
Recommendation 111
    Amaury Bouchra Pilet, Davide Frey, and François Taïani

Blockchain

Blockchain Using Proof-of-Interaction 129
    Jean-Philippe Abegg, Quentin Bramas, and Thomas Noël

LighTx: A Lightweight Proof-of-Bandwidth Transactions Transfer System ... 144
    Imane El Abid, Yahya Benkaouz, and Ahmed Khoumsi

Efficient and Secure TSA for the Tangle   161
    Quentin Bramas

Verification

Separating Map Variables in a Logic-Based Intermediate Verification
Language.   169
    Daniel Dietsch, Matthias Heizmann, Jochen Hoenicke, Alexander Nutz,
    and Andreas Podelski

Petri Net Invariant Synthesis.   187
    Peter Chini and Florian Furbach

Towards Efficient Shape Analysis with Tree Automata   206
    Martin Hruška and Lukáš Holík

Deciding S1S: Down the Rabbit Hole and Through the Looking Glass   215
    Vojtěch Havlena, Ondřej Lengál, and Barbora Šmahlíková

BAM: Efficient Model Checking for Barriers   223
    Michalis Kokologiannakis and Viktor Vafeiadis

Verifying and Optimizing the HMCS Lock for Arm Servers.   240
    Jonas Oberhauser, Lilith Oberhauser, Antonio Paolillo, Diogo Behrens,
    Ming Fu, and Viktor Vafeiadis

Author Index   261

Distributed Systems

On the Impossibility of Confidentiality, Integrity and Accessibility in Highly-Available File Systems

Elena Yanakieva$^{(\boxtimes)}$, Michael Youssef$^{(\boxtimes)}$, Ahmad Hussein Rezae$^{(\boxtimes)}$, and Annette Bieniusa$^{(\boxtimes)}$

TU Kaiserslautern, 67663 Kaiserslautern, Germany
{e_yanakiev15,m_youssef19,a_rezae19,bieniusa}@cs.uni-kl.de

Abstract. Distributed file systems are at the core of many services for sharing data among users. To keep the file contents secure from unauthorized access, such systems make use of custom access control policies similar to the traditional POSIX policies.

In our work, we want to investigate the interdependence of secure access and high-availability. To this end, we formalize the three properties related to data security, namely confidentiality, integrity and accessibility (*CIA*). We proof the CIA impossibility showing that these properties cannot be achieved together in a highly-available partition-tolerant setting. We further discuss a CRDT-based model that implements an access control policy similar to the POSIX one and that guarantees confidentiality and integrity while precluding accessibility only in rare situations.

Keywords: Distributed file system · Access control · Conflict-free replicated data type · POSIX · CRDT

1 Introduction

Distributed file systems have been gaining increasing importance as collaboration tools. In such systems, multiple users can simultaneously create, modify and remove files and directories. Distributed systems are replicated across multiple nodes, typically allowing clients to connect and switch between different nodes at any given time, e.g. to compensate for node failure. Availability and consistency is essential for user satisfaction. To increase the availability of such systems, service providers often offer support for offline operation. However, this poses consistency issues when clients reconnect to the network, e.g. when multiple users have modified the same resource while being offline. Providers approach this problem in different ways. For example, GoogleDrive and Dropbox create copies of the files that have been concurrently modified and escalate the conflict to the user(s) to solve, whereas MicrosoftOneDrive offers a configurable automatic merging policy for Word, Excel, and Visio files.

© Springer Nature Switzerland AG 2021
K. Echihabi and R. Meyer (Eds.): NETYS 2021, LNCS 12754, pp. 3–18, 2021.
https://doi.org/10.1007/978-3-030-91014-3_1

An established approach for guaranteeing high availability and consistency are conflict-free replicated data types (CRDTs) [8,9]. For example, the Inter-Planetary File System (IPFS) provides a file system abstraction on top of a P2P network. The directories are represented in a tree data structure, while the files are structured into blocks. For each block, the different versions are accessible as a directed acyclic graph (DAG). Updates add new versions to the DAG, and branches in the graph can be conflated with new versions or by using CRDTs.

To keep their data secure from unauthorized access, users want to be able to restrict the visibility and modifiability of their data. For example, IPFS handles security by making use of cryptographic encryption of new versions and delegates the accessibility problem to a key management subsystem. While we acknowledge the benefits of using this approach, we argue that meta data conflicts can occur regardless of cryptography. We believe file system security can also be enhanced by providing secure merging semantics on metadata level. To our knowledge, there is little research on what additional security challenges are posed when merging metadata associated to files, such as access control data. In the literature, there is a strong focus on the challenge of guaranteeing the integrity of the file content in the context of text editing, and on the integrity of the directory structure by preventing and resolving conflicts inducing cycles in the tree structure due to concurrent move operations [5,7,10,12]. However, we believe that file systems offer further challenges besides the handling of file content in a distributed setting.

In this paper, we address the problem of defining a merging mechanism for access control policies in a highly available setting. The three main properties of such a mechanism are confidentiality, accessibility, and integrity (*CIA*). Confidentiality and integrity prevent unauthorized access to files and directories, while accessibility ensures that data is accessible to authorized users.

Our main contribution is an impossibility result that shows that CIA cannot be achieved together in a highly-available partition-tolerant setting (Sect. 4). We further discuss the applicability and limitations of CRDTs in a distributed file system model with focus on POSIX-like access control semantics (Sect. 5). To validate our results, we formalize and test the model in Repliss, which depicts non-obvious shortcomings of customary design decisions with automatically generated counterexamples (Sect. 5).

2   Foundations

File systems store, organize and present data in a hierarchical manner. Usually, they organize files and directories in a tree-like structure with a root node being the root directory and the files forming the leaves. Each node has a unique path starting from the root and traversing the directory nodes. In this section, we discuss what it means for a distributed system to be available, consistent and secure. We further summarize the POSIX access control policy as a classical way of defining an authorization scheme for file systems.

2.1   Availability, Consistency, and Partition Tolerance

With the growth of digitalization and online collaboration, distributed file systems have gained greater importance in our private and work lives. To offer customers a satisfactory and reliable user experience, system providers need to guarantee availability and consistency of the users' data. On the one hand, **availability** requires that every request issued to a correct node must result in a response. This implies that every request must terminate. On the other hand, **consistency** requires linearizability of system operations. This implies that there should be a total order on the operations such that it appears as if they were executed sequentially on a single replica [4]. However, following from the CAP theorem [3,4], these two properties cannot be guaranteed together in a distributed setting due to the inevitable network partitioning. Consequently, a system's design needs to provide a trade-off between consistency and availability.

High-availability has a large impact on user satisfaction. Therefore, ongoing research investigates concepts on how to provide high availability coupled with the strongest possible consistency in distributed systems. For this, it is useful to identify to which extent weaker consistency properties are sufficient to guarantee desirable system properties. An example of such a weaker consistency notion that allows to be coupled with high availability is strong eventual consistency [9]. It guarantees that replicas will converge to the same state once they have received the same updates, regardless of their order. One possible approach to support strong eventual consistency in distributed systems is using conflict-free replicated data types (CRDTs) [8,9]. These are a central part to our model and will be further explained in Sect. 5.

2.2   Security in File Systems

Besides providing high availability and data consistency, it is essential that data stored in shared distributed file systems is secure, meaning that unauthorized access is prohibited and authorized access always succeeds. The part of the file system that provides these guarantees is the access control subsystem. File systems implement different policies, which specify the access requirements and can be dynamically adapted. For developing our model, we have studied the POSIX access control policy in detail as it is one of the broadly implemented ones.

The POSIX system design distinguishes between three types of authorization [2] - owner, group and other. The **owner** of a directory or a file is a single user; by default, it is the user that created the entity. The **group** of a file or directory can contain single or multiple users; in most systems, by default, it initially contains only the user that created the entity. **Other** represents every other user that is neither the owner nor a member of the group. Furthermore, there exist three types of access permissions for files and directories. **Read** permission for a file allows the user to open and read its contents. **Write** permission for a file allows the user to modify its contents. **Execute** permission for a file allows users to run it, if it is a program. As we focus on read and write access in our setting, this type is not relevant for our model.

Directories under POSIX semantics act differently than files. For example, having **read** permissions for a directory allows the user to list its contents, but not to interact with them. For the purpose of the paper, we focus on a simplified model and consider only file access policies.

In addition to specifying the permissions, the access control design needs to ensure that the access control behaves correctly even under concurrent policy modifications and that it guarantees that unauthorized access is prohibited and authorized one is allowed. Assuming a total order on all operations can be established, users are authorized to access a resource if, at the time of access, they possess read and/or write permissions for the specific resource. Respectively, a user is unauthorized to access a resource if, at the time of access, they do not posses read and/or write permissions for that specific resource. Furthermore, we define a secure file system as a system that possesses the following three properties - confidentiality, integrity and accessibility[1]. **Confidentiality** ensures that every resource is protected from unauthorized viewing, meaning unauthorized read access will never succeed. **Integrity** protects resources from unauthorized modification, meaning unauthorized write access will never succeed. **Accessibility** guarantees that authorized access is always allowed, meaning that, if an access is requested by an authorized user, the request must terminate with an answer and allow the access.

3   Related Work

In this section, we present the related work that we believe is relevant for our model in Sect. 5. We use a CRDT approach to ensure that our file system is highly available, therefore we present the research so far on CRDT-based distributed file systems. We further discuss access control semantics in distributed information systems as we believe important insights can be gained that are useful for our model.

3.1   CRDT-Based File Systems

The research done in the area of distributed file systems using CRDTs mainly focuses on providing merging semantics for conflicts occurring during the common file system operations, such as adding, modifying, and deleting files and directories [1,5,7,10]. To our knowledge, there is no literature that has set focus on conflict resolution on metadata level, or specifically on access control.

When modeling a distributed file system, the goal is to define merging semantics such that conflicts are not observable by the user, and manual user intervention is needed as little as possible [1]. File systems are typically modeled as a

---

[1] In information systems, this concept is typically named "availability". To prevent confusion, we here use a different term to distinguish it from availability as defined above.

tree structure: there is a dedicated root node, and every other node has a unique path, starting from the root; directories are inner nodes, while files form the leaf nodes [1,5,7,10]. The conflict resolution strategies mostly concentrate on operations that preserve unique names and/or identification of nodes, the structural tree invariant, or state conflicts [7,10]. A naming conflict occurs when two files of the same type and the same name are concurrently added to the same directory. This situation leads to having two files with the same path, which breaks the rule of path uniqueness in the tree structure. Some works propose to merge both files [1,7], others to rename the files and keep both, but merge directories if it occurs on directory level [10]. A state, or an update-remove, conflict occurs when one user concurrently removes a file while another user is modifying it. An information-preserving resolution, in this case, is to preserve the file with its original name [7,10]. This can be achieved by using an add_wins CRDT [7].

Tao et al. [10] identified further indirect conflicts, i.e. operations that do not modify the same data but still result in an anomaly due to invariant violation. Examples of such conflicts are delete-while-edit or moving two directories into each other concurrently. The first one occurs when a directory is concurrently removed while a file inside of the directory is updated. This situation results in the file being removed if it is not acknowledged as a conflict. The second one causes cycles in the file system tree if not handled [7].

Kleppmann et al. [5] propose an algorithm for highly-available move operations in a distributed file system based on operation logs, in addition to CRDTs. They suggest modeling add, move and remove operations as move operations. Removing an element is modeled as a move to the trash node. The trash node serves as a second root node with no parent and no children. Conflict resolution is based on the operation log: when a previous operation arrives at the replica, all future operations are undone and then redone. This situation leads to having high availability because the user waits only for the local operations to be executed.

3.2   Access Control in Distributed Systems

The problem of maintaining dynamically adaptable access control policies under high availability is not restricted to distributed file systems. In his dissertation [11], Matthias Weber discusses highly-available access control semantics in the context of geo-distributed information systems. For a formal model based on causal consistency and highly-available transactions, he proves that data modifications can be guarded by corresponding access control policies. Further, he introduces the term *lower bound* for merging conflicting updates, by defining the sets of combinations with read and write permissions structured as a lattice. When a conflict between two concurrent updates arises, the greatest lower bound of those is the set of the most restrictive combination of permissions. For example, when updating the access rights concurrently to rw and r, the conflict is resolved to r which is the greatest lower bound according to the lattice. Weber shows that taking the greatest lower bound when merging conflicting updates on the access control policy can be expressed as a CRDT. We apply the same notion of lower bound in our model and use a similar CRDT approach.

There are three important insights from his work that we base our model on:

1. **Consistency level:** The consistency level of the application's data store needs to comprise causal consistency, and both data and policy operations subject to a common causality restriction (e.g., by being stored in the same database). For example, using a CRDT store can ensure those requirements. As a result, the protection relation between modifications to data and access control policy can be guaranteed and data leakage can be prevented.
2. **Data type:** The data type representing the policy state needs to have decreasing merge semantics in order to guarantee security. It means that when concurrent updates to the policy happen, the greatest lower bound needs to be taken and hence, its value can only decrease, but never increase.
3. **Decide function:** The decide function that determines whether a data operation is permitted under the current policy can make decisions based on the local state, and no knowledge of the future or other replicas' states is required. This means that merging conflicts can be solved based on the local knowledge and external knowledge is not needed. It is important to know that this can be achieved, because waiting on external knowledge before making a decision is not desirable in any system, because it lowers availability.

4 On the Impossibility of Confidentiality, Integrity and Accessibility in Highly-Available File Systems

Security is of great importance for any file system - local or distributed - as users desire that their data remains secure from unauthorized access, but accessible at all times for authorized users. However, distributed file systems are subject to the same inherent limitations as other distributed systems. In the context of access control, the interplay of consistency, availability, and partition tolerance becomes even more delicate if we assume that access policies are not static, but can be dynamically changed. To highlight the additional impact on the CIA of data under concurrent access control policy changes, we formulated a specialized version of the CAP theorem [3,4].

Before we begin, we want to point out the close relation of confidentiality and integrity to consistency. According to their standard definitions, these three notions require that a total order on operations exists. In the case of consistency, the requirement applies on all system operations; in the case of confidentiality/integrity, the requirement concerns only the access control operations. This relation implies that if confidentiality and/or integrity cannot be met, the system cannot be consistent. Later, we will see that the reverse is also true. A similar relation exists between availability and accessibility. Accessibility is dependent on availability in the sense that unavailability results in the inability to guarantee accessibility. If a request issued by an authorized user U to access some resource R does not receive a response, U would be restricted from accessing resources despite having permission to do so.

Theorem 1. *A partition-tolerant distributed file system that does not provide (strong) consistency, cannot guarantee confidentiality, integrity, and accessibility (CIA).*

Proof. For a contradiction, assume that there exists a partition-tolerant distributed file system that does not provide (strong) consistency, but guarantees accessibility, confidentiality, and integrity.

In the following, we assume that network partitions split the system into two disjoint non-empty sets of nodes: $N = N_1 \uplus N_2$. We further assume that our file system has two users, $User1$ and $User2$, and a file, F. $User1$ is the owner and the only member in the group of F and $User2$ belongs to *Other* authorization type. F has permissions for *Other* set to **rw** (read and write).

Following from that scenario, we now distinguish between three cases. According to the CAP theorem, a distributed system cannot be made available, strongly consistent, and tolerate network partitioning at the same time. If the system does not provide strong consistency, it can guarantee availability. Therefore, in the first case the system provides availability and it always returns the last written permission when an access control request is issued. In the second case, the system provides availability and it responds with the last written permissions when the network is not partitioned and with an "unsuccessful" response when the network is partitioned to any issued access control requests. For the last case, we assume that the system does not provide availability.

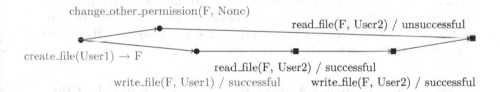

change_other_permission(F, None)

read_file(F, User2) / unsuccessful

create_file(User1) → F

read_file(F, User2) / successful

write_file(F, User1) / successful   write_file(F, User2) / successful

Fig. 1. Confidentiality and Integrity violation.

Case 1. The file system provides availability, always responding with the last locally written permission. For this case, we have depicted a valid scenario as an abstract execution graph in Fig. 1. Operations issued by $User1$ are marked in red and with a circle and operations issued by $User2$ are marked in blue and with a square.

Initially, $User1$ creates file F, which has $User1$ as owner and the only member in its group. Furthermore, by default F has rw (read and write) permission for *Other*. This event is observed by all nodes. After the network partitions into N_1 (top) and N_2 (bottom), $User1$ communicates with a node from N_1 and changes *Other* access for F to *None*. Afterwards, $User1$ communicates with a node from N_2 and successfully writes to F without realizing that they are operating on an earlier version of the file, where others still have full access to its contents. Subsequently, $User2$ communicates with N_2 and issues a *read* request for F, which succeeds, because the last written permission on N_2 for *Other* on F is rw. This, however, violates confidentiality, as the permission was already changed in the system to *None* after the file creation and so *User2* has made an unauthorized read access to F. Furthermore, let us assume that *User2* issues next a write request on F, while communicating with N_2. Analogously, the request succeeds, because the last written permission on N_2 for *Other* on F is rw. This result violates integrity, as the permission was already changed in the system to *None* after the file creation and so *User2* has actually made an unauthorized write access to F.

Continuing on the graph, we see that after the network is not partitioned anymore, all replicas can receive the new updates and synchronize to the same state. This leads to the request of *User2* to read F to be denied (return value is unsuccessful). This scenario serves as a counterexample to the general assumption.

Case 2. The file system provides availability. However, instead of blocking, it does not permit read and write operations until the system is synchronized. In this case, any type of access is prohibited for any user, and the system responds with "unsuccessful" to any read and/or write request. Therefore, it is guaranteed that access from unauthorized users will not be allowed. This fulfills confidentiality and integrity.

Nonetheless, this implies that if an authorized user wants to access the resource, access will not be granted, thus violating accessibility. This scenario is thus another counterexample to the general assumption.

Case 3. Finally, assume the file system does not provide availability. By our prior observation, availability is a precondition for accessibility; thus, accessibility will not be guaranteed in this case. From this, it follows that this case is a further counterexample to the general assumption. □

5   An Access Control Model for Highly-Available File Systems

In the previous section, we showed that confidentiality, integrity and accessibility, as defined in Sect. 2 cannot be guaranteed in a highly-available partition tolerant file system. However, weaker notions may suffice to offer protection of shared data. In the following section, we investigate to which degree weaker notions of the three security properties can be guaranteed, by formalizing a distributed file system model using CRDTs.

Instead of relying on a global total order, we follow the first observation by Weber [11] and establish a common causal order on the access control and data operations. Thus, a user is authorized to execute an operation at a specific point in time if and only if an access control operation that grants the user access rights has been issued and no access control operation that takes away these rights has happened thereafter. The weaker notions of confidentiality, integrity and accessibility follow from this modified definition of authorization. As a reminder, confidentiality does not allow unauthorized read access, integrity does not permit unauthorized write access and accessibility allows access to every authorized user.

We integrate these weaker security properties in a POSIX-like file system model based on CRDTs formalized in Repliss, a testing and verification tool for highly available applications [13]. Our full model can be found in [14]. Before discussing the formalization, we briefly explain what CRDTs are, introduce Repliss and discuss the assumptions and the restrictions on our model.

5.1 Conflict-Free Replicated Data Types

CRDTs extend classical data types with semantics under concurrent updates using datatype-specific conflict-resolution schemes. Common examples include counters, sets (add-wins, remove-wins) and maps (update-wins, delete-wins). Furthermore, CRDTs fulfill strong eventual consistency [8]. This means that two replicas have the same state once they have received the same set of updates.

As shown in Sect. 3, CRDTs have been applied in the context of distributed file systems. However, we believe that their application is not restricted to the files and the directory structure, but also affects their associated metadata such as the access control policies, comprising information on users and their access permissions. In the following model, we observe the challenges emerging when applying CRDTs for access control of data.

5.2 Repliss

Repliss is a testing and verification tool for highly available applications [13]. It provides **invocations** that can simulate access to shared data. The tool's main advantage is that it offers built-in CRDTs such as Set, Map, Counter, and Flag. The **calls** interact with a CRDT data storage based on the data-type specific operations defined by the CRDTs. Repliss also defines a **happens-before**[6] relation between calls or invocations that allows to reason about the application behavior using the history of invocations. To specify a model's safety[2] properties, one can define **invariants** that are assumed to hold at transaction boundaries. The model can be then tested against those invariants. If there are any possible scenarios when the invariants are violated, Repliss constructs counterexamples depicting those scenarios.

In our model from Sect. 5.4, we formalize the weaker notions of the security properties as such invariants.

5.3 Assumptions and Restrictions

The following assumptions on our formal model for the distributed file system are derived from the traditional POSIX semantics:

---

[2] Here, the meaning of safety is as in safety and liveness properties.

- There exists at all times at least one administrator.
- A file has exactly one owner.
- A file has exactly one group.

In addition to these assumptions, we simplify the formal model as follows. As mentioned in Sect. 2, we limit the model to not support directories. While we acknowledge that directories have an effect on access control and operations on such may present further security challenges, we believe that modeling static files captures the most important executions and conflicts related to access control. Simplifying the model also eases the testing process. Therefore, we have chosen to opt them out to narrow the scope.

Next, we have limited the model to support only read and write permissions and to exclude the execute permission commonly found on traditional POSIX file systems. We set our focus on addressing access control for data while the execute permissions are irrelevant for our context.

```
crdt files: Map_dw[NodeId,
    {access_right_owner, FileAccessRights[AccessRight],
     access_right_group, FileAccessRights[AccessRight],
     access_right_other, FileAccessRights[AccessRight],
     file_owner: Register[UserId],
     file_group: Register[GroupId],
     file_data: Register[UserId]}]

crdt groups: Map_dw[GroupId, {group_users: Set_rw[UserId]}}]
crdt users: Map_dw[UserId, {is_admin: Register[Bool]}}]
```

Fig. 2. CRDTs for files, groups and users.

5.4   Formalization

The data model we use for our Repliss model is illustrated in Fig. 2. It consists of three Map CRDTs: files, groups and users. groups contains the current set of groups in the system and contains the user ids of the members for each group. users represents the set of users in the system and whether the user has administrator rights or not. files depicts the set of files currently in the system. Each file has a unique representation NodeId, an owner file_owner, a group file_group and data file_data. We do not model any file content, because for our discussion here it suffices to know whether the file has been accessed to construct scenarios that violate the invariants. Instead, we maintain as file data the id of the user that has last accessed it. Furthermore, each file maintains three different permissions: for owner access_right_owner, for group access_right_group and for all others access_right_other. Each of these permissions is of type FileAccessRights and has a value of type AccessRight, which is depicted in Fig. 3. FileAccessRights is a custom CRDT, which we extended Repliss with. It represents the access rights for a specific file and type of authorization. The corresponding lattice is defined in Fig. 3. To be able to distinguish between concurrent administrator and user actions, the access rights are decorated with an administrator label A which states that they were issued by an administrator, or a user label U which states that the change was issued by a user.

In addition, there are two options when defining lattice and thus the merging semantics as illustrated in Fig. 3. As an example, consider a setting where two concurrent access right operations change a file's owner access rights to Read (UR) and to Write (UW). The restrictive approach (Lattice 1) would be to take the lower bound of the access rights and revoke the owner's write rights by resolving to UNone. The other approach (Lattice 2) would be to take the upper bound of the rights and settle for a Read-Write access for the owner URW. In both cases, admin changes take precedence over user changes. One could argue here that the choice here is whether to favor accessibility over confidentiality and integrity or the other way around. However, it is important to note that in many file system implementations, the administrators usually have separate domain territories where they do not and must not overlap to prevent conflicting actions. In this case, and since the access rights changes in a POSIX-like file system are issued by one of the administrators or the owner of the file, one can opt to favor accessibility since the execution in which conflicting assignments happen is only applicable when two administrators are in conflict.

```
type AccessRight = UNone() | UR() | UW() | URW()
               | ANone() | AR() | AW() | ARW()
```

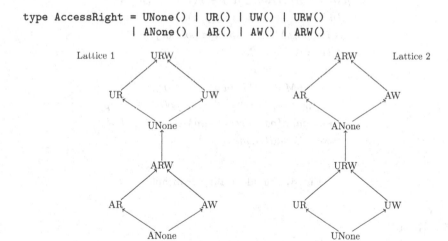

Fig. 3. Access rights.

Besides the CRDT data model, the Repliss model comprises operations for creating users, groups and files, for assigning and removing users to/from a group, for changing a file's ownership and group, and modifying the associated permissions, and for reading and writing a file's data.

Furthermore, we defined invariants to reflect the weaker notions of the security properties to test the model against. We argue that for our purposes confidentiality behaves the same as integrity. In both cases, access is requested from the file.

Figure 4 defines the weak confidentiality property. Assume that **AP** is the set containing successful policy modification invocations, **R** is the set containing successful read invocations, **F** is the set of files, **U** is the set of users and its subset **RU** contains only non-admin users, **P** is the set of permissions as defined previously. Further, **Info** is a function $Invocation \rightarrow InvocationInfo$ that maps the invocation events to the corresponding parameters. **Result** is a predicate that determines whether an invocation was performed successfully. $\rightarrow$ defines the **happens-before** relation.

The invariant in Fig. 4 states the following: Assume there is a read request for file f issued by user u, and there is a policy modification that changes the permission for the user to $Write$ or $None$ (i.e. removing the read access) that happened before the read request. If there doesn't exist another policy modification between this policy change and the read request, then the read request will be denied.

For brevity, we consider `change_policy` to act only on owner rights. The actual policy change event would also modify the group, group permissions and other permissions.

$$
\begin{aligned}
&\forall\, apMod \in AP, read \in R, f \in F, u \in RU, p \in P: \\
&\quad Info(read) = read\_file(f, u) \\
&\quad \wedge\, Info(apMod) = change\_policy(f, u, p) \\
&\quad \wedge\, (p = W \vee p = None) \\
&\quad \wedge\, apMod \rightarrow read \\
&\quad \wedge\, \nexists\, iapMod \in AP, iu \in U, ip \in P: \\
&\qquad Info(iapMod) = change\_policy(f, iu, ip) \\
&\qquad \wedge\, iapMod \rightarrow read \wedge apMod \rightarrow iapMod \\
&\quad \implies \neg Result(read)
\end{aligned}
$$

Fig. 4. Confidentiality invariant.

5.5   Results

When testing the model with Repliss, we obtained several counter examples of executions where the invariants were violated due to event interactions that we had not anticipated[3]. In this section, we discuss these results and the most important insights from testing our model against the described invariants. We start by describing how important it is to define operations and CRDTs in such a way that decisions are made based on the full access control knowledge and therefore write skews can be avoided. Finally, we discuss the difference between taking the lower and the upper bound when merging conflicting updates.

---

[3] The full formal model and the evaluation results are available on Github [12].

Preventing Write Skews in Access Policy Changes. The first result emphasizes what happens if access policy events are represented similarly to their counterpart in a non-distributed POSIX-like file system. Normally, changing a file's owner and changing the file's owner permissions are two separate operations. Modelling the events this way may however lead to an intercepting policy change and result in an unexpected state after converging.

Figure 5 represents an execution where a file has its owner and the owner permissions changed concurrently. The resulting state includes the new owner with the new permissions. From the perspective of the users executing the access policy modification operations, it seems that one desirable outcome would be to have the new owner with the old permissions, the other would be to have the new permissions assigned to the old owner. The actual resulting situation as depicted in Fig. 5 is undesirable by both users.

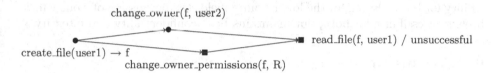

Fig. 5. Conflict between access control modifications.

Unrelated Access Control Semantics. The previous result concludes that if an access policy modification event does not specify the expected state for all access control CRDTs in the same transaction, the outcome can be considered to be undefined by the issuer. To overcome this problem, each access policy modification event can be adjusted to change the state of all CRDTs concerned with the access policy for the file. The desired change is applied alongside re-applying the existing state values for all CRDTs that are not to be changed. For example, if a user wishes to change the owner of a file, the change owner event should not only change the owner, but also confirm the owner permissions. By extending the operations this way, we avoid write skews.

While this change solves the problem, another issue arises. Consider the case where a file's owner permissions are changed to prevent the owner from reading, while concurrently the owner is changed to a new owner. In the case where the new owner with the old permissions is assigned, there are possible executions where the old owner can still read if the file's group contained the old owner alongside read access for that group. One could argue that this is the expected behavior from the perspective that a similar execution could happen on a centralized system, but the pessimistic approach of taking the lower bound of permissions is not functional in such a situation since the old owner must have been completely prevented from reading the file. The argument boils down to the choice of domain for the targeted users in access policies. On one side, one could say that the access policy is defined for the roles owner, group, others irrespective of which users are assigned these roles. The other side of the argument would be a more refined alternative by defining the targeted roles to be the users that are assigned to be the owner, part of a group or simply other users.

While the former is captured by the model, the latter would require an auxiliary CRDT that captures all of the access policy data in one data structure, and thus allows more refined merging procedures. In conclusion, defining the access policy in terms of more than one semantically unrelated CRDTs is not sufficient for the refined model that targets users explicitly instead of roles.

Upper and Lower Bound Semantics. Taking the upper bound favors accessibility over confidentiality and integrity. On the other hand, taking the lower bound favors confidentiality and integrity at the cost of decreased accessibility.

In the event that a permission is changed to read-only on one side, and concurrently write-only on the other. Taking the upper bound of read-write would violate confidentiality for one side and integrity for the other but maintains accessibility for both. Opting for the lower bound yields the permission of None which breaks accessibility for both, but maintains both confidentiality and integrity.

6   Conclusion

Our file-system model presented in this paper underlines the importance of having secure access control mechanisms for distributed file systems that reflect the users' intention and maintain the three central properties - confidentiality, integrity and accessibility. Our paper provides an impossibility result for highly available partition-tolerant distributed file systems guaranteeing all three security properties.

In an attempt to verify whether weaker notions of the security properties can be achieved in the same setting, we devise a CRDT-based file system model formalized in Repliss. Testing the model against the weaker notions of confidentiality and accessibility exposed some corner counter-cases and allowed us to draw important insights. Firstly, it appears that there are non-trivial relations between parts of the access control policy state which need to be considered in the design of operations; otherwise, safety violations can be introduced. For example, if the file's owner and file's owner permission have been changed concurrently, the system will converge to the undesirable state of the file having the new owner with the new permissions. Secondly, having a data model with multiple unrelated CRDTs causes the inability to define merging semantics based on all of the access control-relevant data. Finally, we identify the correlation between confidentiality/integrity and making more restrictive merging semantics.

6.1   Future Work

For the formal analysis presented here, we focused on operations related to files. Directories have different access policy semantics, which introduces more conflict cases. A typical conflict can occur as follows: Owner A of directory D changes D's permissions to exclude user B from accessing D's contents. Concurrently,

B copies a file from D to another directory E, to which B has access. This conflict can be considered an example of confidentiality violation by transitivity because B can now read the copied files' content. We believe that an extension of our model can serve as a formal basis for reasoning about implementations and devise safe conflict resolution strategies.

We further plan to define a custom policy CRDT that combines related access control metadata and applies merge operations that extend to all semantically related CRDT objects. Note that Repliss currently does not support this type of CRDTs.

Finally, it would be interesting to extend an existing distributed file system with the implementation of our model. This is essential to validate our model a practical setup and evaluate the administrative overhead.

References

1. Ahmed-Nacer, M., Martin, S., Urso, P.: File system on CRDT. CoRR abs/1207.5990 (2012). http://arxiv.org/abs/1207.5990
2. Bach, M.J.: The Design of the Unix Operating System. Prentice-Hall, Englewood Cliffs (1986)
3. Brewer, E.A.: Towards robust distributed systems (abstract). In: Proceedings of the Nineteenth Annual ACM Symposium on Principles of Distributed Computing, PODC 2000, p. 7. Association for Computing Machinery, New York (2000). https://doi.org/10.1145/343477.343502
4. Gilbert, S., Lynch, N.: Brewer's conjecture and the feasibility of consistent, available, partition-tolerant web services. SIGACT News **33**(2), 51–59 (2002)
5. Kleppmann, M., Mulligan, D.P., Gomes, V.B., Beresford, A.R.: A highly-available move operation for replicated trees and distributed filesystems (2020)
6. Lamport, L.: Time, clocks, and the ordering of events in a distributed system. Commun. ACM **21**(7), 558–565 (1978)
7. Najafzadeh, M., Shapiro, M., Eugster, P.: Co-design and verification of an available file system. In: Dillig, I., Palsberg, J. (eds.) VMCAI 2018. LNCS, vol. 10747, pp. 358–381. Springer, Cham (2018). https://doi.org/10.1007/978-3-319-73721-8_17
8. Preguiça, N.M., Baquero, C., Shapiro, M.: Conflict-free replicated data types CRDTs. In: Sakr, S., Zomaya, A.Y. (eds.) Encyclopedia of Big Data Technologies. Springer, Heidelberg (2019). https://doi.org/10.1007/978-3-319-63962-8_185-1
9. Shapiro, M., Preguiça, N., Baquero, C., Zawirski, M.: Conflict-free replicated data types. In: Défago, X., Petit, F., Villain, V. (eds.) SSS 2011. LNCS, vol. 6976, pp. 386–400. Springer, Heidelberg (2011). https://doi.org/10.1007/978-3-642-24550-3_29
10. Tao, V., Shapiro, M., Rancurel, V.: Merging semantics for conflict updates in geo-distributed file systems. In: Naor, D., Heiser, G., Keidar, I. (eds.) Proceedings of the 8th ACM International Systems and Storage Conference, SYSTOR 2015, Haifa, Israel, 26–28 May 2015, pp. 10:1–10:12. ACM (2015). https://doi.org/10.1145/2757667.2757683
11. Weber, M., Bieniusa, A., Poetzsch-Heffter, A.: Access control for weakly consistent replicated information systems. In: Barthe, G., Markatos, E., Samarati, P. (eds.) STM 2016. LNCS, vol. 9871, pp. 82–97. Springer, Cham (2016). https://doi.org/10.1007/978-3-319-46598-2_6

12. Youssef, M., Bieniusa, A., Rezae, A.H., Yanakieva, E.: CRDT-filesystem (2021). https://github.com/AntidoteDB/crdt-filesystem
13. Zeller, P., Bieniusa, A., Poetzsch-Heffter, A.: Combining state- and event-based semantics to verify highly available programs. In: Arbab, F., Jongmans, S.-S. (eds.) FACS 2019. LNCS, vol. 12018, pp. 213–232. Springer, Cham (2020). https://doi.org/10.1007/978-3-030-40914-2_11
14. Zeller, P., Youssef, M., Bieniusa, A., Rezae, A.H., Yanakieva, E.: Repliss-filesystem (2021). https://github.com/AntidoteDB/repliss-filesystem

Byzantine Geoconsensus

Joseph Oglio, Kendric Hood, Gokarna Sharma, and Mikhail Nesterenko[✉]

Department of Computer Science, Kent State University, Kent, OH 44242, USA
{joglio,khood5}@kent.edu, {sharma,mikhail}@cs.kent.edu

Abstract. We define and investigate consensus for a set of N processes embedded in the d-dimensional plane, $d \geq 2$, which we call the *Geoconsensus Problem*. The processes have unique coordinates and can communicate with each other through oral messages. Faulty processes are covered by a finite-size convex fault area F. The correct processes know the fault area size but not its location. We prove that the geoconsensus is impossible if all processes may be covered by at most three areas the size of the fault area.

On the constructive side, for $M \geq 1$ fault areas F of arbitrary shape with diameter D, we present a consensus algorithm *BASIC* that tolerates $f \leq N - (2M + 1)$ Byzantine processes provided that there are $9M + 3$ processes with pairwise distance between them greater than D. We present another consensus algorithm *GENERIC* that lifts this distance requirement. For square F with side ℓ, *GENERIC* tolerates $f \leq N - 15M$ Byzantine processes given that all processes are covered by at least $22M$ axis aligned squares of the same size as F. For a circular F of diameter ℓ, *GENERIC* tolerates $f \leq N - 57M$ Byzantine processes if all processes are covered by at least $85M$ circles. We then estimate the tolerance of *GENERIC* for various size combinations of fault and non-fault areas as well as d-dimensional process embeddings, where $d \geq 3$.

1 Introduction

The problem of *Byzantine consensus* [17,24] has been attracting extensive attention from researchers and engineers in distributed systems since its initial statement. The problem has applications in distributed storage [1,2,5,6,16], secure communication [8], safety-critical systems [26], blockchain [21,27,29], and Internet of Things (IoT) [18].

Pease *et al.* [24] defined the problem as follows. Consider a set of N processes with unique identifiers. The processes communicate in synchronous rounds. Each process can communicate with all other processes. Some number $f < N$ of these processes are faulty. The fault is Byzantine which means that the faulty process may behave arbitrarily. The correct processes know the number of the faults f but not the identifiers of the faulty processes. The Byzantine Consensus Problem requires all $N - f$ correct processes to agree on a single value.

Pease *et al.* proved that the maximum number of faults f that can be tolerated by a deterministic algorithm depends on the communication assumptions. Unauthenticated *oral messages* may be modified upon retransmission. If only oral messages are allowed, Pease *et al.* showed that a consensus algorithm may tolerate up $f < N/3$ faults. In case of unforgeable authenticated *written messages*, the consensus is solvable with an arbitrary number of faults $f \leq N$ [24]. It is shown that Byzantine consensus requires

© Springer Nature Switzerland AG 2021
K. Echihabi and R. Meyer (Eds.): NETYS 2021, LNCS 12754, pp. 19–35, 2021.
https://doi.org/10.1007/978-3-030-91014-3_2

20   J. Oglio et al.

at least f rounds of communication [10] and $O(N^2)$ messages [12]. Faster and more efficient solutions are possible if randomized algorithms are allowed [3,14,20].

The original Byzantine Consensus Problem requires the processes to have unique identifiers but does not restrict their location. One way to add some location information to the system is by limiting process communication. In the original problem statement, any pair of processes may communicate directly. Therefore, the communication topology is a complete graph, i.e. a clique. A number of studies relax this connectivity assumption and investigate the problem in arbitrary graphs [24,28] and wireless networks [22]. Several papers study a related problem of *Byzantine broadcast* in incomplete graphs [15,25].

Recently, Lao *et al.* [18] proposed a Byzantine consensus protocol for IoT and blockchain applications, called Geographic-PBFT or simply G-PBFT, which extends a well-known PBFT algorithm [5] to geographic setting. They considered the case of fixed IoT devices embedded in geographic locations for data collection and processing. The location data for these IoT devices can be recorded at deployment, obtained using low-cost GPS receivers or through location estimation algorithms [4,13]. They argued that the fixed IoT devices have more computational power than other mobile IoT devices (e.g., mobile phones and sensors) and are less likely to exhibit Byzantine behavior. Therefore, they exploited the geographical location information of fixed IoT devices to reach consensus. They argued that G-PBFT avoids Sybil attacks [9], reduces the overhead for validating and recording transactions, achieves consensus with high efficiency and low traffic intensity. However, no formal analysis of G-PBFT is given and it is only experimentally validated. Yet, we believe that these developments warrant a study of Byzantine consensus in devices that are aware of their locations.

Our Contribution. In this paper, we formally define and investigate the problem of reaching consensus among processes in fixed geographical locations. We call this variant the *Byzantine Geoconsensus Problem*. We retain all other parameters of the original problem statement. However, if fault locations are not constrained, Geoconsensus differs little from the classic Byzantine consensus: the geographic location of each process can serve as its identifier. Hence, we consider a variant where the faults are constrained geometrically. Specifically, they are limited to a fixed-size *fault area* F. This limitation allows more effective solutions and makes Geoconsensus an interesting problem to study. We are not aware of prior work in Byzantine consensus where processes are embedded in a geometric plane while faulty processes are located in a fixed area.

Let us enumerate the contributions of this paper. Denote by N the number of processes, M the number of fault areas F, D the diameter of F, and f the number of faulty processes. In other words, f is the number of processes covered by fault areas F. Assume that each process can communicate with all other $N-1$ processes and the communication is through oral messages only. Assume that any process covered by a faulty area F may be Byzantine. The correct processes know the size of each faulty area, such as its diameter, number of edges, etc. but do not know their exact locations. In this paper, we make the following five major contributions:

(i) We prove that Geoconsensus is not solvable deterministically if all N processes may be covered by 3 equal size areas F and one of them may be the fault area. This extends to the case of N processes being covered by $3M$ areas F with M

areas being faulty. This is done by adapting the impossibility proof of Pease *et al.* [24] to Geoconsensus.

(ii) We present algorithm *BASIC* that solves Geoconsensus tolerating $f \leq N - (2M + 1)$ Byzantine processes, provided that there are $9M + 3$ processes with pairwise distance between them greater than D. The idea is for each process to deterministically select a leader in each independent coverage area. Once the leaders are selected, any generic Byzantine consensus algorithm can be run. We use the classic algorithm by Pease *et al.* [24]. Non-leader processes accept the result chosen by the leaders.

(iii) We present algorithm *GENERIC* that removes the pairwise distance assumption of *BASIC* and solves Geoconsensus tolerating $f \leq N - 15M$ Byzantine processes, provided that all N processes are covered by $22M$ axis-aligned squares of the same size as the fault area F. For *GENERIC*, we start with covering processes by axis-aligned squares and studying how these squares may intersect with fault areas of various shapes and sizes. We show that determining optimal axis-aligned square coverage is NP-hard and provide constant-ratio approximation algorithms.

(iv) We extend *GENERIC* to circular F tolerating $f \leq N - 57M$ Byzantine processes if all N processes are covered by $85M$ circles of same size as F.

(v) We further extend results of (iii) and (iv) to various shape and alignment combinations of fault and non-fault areas and to d-dimensional process embeddings, $d \geq 3$.

Notice that we considered only square and circular fault areas. However, our results can be immediately extended to more complex shapes as they can be inscribed into simple ones. Providing better bounds on more sophisticated analysis of complex shapes beyond simple inscription is left for future research.

Geoconsensus vs. Generic Byzantine Consensus. Let us contrast the results obtained for Geoconsensus to those of the original Byzantine Consensus Problem. The Geoconsensus provides potentially tighter bounds on the number of faults. The original problem establishes the relationship only between N and f, while Geoconsensus also factors the number of fault areas M. Thus, in the original problem, at most $f < N/3$ faulty processes may be tolerated, whereas our results show that as many as $f \leq N - \alpha M$ faults can be tolerated provided that the processes are placed such that at most βM areas (same size as F) are needed to cover them. Here, α and β are both integers with $\beta \geq c \cdot \alpha$ for some constant c.

Geoconsensus also allows to increase the speed and reduce message complexity of the solution. The original consensus requires at least f consecutive rounds of message exchanges and $O(f \cdot N^2)$ messages. The algorithms presented in this paper rely on the selection of a single leader per coverage area. Since each process knows the location of all other processes, this selection is done without message exchanges. Then, the leaders communicate to achieve consensus. Let N processes be covered by X areas of the same size as fault area F. Then, in one round, at most $O(X^2)$ messages need to be exchanged. To reach consensus the algorithm runs for $O(M)$. Thus, in the worst case, at most $O(M \cdot X^2) < O(f \cdot N^2)$ messages are exchanged.

Pease *et al.* [24] showed that it is impossible to solve consensus through oral messages when $N = 3f$ but provided a solution for $N \geq 3f + 1$. That is, their impossibility bound is tight. In this paper, however, we were able to show that it is impossible to solve consensus if all N processes are covered by $3M$ areas that are the same size as

Table 1. Notation used throughout the paper.

Symbol	Description
$N; \mathcal{P}; (x_i, y_i)$	Number of processes; $\{p_1, \ldots, p_N\}$; planar coordinates of process p_i
$F; D; \mathcal{F}$	Fault area; diameter of F; a set of fault areas F with $|\mathcal{F}| = M$
f	Number of faulty processes
$\mathcal{P}_D$	Processes in $\mathcal{P}$ such that pairwise distance between them is more than D
A (or $A_j(R_i)$); $\mathcal{A}$	Cover area that is of same shape and size as F; a set of cover areas A
$n(F)$	Number of cover areas $A \in \mathcal{A}$ that a fault area F overlaps

F. Yet, for the axis-aligned squares case, the provide the solution where N processes are covered by at least $22M$ areas. Narrowing the gap between the impossible and the achievable is left for further research.

2   Notation, Problem Definition, and Impossibility

Processes. A computer system consists of a set $\mathcal{P} = \{p_1, \ldots, p_N\}$ of N processes. Every process p_i is embedded in the 2-dimensional plane and has unique planar coordinates (x_i, y_i). Each process is aware of coordinates of all the other processes of $\mathcal{P}$ and is capable of sending a message to any of them. The sender of the message may not be spoofed. The communication between processes is through unauthenticated oral messages. This communication is synchronous.

Byzantine Faults. Every process is either permanently correct or faulty. The fault is Byzantine. A faulty process may behave arbitrarily. To simplify the presentation, we assume that all faulty processes are controlled by a unique adversary trying to prevent the system from achieving its task.

Fault Area. The adversary controls the processes as follows. Let the *fault area* F be a finite-size convex area in the plane. Let D be the diameter of F, i.e. the maximum distance between any two points of F. The adversary may place F in any location on the plane. A process p_i is *covered* by F if the coordinates (x_i, y_i) of p_i is either in the interior or on the boundary of F. Any process covered by F may be faulty.

A *fault area set* or just *fault set* is the set $\mathcal{F}$ of identical fault areas F. The size of this set is M, i.e., $|\mathcal{F}| = M$. The adversary controls the placement of all areas in $\mathcal{F}$. Correct processes know the shape and size of the fault areas F. However, correct processes do not know the precise placement of the fault areas $\mathcal{F}$. For example, if $\mathcal{F}$ contains 4 square fault areas F with the side ℓ, then correct processes know that each fault area is of square with side ℓ but do not know where they are located. Table 1 summarizes the notation used in this paper.

Byzantine Geoconsensus. Consider the binary consensus where every correct process is input a value $v \in \{0, 1\}$ and must output an irrevocable decision with the following three properties.

agreement – no two correct processes decide differently;
validity – every correct process outputs a value input to some correct process;
termination – every correct process eventually decides.

Definition 1. *An algorithm solves* the Byzantine Geoconsensus Problem *(or Geocon-sensus for short) for fault area set* $\mathcal{F}$, *if every computation produced by this algorithm satisfies the three consensus properties.*

Impossibility of Geoconsensus. Given a certain set of embedded processes $\mathcal{P}$ and sin-gle area F, the *coverage number* k of $\mathcal{P}$ by F is the minimum number of such areas required to cover each process of $\mathcal{P}$. We show that Geoconsensus is not solvable if the coverage number k is less than 4. When the coverage number is 3 or less, the problem is reducible to the classic consensus with 3 sets of peers where one of the sets is faulty. Pease *et al.* [24] proved the solution for the latter problem to be impossible. The intu-ition is that a group of correct processes may not be able to distinguish which of the other two groups is Byzantine and which one is correct. Hence, the correct groups may not reach consensus.

Theorem 1 (Impossibility of Geoconsensus). *Given a set $\mathcal{P}$ of $N \geq 3$ processes and an area F, there exists no algorithm that solves the Byzantine Geoconsensus Problem if the coverage number k of $\mathcal{P}$ by F is less than 4.*

Proof. Set $N = 3 \cdot \kappa$, for some positive integer $\kappa \geq 1$. Place three areas A on the plane in arbitrary locations. To embed processes in $\mathcal{P}$, consider a bijective placement function $f : \mathcal{P} \to \mathcal{A}$ such that κ processes are covered by each area A. Let v and v' be two distinct input values 0 and 1. Suppose one area A is fault area, meaning that all κ processes in that area are faulty.

This construction reduces the Byzantine Geoconsensus Problem to the impossibil-ity construction for the classic Byzantine consensus problem given in the theorem in Section 4 of Pease *et al.* [24] for the 3κ processes out of which κ are Byzantine. □

3 Geoconsensus Algorithm *BASIC*

In this section, we present the algorithm we call *BASIC* that solves Geoconsensus for up to $f < N - (2M + 1), M \geq 1$ faulty processes located in fault area set $\mathcal{F}$ of size $|\mathcal{F}| = M$ provided that $\mathcal{P}$ contains at least $9M + 3$ processes such that the pairwise distance between them is greater than the diameter D of the fault areas $F \in \mathcal{F}$.

The pseudocode of *BASIC* is shown in Algorithm 1. It contains two parts: the leaders selection and the consensus procedure. Let us discuss the selection of leaders. If the distance between two processes is less than the D, they may be covered by a single fault area F. Therefore, the leaders need to be selected such that, pairwise, they are at least D away from each other. Finding the largest set of such leaders is equivalent to computing the maximum independent set in a unit disk graph. This problem is known to be NP-hard [7]. We, therefore, employ a greedy heuristic.

Denote by $Is(G)$ a distance D maximal independent set of a planar graph G. It is defined as a subset of processes of G such that the distance between any pair of processes of Is is more than D, and every process of G that does not belong to Is is at most D away from a process in Is. That is, $p_i \in Is(G)$ if $\forall p_j \neq p_i \in Is, d(p_i, p_j) > D$ and $\forall p_k \in G \setminus Is, \exists p_m \in Is$ such that $d(p_k, p_m) \leq D$. Denote by $Nb(p_i, D)$, the distance D neighborhood of process p_i. That is, $p_j \in Nb(p_i, D)$ if $d(p_i, p_j) \leq D$. It is known [19, Lemma 3.3] that in every distance D planar graph, there exists a neighborhood whose induced subgraph contains an independent set of size at most 3.

Algorithm 1: Geoconsensus algorithm *BASIC*.

1 **Setting:** A set $\mathcal{P}$ of N processes positioned at distinct coordinates. Each process can
  communicate with all other processes and knows their coordinates. There are $M \geq 1$
  identical fault areas F. The diameter of a fault area is D. The locations of any area F is
  not known to correct processes. Each process covered by any F is Byzantine.
2 **Input:** Each process has initial value either 0 or 1.
3 **Output:** Each correct process outputs decision subject to Geoconsensus.
4 *Procedure for process* $p_k \in \mathcal{P}$
5 // leaders selection
6 Let $P_D \leftarrow \emptyset$, $P_C \leftarrow \mathcal{P}$;
7 **while** $P_C \neq \emptyset$ **do**
8   let $P_3 \subset P_D$ be a set of processes such that $\forall p_j \in P_3$, $Nb(p_j, D)$ has distance D
      independent set of at most 3;
9   let $p_i \in P_3$, located in (x_i, y_i) be the lexicographically smallest process in P_3, i.e.
      $\forall p_j \neq p_i \in P_3$: located in (x_j, y_j) either $x_i < x_j$ or $x_i = x_j$ and $y_i < y_j$;
10   add p_i to P_D;
11   remove p_i from P_C;
12     $\forall p_j \in Nb(p_i, D)$ remove p_j from P_C;
13 // consensus
14 **if** $p_k \in P_D$ **then**
15   run *PSL* algorithm, achieve decision v, broadcast v, output v;
16 **else**
17   wait for messages with identical decision v from at least $2M + 1$ processes from $\mathcal{P}_D$,
      output v;

The set of leaders $\mathcal{P}_D \subset \mathcal{P}$ selection procedure operates as follows. A set P_C of leader candidates is iteratively processed. At first, all processes are candidates. All processes whose distance D neighborhood induces a subgraph with an independent set no more than 3 are found. Among those, the process p_i with lexicographically smallest coordinates, i.e. the process in the bottom left corner, is added to the leader set $\mathcal{P}_D$. Then, all processes in $Nb(p_i, D)$ are removed from the leader candidate set $\mathcal{P}_C$. This procedure repeats until $\mathcal{P}_C$ is exhausted.

The second part of *BASIC* relies on the classic consensus algorithm of Pease *et al.* [24]. We denote this algorithm as *PSL*. The input of *PSL* is the set of $3f + 1$ processes such that at most f of them are faulty as well as the initial value 1 or 0 for each process. As output, the correct processes provide the decision value subject to the three properties of the solution to consensus. *PSL* requires $f + 1$ communication rounds.

The complete *BASIC* operates as follows. All processes select leaders in P_D. Then, the leaders run *PSL* and broadcast their decision. The rest of the correct processes, if any, adopt this decision.

Analysis of *BASIC*. The observation below is immediate since all processes run exactly the same deterministic leaders selection procedure.

Observation 1. *For any two processes $p_i, p_j \in \mathcal{P}$, set P_D computed by p_i is the same as set P_D computed by p_j.*

Lemma 1. *If $\mathcal{P}$ contains at least $3x$ processes such that the distance between any pair of such processes is greater than D, then the size of $\mathcal{P}_D$ computed by processes in BASIC is at least x.*

Proof. In [19, Theorem 4.7], it is proven that the heuristic we use for the leaders selection provides a distance D independent set $\mathcal{P}_D$ whose size is no less than a third of optimal size. Thus, $x \leq |\mathcal{P}_D|$. The lemma follows. $\qquad\square$

Lemma 2. *Consider a fault area F with diameter D. No two processes in $\mathcal{P}_D$ are covered by F.*

Proof. For any two processes $p_i, p_j \in \mathcal{P}_D$, $d(p_i, p_j) > D$. Since any area F has diameter D, no two processes $> D$ away can be covered by F simultaneously. $\qquad\square$

Theorem 2. *Algorithm BASIC solves the Byzantine Geoconsensus Problem for a fault area set $\mathcal{F}$, the size of $M \geq 1$ with fault areas F with diameter D for N processes in $\mathcal{P}$ tolerating $f \leq N - (2M+1)$ Byzantine faults provided that $\mathcal{P}$ contains at least $9M+3$ processes such that their pairwise distance is more than D. The solution is achieved in $M + 2$ communication rounds.*

Proof. If $\mathcal{P}$ contains at least $9M + 3$ processes whose pairwise distance is more than D, then, according to Lemma 1, each process in BASIC selects $\mathcal{P}_D$ such that $|\mathcal{P}_D| \geq 3M + 1$. We have $M \geq 1$ fault areas, i.e., $|\mathcal{F}| = M$. From Lemma 2, a process $p \in \mathcal{P}_D$ can be covered by at most one fault area F. Therefore, if $|\mathcal{P}_D| \geq 3M + 1$, then it is guaranteed that even if M processes in $\mathcal{P}_D$ are Byzantine, $2M + 1$ correct processes in $\mathcal{P}_D$ can reach consensus using PSL algorithm.

In the worst case, the adversary may position fault areas of $\mathcal{F}$ such that all but $2M + 1$ processes in $\mathcal{P}$ are covered. Hence, BASIC tolerates $N - (2M + 1)$ faults.

Let us address the number of rounds that BASIC requires to achieve Geoconsensus. It has two components executed sequentially: leaders selection and PSL. Leaders selection is done independently by all processes and requires no communication. PSL takes $M + 1$ rounds for the $2M + 1$ leaders to arrive at the decision. It takes another round for the leaders to broadcast their decision. Hence, the total number of rounds is $M + 2$. $\qquad\square$

4 Covering Processes

In this section, in preparation for describing the GENERIC Geoconsensus algorithm, we discuss techniques of covering processes by axis-aligned squares and circles. These techniques vary depending on the shape and alignment of the fault area F.

Covering by Squares. The algorithm we describe below covers the processes by square areas A of size $\ell \times \ell$, assuming that the fault areas F are also squares of the same size. Although F may not be axis-aligned, we use axis-aligned areas A to cover processes. Later, we determine the maximum number of such areas A, that non-axis-aligned F may overlap.

Let A be positioned on the plane such that the coordinate of its bottom left corner is (x_1, y_1). The coordinates of its top left, top right, and bottom right corners are respectively $(x_1, y_1 + \ell), (x_1 + \ell, y_1 + \ell)$, and $(x_1 + \ell, y_1)$. Let process p_i be at coordinate (x_i, y_i). We say that p_i is *covered* by A if and only if $x_1 \leq x_i \leq x_1 + \ell$ and $y_1 \leq y_i \leq y_1 + \ell$. We assume that A is *closed*, i.e., process p_i is assumed to be covered by A even if p_i is positioned on the boundary of A.

Let us formally define the problem of covering processes by square areas, which we denote by SQUARE-COVER. Denote by $\mathcal{A}$ a set of square areas A. We say that $\mathcal{A}$ completely covers all N processes if each $p_i \in \mathcal{P}$ is covered by at least one square $A \in \mathcal{A}$.

Definition 2 (The SQUARE-COVER problem). *Suppose N processes are embedded into a 2d-plane such that the coordinates of each process are unique. Given a number $k \geq 1$, is there a set $\mathcal{A}$ of cardinality k composed of identical square areas $A = \ell \times \ell$ that completely covers these N processes?*

Theorem 3. *SQUARE-COVER is NP-Complete.*

Proof. To prove the theorem, we demonstrate that SQUARE-COVER is reducible to the BOX-COVER problem which was shown to be NP-Complete by Fowler *et al.* [11]. BOX-COVER is defined as follows: There is a set of N points on the plane such that each point has unique integer coordinates. A closed box (rigid but relocatable) is set to be a square with side 2 and is axis-aligned. The problem is to decide whether a set of $k \geq 1$ identical axis-aligned closed boxes are enough to completely cover all N points. Fowler *et al.* provided a polynomial-time reduction of 3-SAT to BOX-COVER such that k boxes will suffice if and only if the 3-SAT formula is satisfiable. In this setting, SQUARE-COVER reduces to BOX-COVER for $\ell = 2$. Therefore, the NP-Completeness of BOX-COVER extends to SQUARE-COVER.   □

A Greedy Square Cover Algorithm. Since SQUARE-COVER is NP-Complete, we use an efficient greedy approximation algorithm to find a set $\mathcal{A}$ of k_{greedy} axis-aligned square areas $A = \ell \times \ell$ that completely cover all N processes in $\mathcal{P}$. We prove that $k_{greedy} \leq 2 \cdot k_{opt}$, where k_{opt} is the optimal number of axis-aligned squares in any algorithm to cover those N processes. That is, our heuristic is a 2-approximation of the optimal algorithm. We call this algorithm *GSQUARE*. Each process p_i can run *GSQUARE* independently, because p_i knows all required input parameters for *GSQUARE*.

GSQUARE operates as follows. Suppose the coordinates of process $p_i \in \mathcal{P}$ are (x_i, y_i). Let $x_{min} = \min_{1 \leq i \leq N} x_i, x_{max} = \max_{1 \leq i \leq N} x_i, y_{min} = \min_{1 \leq i \leq N} y_i$, and $y_{max} = \max_{1 \leq i \leq N} y_i$. Let R be an axis-aligned rectangle with the bottom left corner at (x_{min}, y_{min}) and the top right corner at (x_{max}, y_{max}). It is immediate that R is the smallest axis-aligned rectangle that covers all N processes. The width of R is $width(R) = x_{max} - x_{min}$ and the height is $height(R) = y_{max} - y_{min}$. See Fig. 1 for illustration.

Cover rectangle R by a set $\mathcal{R}$ of m *slabs* $\mathcal{R} = \{R_1, R_2, \ldots, R_m\}$. The height of each slab R_i is ℓ, except for possibly the last slab R_m whose height may be less than ℓ. The width of each slab is $width(R)$. That is this width is the same is the width of R.

This slab-covering is done as follows. Place slab R_1 at the bottom of R such that its bottom side aligns with the bottom of R and left and right sides align with the corresponding sides of R. Slide R_1 up so that the bottom-most process $p_{min} = (x_{min}, y_{min}) \in \mathcal{P}$ is on the bottom side of R_1. See Fig. 1 for illustration. Now consider

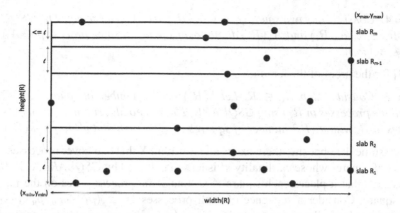

Fig. 1. Selection of axis-aligned smallest enclosing rectangle R covering all N processes in $\mathcal{P}$ and coverage of R by axis-aligned slabs R_i of height ℓ and width $width(R)$. The slabs are selected such that the at least one process is positioned on the bottom side of each slab.

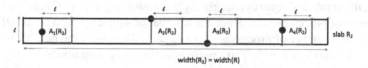

Fig. 2. Selection of axis-aligned areas $A_j(R_2)$ (shown in red) to cover the processes in the slab R_2 of Fig. 1. At least one process is positioned on the left side of each area. (Color figure online)

only the processes in R that are not covered by R_1. Denote this process set by $\mathcal{P}'$. Consider the bottom-most process $y_{min'}$ of $\mathcal{P}'$. Slide the next slab R_2 up so that $p_{min'}$ is on its bottom side. Continue placing slabs over R in this manner until all processes of $\mathcal{P}$ are covered.

We now cover each such slab by axis-aligned square areas $A = \ell \times \ell$. See Fig. 2 for illustration. This square-covering is done similar to slab-covering. Let R_i be a slab to cover. Place the first area A, call it $A_1(R_i)$, on R_i such that the top left corner of A overlaps with the top left corner of slab R_i. Slide $A_1(R_i)$ horizontally to the right until the left-most process in R_i is positioned on the left side of $A_1(R_i)$. Now consider only the processes in R_i not covered by $A_1(R_i)$. Slide the next area A, called $A_2(R_i)$, such that the left-most process in R_i is positioned on the left side of A. Note that there are no uncovered processes between $A_1(R_i)$ and $A_2(R_i)$. Continue to cover all the points in R_i in this manner. The last square may extend past the right side of the slab. Repeat this procedure for every slab of R.

Lemma 3. *Consider any two slabs $R_i, R_j \in \mathcal{R}$ produced by GSQUARE. R_i and R_j do not overlap, i.e., if some process $p \in R_i$, then $p \notin R_j$.*

Proof. It is sufficient to prove this lemma for adjacent slabs. Suppose slabs R_i and R_j are adjacent, i.e., $j = i + 1$. According to algorithm *GSQUARE*, after the location of R_i is selected, only processes that are not covered by the slabs so far are considered for the selection of R_j. The first such process lies above the top (horizontal) side of R_i. Hence, there is a non-empty gap between the top side of R_i and the bottom side of R_j. $\square$

Lemma 4. *Consider any two square areas $A_j(R_i)$ and $A_k(R_i)$ selected by GSQUARE in slab $R_i \in \mathcal{R}$. $A_j(R_i)$ and $A_k(R_i)$ do not overlap, i.e., if some process $p \in A_j(R_i)$, then $p \notin A_k(R_i)$.*

See [23] for the proof of the lemma.

Lemma 5. *Consider slab $R_i \in \mathcal{R}$. Let $k(R_i)$ be the number of squares $A_j(R_i)$ to cover all the processes in R_i using GSQUARE. There is no algorithm that can cover the processes in R_i with $k'(R_i)$ number of squares $A_j(R_i)$ such that $k'(R_i) < k(R_i)$.*

Proof. Assume the opposite: there exists an algorithm X that can cover processes in R_i with a set of squares whose cardinality k' is less than k used by *GSQUARE*. *GSQUARE* operates such that it places each square A so that some process p lies on the left side of this square. Consider a sequence of such processes: $\sigma \equiv \langle p_1 \cdots p_u, p_{u+1} \cdots p_j \rangle$. Consider any pair of subsequent processes p_u and p_{u+1} in σ with respective coordinates (x_u, y_u) and (x_{u+1}, y_{u+1}). *GSQUARE* covers them with non-overlapping squares with side ℓ. Therefore, $x_u + \ell < x_{u+1}$. That is, the distance between consequent processes in σ is greater than ℓ. Any pair of such processes may not be covered by a single square. Therefore, the number of squares required by the posited algorithm X is at least as large as the number of processes in σ. Since the number of squares placed by *GSQUARE* in slab R_i is k, the number of processes in σ is also k. Therefore, the number of squares required by X is no less than k. This contradicts our initial assumption. $\square$

Let $k_{opt}(\mathcal{R})$ be the number of axis-aligned square areas $A = \ell \times \ell$ to cover all N processes in R in the optimal cover algorithm. We now show that $k_{greedy}(\mathcal{R}) \le 2 \cdot k_{opt}(\mathcal{R})$, i.e., *GSQUARE* provides 2-approximation. We divide the slabs in the set $\mathcal{R}$ into two sets $\mathcal{R}_{odd}$ and $\mathcal{R}_{even}$. For $1 \le i \le m$, let

$$\mathcal{R}_{odd} := \{R_i, i \bmod 2 \neq 0\} \text{ and } \mathcal{R}_{even} := \{R_i, i \bmod 2 = 0\}.$$

Lemma 6. *Let $k(\mathcal{R}_{odd})$ and $k(\mathcal{R}_{even})$ be the total number of (axis-aligned) square areas $A = \ell \times \ell$ to cover the processes in the sets $\mathcal{R}_{odd}$ and $\mathcal{R}_{even}$, respectively. Let $k_{opt}(\mathcal{R})$ be the optimal number of axis-aligned squares $A = \ell \times \ell$ to cover all the processes in $\mathcal{R}$. $k_{opt}(\mathcal{R}) \ge \max\{k(\mathcal{R}_{odd}), k(\mathcal{R}_{even})\}$.*

Proof. Consider two slabs R_i and R_{i+2} for $i \ge 1$. Consider a square $A_j(R_i)$ placed by *GSQUARE*. Consider also two processes $p \in R_i$ and $p' \in R_{i+2}$, respectively. The distance between p and p' is $d(p, p') > \ell$. Therefore, if $A_j(R_i)$ covers p, then it cannot cover $p' \in R_{i+2}$. Hence, no algorithm can produce the number of squares $k_{opt}(\mathcal{R})$ less than the maximum between $k(\mathcal{R}_{odd})$ and $k(\mathcal{R}_{even})$. $\square$

Lemma 7. $k_{greedy}(\mathcal{R}) \le 2 \cdot k_{opt}(\mathcal{R})$.

Proof. From Lemma 5, we obtain that *GSQUARE* is optimal for each slab R_i. From Lemma 6, we get that for any algorithm $k_{opt}(\mathcal{R}) \ge \max\{k(\mathcal{R}_{odd}), k(\mathcal{R}_{even})\}$. Moreover, the *GSQUARE* produces the total number of squares $k_{greedy}(\mathcal{R}) = k(\mathcal{R}_{odd}) + k(\mathcal{R}_{even})$. Comparing $k_{greedy}(\mathcal{R})$ with $k_{opt}(\mathcal{R})$, we get

$$\frac{k_{greedy}(\mathcal{R})}{k_{opt}(\mathcal{R})} \le \frac{k(\mathcal{R}_{odd}) + k(\mathcal{R}_{even})}{\max\{k(\mathcal{R}_{odd}), k(\mathcal{R}_{even})\}} \le \frac{2 \cdot \max\{k(\mathcal{R}_{odd}), k(\mathcal{R}_{even})\}}{\max\{k(\mathcal{R}_{odd}), k(\mathcal{R}_{even})\}} \le 2.$$

$\square$

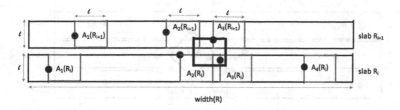

width(R)

Fig. 3. The maximum overlap of an axis-aligned fault area F with the identical axis-aligned cover squares A of same size.

Covering by Circles. Let $\mathcal{A}$ be the set of identical circles of diameter ℓ. We say that $\mathcal{A}$ completely covers all the processes if every process $p_i \in \mathcal{P}$ is covered by at least one of the circles in $\mathcal{A}$. The problem CIRCLE-COVER of completely covering processes by $\mathcal{A}$ may be formally stated similar to SQUARE-COVER in Definition 2. The following theorem, in turn, can be proven similar to Theorem 3 for SQUARE-COVER.

Theorem 4. *CIRCLE-COVER is NP-Complete.*

A Greedy Circle Cover Algorithm. We call this algorithm *GCIRCLE*. Select the square cover set $\mathcal{A}$ as produced by *GSQUARE*. Consider an individual square $A \in \mathcal{A}$. For each side of A, find a midpoint and place a circle of diameter ℓ there. Observe that thus placed four circles completely cover the area of the square A.

Lemma 8. *Let $k^C_{greedy}(\mathcal{R})$ be the number of circles C of diameter ℓ needed to cover all the processes in $\mathcal{P}$ by algorithm GCIRCLE. Let also $k^C_{opt}(\mathcal{R})$ be the minimum number of such circles used by any algorithm. Then, $k^C_{greedy}(\mathcal{R}) \leq 8 \cdot k^C_{opt}(\mathcal{R})$.*

See [23] for the proof of the lemma.

Overlapping Fault Area. The adversary may place the fault area F in any location in the plane. This means that F may not necessarily be axis-aligned. Algorithms *GSQUARE* and *GCIRCLE* produce a cover set $\mathcal{A}$ of axis-aligned squares and circles, respectively. The algorithm we present in the next section needs to know how many areas in $\mathcal{A}$, fault area F overlaps. We now compute the bound for this number. The bound considers both square and circle areas A under various size combinations of fault and non-fault areas. The lemma below is for each $A \in \mathcal{A}$ and F being either squares of side ℓ or circles of diameter ℓ.

Lemma 9. *For the processes of $\mathcal{P}$, consider the cover set $\mathcal{A}$ consisting of the axis-aligned square areas $A = \ell \times \ell$. Place a relocatable square area $F = \ell \times \ell$ in any orientation (not necessarily axis-aligned). F overlaps no more than 7 squares A. If the cover set consists of circles $C \in \mathcal{A}$ of diameter ℓ and F is a circle of diameter ℓ, then F overlaps no more than 28 circles C.*

Proof. Suppose F is axis-aligned. F may overlap at most two squares A horizontally. Indeed, the total width covered by two squares in $\mathcal{A}$ is $> 2\ell$ since the squares do not overlap. Meanwhile, the total width of F is ℓ. Similarly, F may overlap at most two squares vertically. Thus, F may overlap at most 4 distinct axis-aligned areas A. See Fig. 3 for illustration.

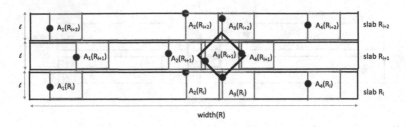

Fig. 4. The maximum overlap of a non-axis-aligned fault area F with the identical axis-aligned cover squares A of the same size.

Consider now that F is not axis-aligned. F can span at most $\sqrt{2}\ell$ horizontally and $\sqrt{2}\ell$ vertically. Therefore, horizontally, F can overlap at most three areas A. Vertically, F can overlap three areas as well. However, not all three areas on the top and bottom rows can be overlapped at once. Specifically, not axis-aligned F can only overlap 2 squares in the top row and 2 in the bottom row. Therefore, in total, F may overlap at most 7 distinct axis-aligned areas. Figure 4 provides an illustration.

Let us consider circular F of size ℓ. It can be inscribed in a square of side ℓ. The square may overlap at most 7 square areas A of side ℓ. Therefore, a circular F can also overlap at most 7 squares. One square area A can be completely covered by 4 circles C. Hence, the circular F may overlap at most $7 \times 4 = 28$ circles C. □

The first lemma below is for each A being an axis-aligned square of side ℓ or a circle of diameter ℓ while F being either a square of side $\ell/\sqrt{2}$ or a circle of diameter $\ell/\sqrt{2}$. The second lemma below considers circular fault area F of diameter $\sqrt{2}\ell$.

Lemma 10. *For the processes in $\mathcal{P}$, consider the cover set $\mathcal{A}$ consisting of the axis-aligned squares $A = \ell \times \ell$. Place a relocatable square area $F = \ell/\sqrt{2} \times \ell/\sqrt{2}$ in any orientation (not necessarily axis-aligned). F overlaps no more than 4 squares A. If the cover set $\mathcal{A}$ consists of circles C of diameter ℓ each, and F is a circle of diameter $\ell/\sqrt{2}$, then F overlaps no more than 16 circles C.*

See [23] for the proof of the lemma.

Lemma 11. *For the processes in $\mathcal{P}$, consider the cover set $\mathcal{A}$ consisting of the axis-aligned square areas $A = \ell \times \ell$. Place a relocatable circular fault area F of diameter $\sqrt{2}\ell$. F overlaps no more than 8 squares A. If $\mathcal{A}$ consists of circles C of diameter ℓ, then circular F of diameter $\sqrt{2}\ell$ overlaps no more than 32 circles C.*

See [23] for the proof of the lemma.

5   Geoconsensus Algorithm *GENERIC*

We are now ready to present an algorithm for solving Geoconsensus that we call *GENERIC*. *GENEREC* follows the same logic as *BASIC* but uses the *GSQUARE* or *GCIRCLE* algorithms, described in the previous section, to obtain the coverage of

---

Algorithm 2: Geoconsensus algorithm *GENERIC*.

---

1 **Setting:** A set $\mathcal{P}$ of N processes positioned at distinct planar coordinates. Each process can communicate with all other processes and knows the coordinates of all other processes. The processes covered by a fault area F at unknown location may be Byzantine. There are $M \geq 1$ of identical fault areas F and processes know M.

2 **Input:** Each process has initial value either 0 or 1.

3 **Output:** Each correct process outputs decision subject to Geoconsensus

4 *Procedure for process p_k*

5 // leaders selection

6 compute the set $\mathcal{A}$ of covers $A_j(R_i)$ using either *GSQUARE* or *GCIRCLE*;

7 **for** every cover $A_j(R_i) \in \mathcal{A}$ **do**

8     $\mathcal{P}_{min} \leftarrow$ a set of processes with minimum y-coordinate among covered by $A_j(R_i)$;

9 **if** $|\mathcal{P}_{min}| = 1$ **then**

10     $l_j(A_j(R_i)) \leftarrow$ the only process in $\mathcal{P}_{min}$;

11 **else**

12     $l_j(A_j(R_i)) \leftarrow$ the process in $\mathcal{P}_{min}$ with minimum x-coordinate;

13 // consensus

14 Let P_L be the set of leaders, one for each $A_j(R_i) \in \mathcal{A}$;

15 **if** $p_k \in \mathcal{P}_L$ **then**

16   run *PSL* algorithm, achieve decision v, broadcast v, output v

17 **else**

18   wait for messages with identical decision v from at least $2M + 1$ processes from $\mathcal{P}_L$, output v

---

processes in $\mathcal{P}$ by a set $\mathcal{A}$. $\mathcal{A}$ is a set of axis-aligned squares separated such that at most a bounded number of them can be covered by a fault area. A single process per square then participates in the classic consensus.

The pseudocode for *GENERIC* is given in Algorithm 2. The algorithm operates as follows. Each process p_k computes a set $\mathcal{A}$ of covers $A_j(R_i)$ that are of same size as F. Then p_k determines the leader $l_j(A_j(R_i))$ in each cover $A_j(R_i)$. The process in $A_j(R_i)$ with smallest y-coordinate is selected as a leader. If there exist two processes with the same smallest y-coordinate, then the process with the smaller x-coordinate between them is picked. If p_k is selected leader, it participates in running *PSL* [24] (or any other Byzantine consensus algorithm). The leaders run *PSL* then broadcast the achieved decision. The non-leader processes adopt it.

Analysis of *GENERIC*. Let us discuss the correctness and fault-tolerance guarantees of *GENERIC*. In all theorems of this section, *GENERIC* achieves the solution in $M + 2$ communication rounds. The proof for this claim is similar to that for *BASIC* in Theorem 2. Let the fault area $F = \ell \times \ell$ be a, not necessarily axis-aligned, square.

Theorem 5. *Given a set $\mathcal{P}$ of N processes and one square area F positioned at an unknown location such that any process of $\mathcal{P}$ covered by F may be Byzantine. Algorithm GENERIC solves Geoconsensus with the following guarantees:*

- If $F = \ell \times \ell$ and not axis-aligned and $A = \ell \times \ell$, $f \leq N - 15$ faulty processes can be tolerated given that $|\mathcal{A}| \geq 22$.
- If $F = \ell \times \ell$ and axis-aligned and $A = \ell \times \ell$, $f \leq N - 9$ faulty processes can be tolerated given that $|\mathcal{A}| \geq 13$.
- If $F = \ell/\sqrt{2} \times \ell/\sqrt{2}$ but $A = \ell \times \ell$, then even if F is not axis aligned, $f \leq N - 9$ faulty processes can be tolerated given that $|\mathcal{A}| \geq 13$.

Proof. We start by proving the first case. *GSQUARE* produces the cover set $\mathcal{A}$ of at least $|\mathcal{A}| = 22$ areas. From Lemma 9, we obtain that a square fault area $F = \ell \times \ell$, regardless of orientation and location, can overlap at most $n(F) = 7$ axis-aligned squares $A = \ell \times \ell$. *GENERIC* runs *PSL* algorithm using the single leader process in each area A. For its correct operation, *PSL* requires the number of correct processes to be more than twice the number of faulty ones. This is guaranteed since at least $2 \cdot |\mathcal{A}|/3 + 1 = 2 \cdot 22/3 + 1 \geq 2 \cdot n(F) + 1 = 2 \cdot 7 + 1$ leader processes are correct and they can reach consensus using *PSL*.

Let us address the second case. An axis-aligned square F can overlap at most $n(F) = 4$ axis-aligned squares A. Therefore, when $|\mathcal{A}| \geq 13$, we have that $|\mathcal{A}| - 9 \geq 2 \cdot n(F) + 1$ leader processes are correct and they can reach consensus. In this case, $f \leq N - 9$ processes can be covered by F and still they all can be tolerated.

Let us now address the third case, when $F = \ell/\sqrt{2} \times \ell/\sqrt{2}$ but $A = \ell \times \ell$. Regardless of its orientation, F can overlap at most $n(F) = 4$ squares A. Therefore, $|\mathcal{A}| \geq 13$ is sufficient for consensus and total $f \leq N - 9$ processes can be tolerated. □

For the set $\mathcal{F}$ of multiple fault areas F with $|\mathcal{F}| = M$, Theorem 5 extends as follows.

Theorem 6. *Given a set $\mathcal{P}$ of N processes and a set of $M \geq 1$ of square areas F positioned at unknown locations such that any process of $\mathcal{P}$ covered by any F may be Byzantine. Algorithm GENERIC solves Geoconsensus with the following guarantees:*

- *If each $F = \ell \times \ell$ and not axis-aligned and $A = \ell \times \ell$, $f \leq N - 15M$ faulty processes can be tolerated given that $|\mathcal{A}| \geq 22M$.*
- *If each $F = \ell \times \ell$ and axis-aligned and $A = \ell \times \ell$, $f \leq N - 9M$ faulty processes can be tolerated given that $|\mathcal{A}| \geq 13M$.*
- *If each $F = \ell/\sqrt{2} \times \ell/\sqrt{2}$ but $A = \ell \times \ell$, then even if F is not axis-aligned, $f \leq N - 9M$ faulty processes can be tolerated given that $|\mathcal{A}| \geq 13M$.*

Proof. The proof for the case of $M = 1$ extends to the case of $M > 1$ as follows. Theorem 5 gives the bounds $f \leq N - \gamma$ and $|\mathcal{A}| \geq \delta$ for one fault area for some positive integers γ, δ. For M fault areas, M separate $|\mathcal{A}|$ sets are needed, with each set tolerating a single fault area F. Therefore, the bounds of Theorem 5 extend to multiple fault areas with a factor of M, i.e., *GENERIC* needs $M \cdot \delta$ covers and $f \leq N - M \cdot \gamma$ faulty processes can be tolerated. Using the appropriate numbers from Theorem 5 provides the claimed bounds. □

We have the following theorem for the case of circular fault set $\mathcal{F}$, $|\mathcal{F}| = M \geq 1$.

Theorem 7. *Given a set $\mathcal{P}$ of N processes and a set of $M \geq 1$ circles F positioned at unknown locations such that any process of $\mathcal{P}$ covered by F may be Byzantine. Algorithm GENERIC solves Geoconsensus with the following guarantees:*

- *If each F and A are circles of diameter ℓ, $f \leq N - 57M$ faulty processes can be tolerated given that $|\mathcal{A}| \geq 85M$.*
- *If each F is a circle of diameter $\sqrt{2}\ell$ and A is a circle of diameter ℓ, $f \leq N - 65M$ faulty processes can be tolerated given that $|\mathcal{A}| \geq 97M$.*
- *If each F is a circle of diameter $\ell/\sqrt{2}$ and A is a circle of diameter ℓ, $f \leq N - 33M$ faulty processes can be tolerated given that $|\mathcal{A}| \geq 49M$.*

See [23] for the proof of the theorem.

6 Extensions to Higher Dimensions

Our approach can be extended to solve Geoconsensus in d-dimensions, $d \geq 3$. *BASIC* extends as is. *GENERIC* runs correctly so long as we determine (i) the cover set $\mathcal{A}$ of appropriate dimension and (ii) the overlap bound – the maximum number of d-dimensional covers A that the fault area F may overlap. The bound on f then depends on M and the cover set size $|\mathcal{A}|$. In what follows, we discuss 3-dimensional space. The still higher dimensions can be treated similarly.

If $d = 3$, the objective is to cover the embedded processes of $\mathcal{P}$ by cubes of size $\ell \times \ell \times \ell$ or spheres of diameter ℓ. It can be shown that the greedy cube (sphere) cover algorithm, let us call it *GCUBE* (*GSPHERE*), provides $2^{d-1} = 4$ (16) approximation of the optimal cover. The idea is to appropriately extend the 2-dimensional slab-based division and axis-aligned square-based covers discussed in Sect. 4 to 3-dimensions with rectangular cuboids and cube-based covers. See [23] for detailed discussion. We summarize the results for cubic covers and cubic fault areas in Theorem 8.

Theorem 8. *Given a set $\mathcal{P}$ of N processes embedded in 3-d space and a set of $M \geq 1$ of cubic areas F at unknown locations, such that any process of $\mathcal{P}$ covered by F may be Byzantine. Algorithm GENERIC solves Geoconsensus with the following guarantees:*

- *If F is a cube of side ℓ and not axis-aligned and A is also a cube of side ℓ, $f \leq N - 55M$ faulty processes can be tolerated given that the cover set $|\mathcal{A}| \geq 82M$.*
- *If F is a cube of side ℓ and axis-aligned and A is also a cube of side ℓ, $f \leq N - 17M$ faulty processes can be tolerated given that $|\mathcal{A}| \geq 25M$.*
- *If F is a sphere of diameter ℓ and A is a sphere of diameter ℓ, $f \leq N - 433M$ faulty processes can be tolerated given that $|\mathcal{A}| \geq 649M$.*

7 Concluding Remarks

In light of the recent development of location-based consensus protocols, such as G-PBFT [18], we have formally defined and studied the consensus problem of processes that are embedded in a d-dimensional plane, $d \geq 2$, on fixed locations known to every other process. We have explored both the possibility as well bounds for a solution to this

Geoconsensus. Our results establish trade-offs between the three parameters N, M, and f, in contrast to the trade-off between only two parameters N and f in the Byzantine consensus literature. Our results also show the dependency of the tolerance guarantees on the shapes and alignment of the fault areas.

Acknowledgement. This research was supported in part by National Science Foundation under Grants No. CCF-1936450 and CAREER CNS-2045597.

References

1. Abd-El-Malek, M., Ganger, G.R., Goodson, G.R., Reiter, M.K., Wylie, J.J.: Fault-scalable Byzantine fault-tolerant services. ACM SIGOPS Oper. Syst. Rev. **39**(5), 59–74 (2005)
2. Adya, A., et al.: Farsite: federated, available, and reliable storage for an incompletely trusted environment. ACM SIGOPS Oper. Syst. Rev. **36**(SI), 1–14 (2002)
3. Ben-Or, M., Kelmer, B., Rabin, T.: Asynchronous secure computations with optimal resilience. In: Proceedings of the Thirteenth Annual ACM Symposium on Principles of Distributed Computing, pp. 183–192 (1994)
4. Bulusu, N., Heidemann, J., Estrin, D., Tran, T.: Self-configuring localization systems: design and experimental evaluation. ACM Trans. Embed. Comput. Syst. (TECS) **3**(1), 24–60 (2004)
5. Castro, M., Liskov, B.: Practical Byzantine fault tolerance and proactive recovery. ACM Trans. Comput. Syst. (TOCS) **20**(4), 398–461 (2002)
6. Castro, M., Rodrigues, R., Liskov, B.: Base: using abstraction to improve fault tolerance. ACM Trans. Comput. Syst. (TOCS) **21**(3), 236–269 (2003)
7. Clark, B.N., Colbourn, C.J., Johnson, D.S.: Unit disk graphs. Discret. Math. **86**(1–3), 165–177 (1990)
8. Cramer, R., Gennaro, R., Schoenmakers, B.: A secure and optimally efficient multi-authority election scheme. Eur. Trans. Telecommun. **8**(5), 481–490 (1997)
9. Douceur, J.R.: The Sybil attack. In: Druschel, P., Kaashoek, F., Rowstron, A. (eds.) IPTPS 2002. LNCS, vol. 2429, pp. 251–260. Springer, Heidelberg (2002). https://doi.org/10.1007/3-540-45748-8_24
10. Fischer, M.J., Lynch, N.A.: A lower bound for the time to assure interactive consistency. Inf. Process. Lett. **14**(4), 183–186 (1982)
11. Fowler, R.J., Paterson, M., Tanimoto, S.L.: Optimal packing and covering in the plane are NP-complete. Inf. Process. Lett. **12**(3), 133–137 (1981)
12. Hadzilacos, V., Halpern, J.Y.: Message-optimal protocols for Byzantine agreement. Math. Syst. Theory **26**(1), 41–102 (1993)
13. Hightower, J., Borriello, G.: Location systems for ubiquitous computing. Computer **34**(8), 57–66 (2001)
14. King, V., Saia, J.: Breaking the $O(n^2)$ bit barrier: scalable Byzantine agreement with an adaptive adversary. J. ACM (JACM) **58**(4), 1–24 (2011)
15. Koo, C.Y.: Broadcast in radio networks tolerating Byzantine adversarial behavior. In: PODC, pp. 275–282 (2004)
16. Kubiatowicz, J., et al.: OceanStore: an architecture for global-scale persistent storage. ACM SIGOPS Oper. Syst. Rev. **34**(5), 190–201 (2000)
17. Lamport, L., Shostak, R., Pease, M.: The Byzantine generals problem. ACM Trans. Program. Lang. Syst. **4**(3), 382–401 (1982)
18. Lao, L., Dai, X., Xiao, B., Guo, S.: G-PBFT: a location-based and scalable consensus protocol for IoT-blockchain applications. In: IPDPS, pp. 664–673 (2020)

19. Marathe, M.V., Breu, H., Hunt, H.B., III., Ravi, S.S., Rosenkrantz, D.J.: Simple heuristics for unit disk graphs. Networks **25**(2), 59–68 (1995)
20. Martin, J.P., Alvisi, L.: Fast Byzantine consensus. IEEE Trans. Dependable Secure Comput. **3**(3), 202–215 (2006)
21. Miller, A., Xia, Y., Croman, K., Shi, E., Song, D.: The honey badger of BFT protocols. In: Proceedings of the 2016 ACM SIGSAC Conference on Computer and Communications Security, pp. 31–42 (2016)
22. Moniz, H., Neves, N.F., Correia, M.: Byzantine fault-tolerant consensus in wireless ad hoc networks. IEEE Trans. Mob. Comput. **12**(12), 2441–2454 (2012)
23. Oglio, J., Hood, K., Sharma, G., Nesterenko, M.: Byzantine geoconsensus. Technical report. 2010.02436 [cs.DC], arXiv, October 2020. http://arxiv.org/abs/2010.02436
24. Pease, M., Shostak, R., Lamport, L.: Reaching agreement in the presence of faults. J. ACM **27**(2), 228–234 (1980). https://doi.org/10.1145/322186.322188
25. Pelc, A., Peleg, D.: Broadcasting with locally bounded Byzantine faults. Inf. Process. Lett. **93**(3), 109–115 (2005). https://doi.org/10.1145/322186.322188
26. Rushby, J.: Bus architectures for safety-critical embedded systems. In: Henzinger, T.A., Kirsch, C.M. (eds.) EMSOFT 2001. LNCS, vol. 2211, pp. 306–323. Springer, Heidelberg (2001). https://doi.org/10.1007/3-540-45449-7_22
27. Sousa, J., Bessani, A., Vukolic, M.: A Byzantine fault-tolerant ordering service for the hyperledger fabric blockchain platform. In: DSN, pp. 51–58. IEEE (2018)
28. Vaidya, N.H., Tseng, L., Liang, G.: Iterative approximate Byzantine consensus in arbitrary directed graphs. In: PODC, pp. 365–374 (2012)
29. Zamani, M., Movahedi, M., Raykova, M.: Rapidchain: scaling blockchain via full sharding. In: CCS, pp. 931–948 (2018)

Loosely-self-stabilizing Byzantine-Tolerant Binary Consensus for Signature-Free Message-Passing Systems

Chryssis Georgiou[1], Ioannis Marcoullis[1(✉)], Michel Raynal[2,3], and Elad M. Schiller[4]

[1] Computer Science, University of Cyprus, Nicosia, Cyprus
{chryssis,imarcoullis}@cs.ucy.ac.cy
[2] IRISA, Univ. Rennes 1, Rennes, France
michel.raynal@irisa.fr
[3] Polytechnic University, Kowloon, Hong Kong
[4] Computer Science and Engineering, Chalmers University of Technology, Gothenburg, Sweden
elad@chalmers.se

Abstract. At PODC 2014, A. Mostéfaoui, H. Moumen, and M. Raynal presented a new and simple randomized signature-free binary consensus algorithm (denoted here as MMR) that copes with the net effect of asynchrony and Byzantine behaviors. Assuming message scheduling is fair and independent from random numbers, MMR is optimal in several respects: it deals with up to t Byzantine processes, where $t < n/3$, n being the number of processes, $\mathcal{O}(n^2)$ messages, and $\mathcal{O}(1)$ expected time. The present article presents a non-trivial extension of MMR to an even more fault-prone context, namely, in addition to Byzantine processes, it considers also that the system can experience transient failures. To this end it considers self-stabilization techniques to cope with communication failures and arbitrary transient faults, *i.e.*, any violation of the assumptions according to which the system was designed to operate.

The proposed algorithm is the first loosely-self-stabilizing Byzantine fault-tolerant binary consensus algorithm suited to asynchronous message-passing systems. This is achieved via an instructive transformation of MMR to a loosely-self-stabilizing solution that can violate safety requirements with probability $\Pr = O(1/(2^M))$, where M is a predefined constant that can be set to any positive integer at the cost of $3Mn+\log M$ bits of local memory. In addition to making MMR resilient to transient faults, the obtained loosely-self-stabilizing algorithm preserves its properties of optimal resilience and termination, *i.e.*, $t < n/3$ and $\mathcal{O}(1)$ expected time. Furthermore, it only requires a bounded amount of memory.

Keywords: Binary consensus · Byzantine fault-tolerance · Self-stabilization

The work of M. Raynal was partially supported by the French ANR project ByBLoSS (20-CE25-0002-01). The work of I. Marcoullis was funded by the ONISILLOS postdoctoral funding scheme of the University of Cyprus.

K. Echihabi and R. Meyer (Eds.): NETYS 2021, LNCS 12754, pp. 36–53, 2021.
https://doi.org/10.1007/978-3-030-91014-3_3

1   Introduction

En route to constructing robust distributed systems, rose the need for different (possibly geographically dispersed) computational entities to take common decisions. Of past and recent contexts in which the need for agreement appeared, one can cherry-pick applications, such as service replication, cloud computing, load balancing, and distributed ledgers. In distributed computing, the problem of agreeing on a single value after the proposal of values by computational entities, called *nodes*, is called *consensus* [5,34]. The most basic form of the consensus problem is for nodes to decide between two possible values, *e.g.*, zero or one. This version of the problem is called *binary consensus* [43, Ch. 14]. This work aims to fortify consensus protocols with fault-tolerance guarantees that are more powerful than any existing known solution. Such solutions are imperative for many distributed systems that run in hostile environments, such as Blockchain.

1.1   Problem Definition

The problem of letting all nodes to uniformly select a single value among all the values that they propose is called consensus. When the set, V, of values that can be proposed, includes just two values, *i.e.*, $V = \{0,1\}$, the problem is called binary consensus, see Definition 1. Otherwise, it is called multivalued consensus.

Definition 1. *Every node p_i has to propose a value $v_i \in \{0,1\}$, via an invocation of the* propose$_i(v_i)$ *operation. Let Alg be an algorithm that solves binary consensus. Alg has to satisfy* safety, *i.e., BC-validity and BC-agreement, and* liveness, *i.e., BC-termination, requirements.*

– **BC-validity.**   *The value $v \in \{0,1\}$ decided by a correct node is a value proposed by a correct node.*
– **BC-agreement.**   *Any two correct nodes that decide, do so with identical decided values.*
– **BC-termination.**   *All correct nodes decide.*

Starting from the algorithm of Mostéfaoui, Moumen, and Raynal [41], from now on MMR, this study proposes an even more fault-tolerant consensus algorithm, which is a variant on MMR. Note that MMR provides randomized liveness guarantees. The termination of the proposed solution satisfies BC-termination within a constant time that depends on a predefined parameter $M \in \mathbb{Z}^+$. However, it satisfies safety with the probability of $1 - \mathcal{O}(2^{-M})$. Since the number of bits that each node needs to store is $3nM + \lceil \log M \rceil$, we note that the probability for violating safety can be made, in practice, to be extremely small, where n is the number of nodes in the system.

1.2   Fault Model

We study asynchronous solutions for message-passing systems where the algorithm cannot explicitly access the local clock or have guarantees on the communication delay. We model a broad set of benign failures that can occur to computers and networks, *e.g.*, due to procrastination, equivocation, selfishness, hostile (human) interference, deviation from the program code, etc. Specifically, our fault model includes (*i*) communication failures, such as packet omission, duplication, and reordering, as well as (*ii*) up to t node failures, *i.e.*, crashed or Byzantine. In detail, a faulty node runs the algorithm correctly but the adversary completely controls the messages that the algorithm sends, *i.e.*, it can modify the content of a message, delay the delivery of a message, or omit it altogether. The adversary's control can challenge the algorithm by creating failure patterns in which a fault occurrence appears differently to different system components. Moreover, the adversary is empowered with the unlimited ability to compute and coordinate the most severe failure patterns. We assume a known maximum number, t, of nodes that the adversary can capture. We also restrict the adversary from letting a captured node impersonate a non-faulty one. In addition, we limit the adversary's ability to impact message delivery between any two non-faulty processes by assuming fair scheduling of message arrivals.

In addition to the failures captured by our model, we also aim to recover from *arbitrary transient faults*, *i.e.*, any temporary violation of assumptions according to which the system and network were designed to operate. This includes the corruption of control variables, such as the program counter, packet payload, and indices, *e.g.*, sequence numbers, which are responsible for the correct operation of the studied system, as well as operational assumptions, such as that at least a distinguished majority of nodes never fail. Since the occurrence of these failures can be arbitrarily combined, we assume that these transient faults can alter the system state in unpredictable ways. In particular, when modeling the system, Dijkstra [17] assumes that these violations bring the system to an arbitrary state from which a *self-stabilizing system* should recover, see [3,18] for details. Dijkstra requires recovery after the last occurrence of a transient fault and once the system has recovered, it must never violate the task specification.

For the case of the studied problem, there are currently no known ways to meet Dijkstra's self-stabilizing design criteria. *Loosely-self-stabilizing systems* [45] require that, once the system has recovered, only rarely and briefly can it violate the safety specifications. Although it is a weaker design criterion than the one defined by Dijkstra, the violation occurrence can be made to be so rare, that the risk of breaking the safety requirements of Definition 1 becomes negligible.

1.3   Related Work

Impossibilities and Lower-Bounds. The FLP impossibility result [27] concluded that consensus is impossible to solve deterministically in asynchronous settings in the presence of even a single crash failure. In [26] it was shown that a lower bound of $t + 1$ communication steps are required to solve consensus deterministically in both synchronous and asynchronous environments. The proposed solution is a randomized one. In the presence of asynchrony, transient-faults, and (non-Byzantine) crash failures, there are known problems such as leader election and counting the number of nodes in the system, for which there are no (randomized) self-stabilizing solutions [2,4]. In this work, we consider weaker design criteria than Dijkstra's self-stabilization.

In the presence of Byzantine faults, consensus is not solvable if a third or more of the nodes are faulty [34]. Thus, optimally resilient Byzantine consensus algorithms, such as the one we present, tolerate $t < n/3$ faulty nodes. The task is also impossible if a node can impersonate some other node in its communication with the other entities [5]. We assume the absence of spoofing attacks and similar means of impersonation. In the presence of asynchrony, transient faults, and Byzantine failures, the task of unison is known to be unsolvable (unless the strongest fairness assumptions are made) [24,25]. As indicated by the above impossibility results, the studied problem remains challenging even under randomization and fairness assumptions during the recovery period.

Self-stabilizing non-Byzantine fault-tolerant solutions Lundström, Raynal, and Schiller [38] presented the first self-stabilizing solution for the problem of binary consensus for message-passing systems where nodes may fail by crashing. They ensure a line of self-stabilizing solutions [28,29,36,37,39]. This line follows the approach proposed by Dolev, Petig, and Schiller [30,31] for self-stabilization in the presence of seldom fairness. Namely, in the absence of transient-faults, these self-stabilizing solutions are wait-free and no assumptions are made regarding the system's synchrony or fairness of its scheduler. However, the recovery from transient faults does require fair execution, *e.g.,* to perform a global reset, see [28], but only during the recovery period. Our work does not assume execution fairness either in the presence or absence of arbitrary transient-faults. As in MMR, our loosely-self-stabilizing Byzantine fault-tolerant solution assumes fair scheduling of message arrivals and the accessibility to an independent common coin service.

We note the existence of other approaches for recovering from transient faults without assuming execution fairness during the recovery period [1,19, 44]. However, none of these results consider both Byzantine fault-tolerance and self-stabilization.

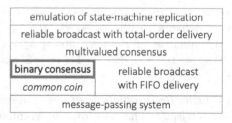

Fig. 1. The hybrid architecture of asynchronous and synchronous components

Self-stabilizing Byzantine Fault-Tolerant Solutions. In the context of this dual design criteria, there are solutions for topology discovery [21], storage [9–13], clock synchronization [22,33,35], approximate agreement [14], asynchronous unison [25] to name a few. The most relevant work is the one by Binun *et al.* [7,8] and Dolev *et al.* [20] for a deterministic Byzantine fault-tolerant emulation of state-machine replication. Binun *et al.* present the first self-stabilizing solution for synchronous message-passing systems and Dolev *et al.* present the first practically-self-stabilizing solution for partially-synchronous settings. We study another problem, which is binary consensus.

Applications. Binary consensus is a fundamental component of total order reliable broadcast, *e.g.,* [15,16]. In what appears as a revival of the topic, several Blockchain consensus protocols are also using similar approaches.

HoneyBadger [40] was the first randomized BFT protocol for Blockchain. They employ MMR as their binary consensus protocol. The BEAT [23] suite of protocols for blockchain consensus also uses the MMR.

1.4 Our Architecture of Asynchronous and Synchronous Components

A Blockchain can be seen as a replication service for state-machine emulation in extremely hostile environments. The stacking of reliable broadcast protocols can facilitate this emulation, see Fig. 1. Specifically, the order of all state transitions of the automaton can be agreed by using total order reliable broadcast. The order of the broadcasts is agreed via multivalued consensus. Whenever multivalued consensus is called, the latter invokes binary consensus several times.

Using Both Asynchronous and Synchronous Components. Existing solutions for binary consensus use either randomization techniques or synchrony assumptions in order to circumvent FLP. The system as a whole can avoid communication-related bottlenecks by making design choices that prefer weaker synchrony assumptions for the components that are more communication demanding. Binary consensus protocols are inherently communication-intensive. Therefore, we select to study the non-self-stabilizing probabilistic MMR algorithm [41] for solving binary consensus in asynchronous message-passing-systems. MMR assumes access to a common coin service. Ben-Or, Dolev, and Hoch [6], in short BDH, presented a self-stabilizing synchronous solution for common coin provision.

Enhancing the Computation Model with Common Coins. A common coin service delivers to all nodes, via the operation randomBit$(r) : r \in \mathbb{Z}^+$, identical sequences of random bits $b_1, b_2, \ldots, b_r, \ldots : b_r \in \{0, 1\}$. We assume that $\Pr(b_r = 0) = \Pr(b_r = 1) = 1/2$ and that b_r is independent of $b_{r'}$, where $r, r' \in \mathbb{Z}^+$.

Periodic Re-installation of the Common Seed and Initialization of Consensus Objects. The use of the common coin service can aim at devising a random seed that is long enough to support plenty of invocations of binary consensus. In detail, let s_t be a common seed that BDH renewed at time τ. A pseudo-random generator can use s_τ for generating the sequence $bits_{\tau,1}, bits_{\tau,2}, \ldots, bits_{\tau,x}$ of unique M-bits integers, where $M \in \mathbb{Z}^+$ is a bound on the number of pseudo-random bits each invocation of binary consensus might use and $x \in \mathbb{Z}^+$ is the highest value that guarantees that the sequence $bits_{\tau,1}, bits_{\tau,2}, \ldots, bits_{\tau,x}, bits_{\tau+1,1}, bits_{\tau+1,2}, \ldots, bits_{\tau+1,x}, \ldots$ satisfies the common coin requirements over time. We note that high x values can mitigate the effect of BDH's synchrony assumption on the benefits gained from selecting an asynchronous algorithm for solving binary consensus.

Also, it is imperative to re-install the common seed repeatedly due to the need to tolerate corruption of the common seed, after the occurrence of a transient fault. This is also imposed by the properties of pseudo-random generator functions, which eventually cannot avoid repeating the same sequences of "random" bits. Note that one can use the events of common seed re-installation also for the initialization of consensus objects. This can help to simplify the correctness proof since it implies that recovery from transient faults depends only on the termination of all operations after the occurrence of the last transient fault. Our hybrid architecture and the assumptions made above help us to circumvent the aforementioned impossibility results.

1.5 Our Contribution

We present a fundamental module for dependable distributed systems: a loosely-self-stabilizing asynchronous algorithm for binary consensus for message-passing systems that are prone to Byzantine node failures. We obtain this new algorithm via a transformation of the non-self-stabilizing probabilistic MMR. The proposed algorithm preserves MMR's elegant properties, such as optimal reliance and termination within a constant expected time (without assuming fair execution). After the occurrence of the last transient fault, the system recovers within a bounded time (while assuming fair execution). Unlike in MMR, each node uses a bounded amount of memory of $3nM + \lceil \log M \rceil$, where $M \in \mathbb{Z}^+$ and n is the number of nodes. This implies that with a probability in $\mathcal{O}(2^{-M})$ the safety requirement of Definition 1 can be violated. However, by selecting a sufficiently large value of M, the risk of violating the safety requirements becomes negligible at affordable costs.

   In the absence of transient-faults, our solution achieves consensus without assuming execution fairness. After the occurrence of any finite number of arbitrary transient-faults, the system recovers within a finite time (while assuming fairness).

   To the best of our knowledge, it is *the first* loosely-self-stabilizing Byzantine fault-tolerant asynchronous algorithm for solving binary consensus in message-passing systems. As such, there is a long line of distributed applications, such as service replication and Blockchain, that our contribution can facilitate solutions that are more fault-tolerant than the existing implementations since they cannot recover after the occurrence of the last transient fault.
Due to the page limit, some details are omitted and can be found in the complementary technical report [32].

2 System Settings

We consider an asynchronous message-passing system that has no guarantees on the communication delay or access to clock-related mechanisms, *e.g.*, timeouts. The system consists of a set, $\mathcal{P}$, of n fail-prone nodes with unique identifiers. Any pair of nodes $p_i, p_j \in \mathcal{P}$ has access to a bidirectional communication channel, $channel_{j,i}$, that, at any time, has at most channelCapacity $\in \mathbb{Z}^+$ packets on transit from p_j to p_i (this is due to an impossibility [18, Chapter 3.2]).

   In the *interleaving model* [18], the node's program is a sequence of *(atomic) steps*. Each step starts with an internal computation and finishes with a single communication operation, *i.e.*, a message *send* or *receive*. The state, s_i, of node $p_i \in \mathcal{P}$ includes all of p_i's variables and $channel_{j,i}$. The term *system state* (or configuration) refers to the tuple $c = (s_1, s_2, \cdots, s_n)$. We define an *execution* *(or run)* $R = c[0], a[0], c[1], a[1], \ldots$ as an alternating sequence of system states $c[x]$ and steps $a[x]$, such that each $c[x+1]$, except for the starting one, $c[0]$, is obtained from $c[x]$ by $a[x]$'s execution.

2.1   Task Specifications

Definition 1 considers the propose(v) operation. We refine the definition of propose(v) by specifying how the decided value is retrieved. This value is either returned by the propose() operation, as in MMR, or via the returned value of the result() operation, as in the proposed solution. In the latter case, the symbol $\perp$ is returned as long as no value was decided. Also, the symbol Ψ indicates a (transient) error that occurs only when the proposed algorithm exceeds the bound on the number of iterations that it may take.

Legal Executions. The set of *legal executions* (*LE*) refers to all the executions in which the requirements of the task T hold. In this work, T_{binCon} denotes the task of binary consensus, which Definition 1 specifies, and LE_{binCon} denotes the set of executions in which the system fulfills T_{binCon}'s requirements.

2.2   The Fault Model and Self-stabilization

Communication Failures and Fairness. We consider solutions that are oriented towards asynchronous message-passing systems and thus they are oblivious to the time at which the packets arrive and depart. The communication channels are prone to packet failures, such as omission, duplication, and reordering. However, if p_i sends a message infinitely often to p_j, node p_j receives that message infinitely often. We refer to the latter as the *fair communication* assumption. We also follow the assumption of MMR regarding the fair scheduling of message arrivals (also in the absence of transient faults) that does not depend on the current coin's value.

Arbitrary Node Failures. Byzantine faults model any fault in a node including crashes, arbitrary behavior, and malicious behaviors. Here the adversary lets each node receive the arriving messages and calculate its state according to the algorithm. However, once a node (that is captured by the adversary) sends a message, the adversary can modify the message in any way, delay it for an arbitrarily long period or even remove it from the communication channel. Note that the adversary has the power to coordinate such actions without any limitation about his computational or communication power. For the sake of solvability [34, 42, 46], the number, t, of Byzantine failure needs to be less than one-third of the number, n, of nodes in the system, *i.e.*, $3t+1 \leq n$. The set of non-faulty nodes is denoted by *Correct* and called the set of correct nodes.

Arbitrary Transient Faults. We consider any temporary violation of the assumptions according to which the system was designed to operate. We refer to these violations and deviations as *arbitrary transient faults* and assume that they can corrupt the system state arbitrarily (while keeping the program code intact). The occurrence of an arbitrary transient fault is rare. Thus, our model assumes that the last arbitrary transient fault occurs before the system execution starts [18]. Also, it leaves the system to start in an arbitrary state.

Dijkstra's Self-stabilization. An algorithm is *self-stabilizing* with respect to the task of LE, when every (unbounded) execution R of the algorithm reaches within a finite period a suffix $R_{legal} \in LE$ that is legal. Namely, Dijkstra [17] requires $\forall R : \exists R' : R = R' \circ R_{legal} \wedge R_{legal} \in LE \wedge |R'| \in \mathbb{Z}^+$, where the operator $\circ$ denotes that $R = R' \circ R''$ is the concatenation of R' with R''. The part of the proof that shows the existence of R' is called the *convergence* (or recovery) proof, and the part that shows that $R_{legal} \in LE$ is called the *closure* proof.

Execution Fairness and Wait-Free Guarantees. We say that a system execution is *fair* when every step of a correct node that is applicable infinitely often is executed infinitely often and fair communication is kept. Self-stabilizing algorithms often assume that their executions are fair [18]. Wait-free algorithms guarantee that operations (that were invoked by non-failing nodes) always terminate in the presence of asynchrony and any number of node failures. This work assumes execution fairness during the period in which the system recovers from the occurrence of the last arbitrary transient fault. In other words, the system is wait-free only during legal executions, which are absent from arbitrary transient-faults. Moreover, the system recovery from arbitrary transient-faults is not wait-free, but this bounded recovery period occurs only once throughout the system execution.

Loosely-Self-stabilizing Systems. Satisfying the design criteria of Dijkstra's self-stabilizing systems is non-trivial since it is required to eventually satisfy strictly always the task's specifications. These severe requirements can lead to some impossibility conditions, as in our case of solving binary consensus without synchrony assumptions [2, 25, 26].

   To circumvent such challenges, Sudo et al. [45] proposed the design criteria for loosely-self-stabilizing systems, which relaxes Dijkstra's criteria by requiring that, starting from any system state, the system (i) reaches a legal execution within a relatively short period, and (ii) remains in the set of legal for a relatively long period. The definition of loosely-self-stabilizing systems by Sudo *et al.* considers the task of leader election, which any system state may, or may not, satisfy. This paper focuses on an operation-based task that has both safety and liveness requirements. Only at the end of the task execution, can one observe whether the safety requirements were satisfied. Thus, Definition 3 presents a variation of Sudo *et al.*'s definition that is operation-based and requires criterion (i) to hold within a finite time rather than within 'a short period'.

   To that end, Definition 2 says what it means for a system S that implements operation op() to satisfy task $T_{op()}$'s safety requirements with a probability p_S. Definition 2 uses the term correct invocation of operation op(). Recall that Sect. 2.1, defines what a correct invocation of binary consensus is, *i.e.*, it is required that all correct processors invoke the propose() operation exactly once during any execution that is in LE_{binCon}.

Definition 2 (Probabilistic satisfaction of repeated invocations of operation op()). *For a given system S that aims at satisfying task $T_{op()}$ in a probabilistic manner, denote by $IE_S(LE_{op()})$ the set of all infinite executions that system S can run, such that for any $R \in IE_S(LE_{op()})$ it holds that $R = R_1 \circ R_2 \circ, \ldots$ is an infinite composition of finite executions, $R_1, R_2, \ldots \in LE_{op()}$. Moreover, each $R_x : x \in \mathbb{Z}^+$ includes the correct invocation of op() that always satisfies $T_{op()}$'s liveness requirements.*

We say that R satisfies task $T_{op()}$'s safety requirements with the probability $\Pr_R$ if (i) for any $x \in \mathbb{Z}^+$ it holds that $R_x \in LE_{op()}$ with probability $\Pr_{R_x} \leq \Pr_R$ and (ii) for any $x, y \in \mathbb{Z}^+$ the event of $R_x \in LE_{op()}$ and $R_y \in LE_{op()}$ are independent. Furthermore, we say system S satisfies task $T_{op()}$ with the probability $\Pr_S$ if $\forall R \in IE_S(LE_{op()}) : \Pr_R \leq \Pr_S$.

Definition 3 (Probabilistic (operation-based) eventually-loosely-self-stabilizing systems). *Let S be a system that implements a probabilistic solution for task $T_{op()}$. Let R be any unbounded execution of S, which includes repeated sequential and correct invocations of op(), such that task $T_{op()}$ terminates within a period of ℓ_S steps in R. Suppose that within a finite number of steps in R, the system S reaches a suffix of R that satisfies $T_{op()}$'s safety requirements with the probability $\Pr_S = 1 - p : p \in o(\ell_S)$. In this case, we say that system S is eventually-loosely-self-stabilizing, where ℓ_S is the complexity measure.*

Definition 3 says that any eventually-loosely-self-stabilizing system recovers within a finite period. After that period, the probability to violate safety-requirement is exponentially small. This work shows that the studied algorithm has an eventually-loosely-self-stabilizing variation for which the probability to violate safety can be made so low that it becomes negligible.

3   The Studied Non-self-stabilizing Solution

We present a non-self-stabilizing algorithm that serves as a stepping stone to the proposed algorithm (Sect. 4). Algorithm 1 (Algorithm 2 in the complementary technical report [32]) has a bound, M, on the number of communication iterations; this algorithm is a bounded version of the studied algorithm MMR. A detailed review of MMR can be found in [32] (given as Algorithm 1 there).

Variables. Algorithm 1 uses variable r (initialized to zero) for counting the number of iterations of the do-forever loop (lines 9 to 19), which we refer to as *asynchronous (communication) round*. During round r, every node $p_i \in \mathcal{P}$ stores in the set $est_i[r][i]$ its estimated decision values, where $est_i[0][i] = \{v\}$ stores its own proposal and $est_i[M+1][i]$ aims to hold the decided value. Since nodes exchange these estimates, $est_i[r][j]$ stores the last estimate that p_i received from p_j. Note that $est_i[r][j] \subseteq \{0, 1\}$ holds a set of values and it is initialized by the empty set, $\emptyset$. At the end of round r, node $p_i \in \mathcal{P}$ tests whether it is ready to decide after it selects a single value $w \in est_i[r][i]$ to be exchanged with other nodes. In order to ensure reliable broadcast in the presence of packet loss, there is a need to store w in auxiliary storage, $aux_i[r][i]$, so that p_i can retransmit w. Note that all entries in $aux[][]$ are initialized to $\perp$.

Algorithm 1: Non-stabilizing Byzantine-tolerant binary consensus using M iterations and $o(1)$ safety violation probability; code for p_i.

---

1 **operations:** propose(v) **do** $\{(est[0][i], aux[0][i] \leftarrow (\{v\}, \bot)\}$;

2 result() **do** $\{$**if** $(est[M+1][i] = \{v\})$ **then return** v **else if** $(r \geq M \land$ infoResult() $\neq \emptyset)$ **then return** Ψ**else return** $\bot$;$\}$

3 **macros:** $binValues(r, x)$ **return**
   $\{y \in \{0,1\} : \exists s \subseteq \mathcal{P} : |\{p_j \in s : y \in est[r][j]\}| \geq x\}$;

4 infoResult() **do** $\{$**if** $(\exists s \subseteq \mathcal{P} : n{-}t \leq |s| \land (\forall p_j \in s : aux[r][j] \in binValues(r, 2t+1)))$ **then return** $\{aux[r][j]\}_{p_j \in s}$ **else return** $\emptyset\}$;

5 **functions:** decide(x) $\{$**if** $(est[M+1][i] = \emptyset \lor aux[M+1][i] = \bot)$ **then** $(est[M+1][i], aux[M+1][i]) \leftarrow (\{x\}, x)\}$

6 tryToDecide($values$) **begin**

7 $\quad$ **if** $(values \neq \{v\})$ **then** $est[r][i] \leftarrow \{$**randomBit**$(r)\}$;

8 $\quad$ **else** $\{est[r][i] \leftarrow \{v\}$; **if** $(v = $ **randomBit**$(r))$ **then** decide(v)$\}$;

9 **do forever begin**

10 $\quad$ **if** $(est[0][i] \neq \emptyset)$ **then**

11 $\quad\quad$ $r \leftarrow \min\{r+1, M\}$;

12 $\quad\quad$ **repeat**

13 $\quad\quad\quad$ **foreach** $p_j \in \mathcal{P}$ **do send**
   EST(True, r, $est[r{-}1][i] \cup binValues(r, t+1)$) **to** p_j

14 $\quad\quad\quad$ **if** $(\exists w \in binValues(r, 2t+1))$ **then** $aux[r][i] \leftarrow w$;

15 $\quad\quad$ **until** $aux[r][i] \neq \bot$;

16 $\quad\quad$ **repeat**

17 $\quad\quad\quad$ **foreach** $p_j \in \mathcal{P}$ **do send** AUX(True, r, $aux[r][i]$) **to** p_j

18 $\quad\quad$ **until** infoResult() $\neq \emptyset$;

19 $\quad\quad$ tryToDecide(infoResult());

20 **upon** EST(aJ, rJ, vJ) **arrival from** p_j **do begin**

21 $\quad$ $est[rJ][j] \leftarrow est[rJ][j] \cup vJ$;

22 $\quad$ **if** (aJ) **then send** EST(False, rJ, $est[rJ{-}1][i]$) **to** p_j;

23 **upon** AUX(aJ, rJ, vJ) **arrival from** p_j **do begin**

24 $\quad$ **if** $(vJ \neq \bot)$ **then** $aux[rJ][j] \leftarrow vJ$;

25 $\quad$ **if** (aJ) **then send** AUX(False, rJ, $aux[rJ][i]$) **to** p_j;

Detailed Description. Algorithm 1 includes the following three stages.

1. *Invocation.* An invocation of operation propose(v) (line 1) initializes $est_i[0][i]$ with the estimated value v. No communication or decision occurs before such an invocation occurs. These actions are only possible through the lines enclosed in the do forever loop (lines 9 to 19). These lines are not accessible before such an invocation, because of the condition of line 10. Each iteration of the do forever loop is initiated with a round increment (line 11); this line ensures that r is bounded by M.

2. *Communication.* The first communication phase, which queries the estimated binary values, is implemented in the repeat-until loop of lines 12–15. The receiver's side of this communication is given in the code of lines 20–22. Similarly, the second communication phase, which informs about the query results through the use of auxiliary messages, is given in the repeat-until loop of lines 20–22. Lines 23–25 are the receiver side's actions for this phase.

3. *Decision.* The decision phase (line 19) is a call to tryToDecide(). Lines 6 to 8 are the implementation of tryToDecide(). This exactly maps the *Try-to-decide* phase of MMR: (i) If the *values* set that was composed of the auxiliary messages that were received is a single value, then this is the estimate of the next round. (ii) If this is also the output of randomBit() then this is the value to be decided. (iii) If *values* is not a single value then the estimate for the next round is the randomBit() output. The actual decision action (line 5) is for both $est[M + 1][i]$ and $aux[M + 1][i]$ to be assigned the decided value.

As specified in Sect. 2.1, result() (line 2) aims to return the decided value. However, the $\perp$ symbol is returned when no value was decided. Also, it indicates whether r has exceeded the limit M, in which case it returns the error symbol, Ψ, laying the ground for the proposed algorithm presented in Sect. 4 (Algorithm 2).

Bounding the Number of Iterations. Algorithm 1 preallocates $\mathcal{O}(M)$ of memory space for every node in the system, where $M \in \mathbb{Z}^+$ is a predefined constant that bounds the maximum number of iterations that Algorithm 1 may take. Lemma 1 shows that Algorithm 1 may exceed the limit M with a probability that is in $\mathcal{O}(2^{-M})$. Once that happens, Definition 1's safety requirements can be violated.

Lemma 1. *By the end of round r, with probability $\Pr(r) = 1 - (1/2)^r$, we have* result$_i() \in \{0, 1\} : p_i \in \mathcal{P} : i \in Correct.$

Proof Sketch of Lemma 1. We show that $\exists v \in \{0, 1\} : \forall i \in Correct$: $est_i[r][i] = \{v\}$ holds with the probability $\Pr(r) = 1 - (1/2)^r$. Let $values_i^r$ be the parameter that p_i passes to tryToDecide() (line 6) on round r. If $\forall k \in Correct : values_i^r = \{0, 1\}$ or $\forall k \in Correct : values_i^r = \{v_k(r)\}$ hold, p_k assigns the same value to $est_k[r][k]$, which is $\{$randomBit$_k(r)\}$, and respectively, $v_k(r)$. The remaining case is when some correct nodes assign $\{v_k(r)\}$ to $est_k[r][k]$ (line 8), whereas others assigns $\{$randomBit$_k(r)\}$ (line 7).

Recall the assumption that the Byzantine nodes have no control over the network or its scheduler. Due to the common coin properties, randomBit$_k(r)$ and randomBit$_k(r')$ are independent, where $r \neq r'$. The assignments of $\{v_k(r)\}$ and $\{$randomBit$_k(r)\}$ are equal with the probability of $\frac{1}{2}$. Thus, $\Pr(r)$ is the probability that $[\exists r' \leq r : $ randomBit$(r) = v(r)] = \frac{1}{2} + (1 - \frac{1}{2})\frac{1}{2} + \cdots + (1 - \frac{1}{2})^{r-1}\frac{1}{2} = 1 - (\frac{1}{2})^r$.

The complete proof shows that the repeat-until loop in lines 12 to 15 cannot block forever and all the correct nodes p_i keep their estimate value $est_i = \{v\}$ and consequently the predicate ($values_i^{r'} = \{v\}$) at line 7 holds for round r', where $values_i^{r'} = \cup_{j \in s}\{aux_i[r][j]\}$. With probability $\Pr(r) = 1 - (1/2)^r$, by round r, randomBit$(r) = v$ holds. Then, the if-statement condition of line 7 does not hold and the one in line 8 does hold. Thus, all the correct nodes decide v. $\square_{Lemma\ 1}$

Algorithm 2: Loosely-stabilizing Byzantine-tolerant binary consensus using M iterations and $o(1)$ safety violation probability; code for p_i.

---

26 **constants:** $initState := (0, [[\emptyset, \ldots], \ldots, [\emptyset, \ldots]], [[\bot, \ldots], \ldots, [\bot, \ldots]]);$

27 **operations:** propose(v) **do** $\{(r, est, aux) \leftarrow initState; est[0][i] \leftarrow \{v\}\};$

28 result() **do** {**if** $(est[M{+}1][i] = \{v\})$ **then return** v **else if** $(r \geq M \wedge$ infoResult$() \neq \emptyset)$ **then return**Ψ**else return** $\bot;$}

29 **macros:** $binValues(r, x)$ **return**
   $\{y \in \{0,1\} : \exists s \subseteq \mathcal{P} : |\{p_j \in s : y \in est[r][j]\}| \geq x\};$

30 infoResult() **do** {**if** $(\exists s \subseteq \mathcal{P} : n{-}t \leq |s| \wedge (\forall p_j \in s : aux[r][j] \in binValues(r, 2t{+}1)))$ **then return** $\{aux[r][j]\}_{p_j \in s}$ **else return** $\emptyset;$}

31 **functions:** decide(x) {**foreach**$r' \in \{r, \ldots, M{+}1\}$ **do if**
   $(est[r'][i] = \emptyset \vee aux[r'][i] = \bot)$ **then** $(est[r'][i], aux[r'][i]) \leftarrow (\{x\}, x)\}$

32 tryToDecide$(values)$ **begin**

33 | **if** $(values \neq \{v\})$ **then** $est[r][i] \leftarrow \{\textbf{randomBit}(r)\};$

34 | **else** $\{est[r][i] \leftarrow \{v\};$ **if** $(v = \textbf{randomBit}(r))$ **then** decide$(v)\};$

35 **do forever begin**

36 | **if** $((r, est, aux) \neq initState)$ **then**

37 | $r \leftarrow \min\{r{+}1, M\};$

38 | **repeat**

39 | **if** $(est[0][i] \neq \{v\})$ **then** $est[0][i] \leftarrow \{w\} : \exists w \in est[0][i];$

40 | **foreach** $r' \in \{1, \ldots, r{-}1\} : est[r'][i] = \emptyset \vee aux[r'][i] = \bot$ **do**

41 | $(est[r'][i], aux[r'][i]) \leftarrow (est[0][i], x) : x \in est[0][i];$

42 | **if** $((\exists w \in binValues(r, 2t{+}1) \wedge (aux[r][i] = \bot \vee aux[r][i] \notin binValues(r, 2t{+}1)))$ **then** $aux[r][i] \leftarrow w;$

43 | **foreach** $p_j \in \mathcal{P}$ **do send**
   EST(True, $r, est[r{-}1][i] \cup binValues(r, t{+}1), aux[r][i])$ **to** $p_j;$

44 | **until** infoResult$() \neq \emptyset;$

45 | tryToDecide(infoResult());

46 **upon** EST(aJ, rJ, vJ, uJ) **arrival from** p_j **begin**

47 | $est[rJ][j] \leftarrow est[rJ][j] \cup vJ; aux[rJ][j] \leftarrow uJ;$

48 | **if** aJ **then send** EST(False, $rJ, est[rJ{-}1][i], aux[r][i])$ **to** $p_j;$

---

4   The Proposed Self-stabilizing Solution

Algorithm 2 (Algorithm 3 in [32]) presents a solution that can recover from transient faults. We demonstrate its correctness in Sect. 5. The main concern that we have when designing a loosely-self-stabilizing version of MMR is to make sure that no transient fault can cause the algorithm to not terminate, *e.g.*, block forever in one of the repeat-until loops in lines 13 to 15 and 17 to 18 of Algorithm 1.

Recall that Algorithm 1 is a code transformation of MMR [41] that runs for M iterations and violates Definition 1's safety requirement with a probability that is in $\mathcal{O}(2^{-M})$. The proposed solution appears in Algorithm 2. We obtain this solution via code transformation from Algorithm 1. The latter transformation aims to offer recovery from transient faults.

Note that a transient fault can corrupt the state of $p_i \in \mathcal{P}$ by, *e.g.*, setting $est_i[i]$ with $\{0, 1\}$. Line 39 addresses this concern. Another case of state corruption is when the round counter, r_i, equals to r, but there is $r' < r$ and entries $est_i[r']$ or $aux_i[r']$ that point to their initial values *i.e.*, $\exists r' \in \{1, \dots, r-1\}$: $est_i[r'][i] = \emptyset \vee aux_i[r'][i] = \bot$. Line 39 addresses this concern. Since we wish not that the for-each condition in line 40 to hold when a correct node decides, line 31 makes sure that all entries of $est[r']$ and $aux[r']$ store the decided value, where r' is any round number that is between the current round number, r, and $M{+}1$, which is the entry that stores the decided value.

The last concern that Algorithm 2 needs to address is the fact that the repeat-until loop in lines 17 to 18 of Algorithm 1 depends on the assumption that $aux_i[r][i] \neq \bot$, which is supposed to be fulfilled by the repeat-until loop in lines 13 to 15. However, a transient fault can place the program counter to point at line 17 without ever satisfying the requirement of $aux_i[r][i] \neq \bot$. Therefore, Algorithm 2 combines in lines 42 to 43 the repeat-until loops of lines 13 to 15 and 17 to 18 of Algorithm 1. A similar combination occurs at the upon events.

5   Correctness

We show that Algorithm 2 cannot block forever even in the presence of transient faults and that the safety requirements of Definition 1 hold with probability $\Pr(r) = 1 - (1/2)^{-M}$. Due to the page limit, the detailed version of the proof can be found in the complementary technical report [32].

5.1   Transient Fault Recovery

We say that a system state c is *resolved* if $\forall i \in Correct : \big|est_i[0][i]\big| \in \{0, 1\} \wedge \nexists r' \in \{1, \dots, r-1\}$: $est_i[r'][i] = \emptyset \vee aux_i[r'][i] = \bot$. Suppose that during execution R, every $p_i \in \mathcal{P} : i \in Correct$ invokes $\mathsf{propose}_i()$ exactly once. In this case, we say that R includes a *complete invocation* of binary consensus. Theorem 1 shows recovery to resolved system states and termination during executions that include a complete invocation of binary consensus. The statement of Theorem 1 uses the term *active* for node $p_i \in \mathcal{P}$ when referring to the case of $est_i[0][i] \neq initState$.

Theorem 1 shows that the system recovers within a finite time after the occurrence of the last arbitrary transient-fault. Recall that we assume execution fairness after the occurrence of that fault (Sect. 2.2).

Theorem 1 (Convergence). *Let R be an execution of Algorithm 2. (i) Within one complete asynchronous round, the system is resolved. Also, suppose that throughout R all correct nodes are active. (ii) Eventually, for every $i \in Correct$, it holds that the operation $\mathsf{result}_i()$ returns $v \in \{0, 1, \Psi\}$.*

Proof Sketch of Theorem 1. Lines 39 to 41 imply that Invariant (i) holds.

Lemma 2. *The repeat-until loop in lines 42 to 44 cannot block forever.*

Proof Sketch of Lemma 2. The proof of the lemma is by contradiction, which Argument 4 demonstrates since line 44's exist condition eventually holds. **Argument 1:** *Eventually $aux_i[r][i] \in binValues_i(r, 2t+1)$ holds.* Suppose that in R's starting system state, $\neg(aux[r][i] \neq \bot \wedge aux[r][i] \in binValues(r, 2t+1))$ holds (otherwise the proof is done). There are $n{-}t \geq 2t{+}1 = (t{+}1){+}t$ correct nodes and each broadcasts $\mathrm{EST}(\bullet, rnd = r, est = \{w, \bullet\}, \bullet) : w \in \{0,1\}$ (line 43). Therefore, $\exists v \in \{0,1\}$ where at least $(t{+}1)$ correct nodes broadcast $\mathrm{EST}(\bullet, rnd = r, est = \{v, \bullet\}, \bullet)$. Since every node receives $\mathrm{EST}(\bullet, rnd = r, est = \{v, \bullet\}, \bullet)$ from at least $(t{+}1)$ nodes (line 48), eventually every correct node relays v via $\mathrm{EST}(\bullet, rnd = r, est = \{v, \bullet\}, \bullet)$ that line 43 sends because $v \in binValues_i(r, t{+}1)$.

Since $n{-}t \geq 2t{+}1$ holds, $(\exists w \in binValues(r, 2t{+}1))$ in the if-statement condition at line 42 is eventually satisfied for any $i \in Correct$. Thus, if $(aux[r][i] = \bot \vee aux[r][i] \notin binValues(r, 2t{+}1))$ does not hold, line 42 makes sure it does.

Argument 2: *Eventually $\exists i \in Correct : w \in binValues_i(r_i, 2t{+}1) \implies \exists s \subseteq Correct : t{+}1 \leq |s| \wedge \forall k \in s : w \in est_k[k]$.* We prove the argument by contradiction. Specifically, suppose that $\exists i \in Correct : w \in binValues_i(r_i, 2t{+}1)$ holds throughout R and yet $\forall s \subseteq Correct : t{+}1 \leq |s| : \exists k \in s : w \notin est_k[k]$.

By lines 43 and 47, the only way in which $w \in binValues_i(r_i, 2t{+}1)$ hold in every $c' \in R$, is if there is $c \in R$ that appears before c', such that $\exists s \subseteq Correct : t{+}1 \leq |s| : \forall k \in s : w \in est_k[k]$. Thus, a contradiction and the argument is true.

Argument 3: *Eventually $c' \in R$ is reached in which $\exists s \subseteq Correct : t{+}1 \leq |s| \wedge \forall p_k \in s : w \in est_k[k] \implies \forall i \in Correct : w \in binValues_i(r_i, 2t{+}1)$.* By line 43, there are $(t{+}1)$ correct nodes that broadcast $\mathrm{EST}(\bullet, rnd = r, est = \{w, \bullet\}, \bullet)$. Since every correct node receives w from at least $(t{+}1)$ nodes (line 47), every correct node eventually reply w via the message $\mathrm{EST}(\bullet, rnd = r, est = \{w, \bullet\}, \bullet)$ at lines 43 and 48 because $w \in binValues_i(r, t{+}1)$. Since $n{-}t \geq 2t{+}1$ holds, $(\exists w \in binValues(r, 2t{+}1))$ holds and the argument is true.

Argument 4: *Suppose that the condition $cond(i) := \mathsf{infoResult}_i() \neq \emptyset : i \in Correct$ does not hold in R's starting state. Eventually, $c'' \in R$, in which $cond(i) : i \in Correct$ holds.* We prove the argument by contradiction. Specifically, suppose that $cond(i)$ never holds, i.e., $\nexists c'' \in R$. We note that $cond(i)$ must hold if $binValues_i(r_i, 2t{+}1) = \{0,1\}$ or $binValues_i(r_i, 2t{+}1) = \{v\} \wedge \exists s \subseteq \mathcal{P} : n{-}t \leq |s| \wedge (\cup_{p_k \in s}\{aux_i[r][k]\}) = \{w\} \wedge w = v$. Suppose that $binValues_i(r_i, 2t{+}1) = \{v\} \subsetneq \{0,1\}$ and $\forall s \subseteq \mathcal{P} : n{-}t \leq |s| \implies w \in (\cup_{p_k \in s}\{aux_i[r][k]\}) : w \neq v$ throughout R. A contradiction is reached by showing that eventually $w \in binValues_i(r_i, 2t{+}1)$. By lines 43 and 47, the only way in which $w \in (\cup_{p_k \in s}\{aux_i[r][k]\})$ holds in every system state $c' \in R$, is if there is a system state c that appear in R before c', such that $\exists p_k \in \mathcal{P} : aux_k[r][k] = w$. Note that c' and c can be selected such that the following is true. By Argument 1,

$aux_k[r][k] \in bin Values_k(r, 2t+1)$. By Argument 2, $w \in bin Values_k(r_i, 2t+1) \implies \exists s \subseteq Correct : t+1 \leq |s| \wedge \forall p_k \in s : w \in est_k[k]$ in c. By Argument 3, $\exists s \subseteq Correct : t+1 \leq |s| \wedge \forall k \in s : w \in est_k[k] \implies \forall i \in Correct : w \in bin Values_i(r_i, 2t+1)$ in c. Thus, a contradiction is reached (w.r.t. argument), which implies that the argument is true. $\qquad \Box_{Lemma}$ 2

Lemma 3. *Invariant (ii) holds.*

Proof of Lemma 3. Lemma 2 says that the repeat-until loop in lines 42 to 44 does not block. By line 37, every iteration of the do-forever loop (lines 36 to 45) can be associated with at most one asynchronous round. Thus, line 28 and Argument (4) of the proof of Lemma 2 imply that $(r_i \geq M \wedge \mathsf{infoResult}_i() \neq \emptyset)$ holds eventually and $\mathsf{result}_i()$ returns a non-$\perp$ value.

5.2 Satisfying the Task Specifications

Theorems 2 and 3 show that Algorithm 2 satisfies the task requirements (Sect. 2.1). We say that system state $c \in R$ is *well-initialized* if $\forall i \in Correct : (r_i, est_i, aux_i) := initState$ holds. The proofs of theorems 2 and 3 appear in [32]. Recall that in the absence of transient-faults, our solution achieves consensus without assuming execution fairness.

Theorem 2 (Closure). *Let R be an execution of Algorithm 2 that starts from a well-initialized system state and includes a complete invocation of binary consensus. Within $\mathcal{O}(r) : r \leq M$ asynchronous rounds, with probability $\Pr(r) = 1 - (1/2)^r$, and for any $p_i \in \mathcal{P} : i \in Correct$, $\mathsf{result}_i() \in \{0, 1\}$.*

Theorem 3. *Let R be an execution of Algorithm 2 that starts in a well-initialized system state and during which every correct node $p_i \in \mathcal{P}$ invokes $\mathsf{propose}_i()$ exactly once. R implements a loosely-self-stabilizing solution for binary consensus that can tolerate up to t Byzantine nodes, where $n \geq 3t + 1$. Also, within four asynchronous rounds, all correct nodes are expected to decide.*

6 Discussion

We have presented a new loosely-self-stabilizing variation on the MMR algorithm [41] for solving binary consensus in the presence of Byzantine failures in message-passing systems. The proposed solution preserves the following properties of the studied algorithm: it does not require signatures, it offers optimal fault-tolerance, and the expected time until termination is the same as the studied algorithm. The proposed solution is able to achieve this using a new application of the design criteria of loosely-self-stabilizing systems, which requires the satisfaction of safety property with a probability in $\mathcal{O}(1 - 2^{-M})$. For any practical purposes and in the absence of transient faults, one can select M to be sufficiently large so that the risk of violating safety is negligible. We believe that this work is preparing the groundwork needed to construct self-stabilizing (Byzantine fault-tolerant) algorithms for distributed systems, such as Blockchains.

References

1. Alon, N., Attiya, H., Dolev, S., Dubois, S., Potop-Butucaru, M., Tixeuil, S.: Practically stabilizing SWMR atomic memory in message-passing systems. J. Comput. Syst. Sci. **81**(4), 692–701 (2015)
2. Anagnostou, E., Hadzilacos, V.: Tolerating transient and permanent failures (extended abstract). In: Schiper, A. (ed.) WDAG 1993. LNCS, vol. 725, pp. 174–188. Springer, Heidelberg (1993). https://doi.org/10.1007/3-540-57271-6_35
3. Altisen, K., Devismes, S., Dubois, S., Petit, F.: Introduction to Distributed Self-Stabilizing Algorithms. Synthesis Lectures on Distributed Computing Theory. Morgan & Claypool Publishers, San Rafael (2019)
4. Beauquier, J., Kekkonen-Moneta, S.: Fault-tolerance and self-stabilization: impossibility results and solutions using self-stabilizing failure detectors. Int. J. Syst. Sci. **28**(11), 1177–1187 (1997)
5. Ben-Or, M.: Another advantage of free choice: completely asynchronous agreement protocols. In: Second Annual ACM SIGACT-SIGOPS Symposium on Principles of Distributed Computing, pp. 27–30. ACM (1983)
6. Ben-Or, M., Dolev, D., Hoch, E.N.: Fast self-stabilizing Byzantine tolerant digital clock synchronization. In: Twenty-Seventh Annual ACM Symposium on Principles of Distributed Computing, PODC'08, pp. 385–394. ACM (2008)
7. Binun, A., et al.: Self-stabilizing Byzantine-tolerant distributed replicated state machine. In: Bonakdarpour, B., Petit, F. (eds.) SSS 2016. LNCS, vol. 10083, pp. 36–53. Springer, Cham (2016). https://doi.org/10.1007/978-3-319-49259-9_4
8. Binun, A., Dolev, S., Hadad, T.: Self-stabilizing Byzantine consensus for blockchain. In: Dolev, S., Hendler, D., Lodha, S., Yung, M. (eds.) CSCML 2019. LNCS, vol. 11527, pp. 106–110. Springer, Cham (2019). https://doi.org/10.1007/978-3-030-20951-3_10
9. Bonomi, S., Dolev, S., Potop-Butucaru, M., Raynal, M.: Stabilizing server-based storage in Byzantine asynchronous message-passing systems. In: ACM Symposium on Principles of Distributed Computing, PODC'15, pp. 471–479. ACM (2015)
10. Bonomi, S., Potop-Butucaru, M., Tixeuil, S.: Stabilizing Byzantine-fault tolerant storage. In: 2015 IEEE International Parallel and Distributed Processing Symposium, IPDPS'15, pp. 894–903. IEEE Computer Society (2015)
11. Bonomi, S., Pozzo, A.D., Potop-Butucaru, M., Tixeuil, S.: Optimal mobile Byzantine fault tolerant distributed storage. In: ACM Symposium on Principles of Distributed Computing, PODC'16, pp. 269–278. ACM (2016)
12. Bonomi, S., Pozzo, A.D., Potop-Butucaru, M., Tixeuil, S.: Optimal storage under unsynchronized mobile Byzantine faults. In: 36th IEEE Symposium on Reliable Distributed Systems, SRDS'17, pp. 154–163. IEEE Computer Society (2017)
13. Bonomi, S., Del Pozzo, A., Potop-Butucaru, M., Tixeuil, S.: Brief announcement: optimal self-stabilizing mobile Byzantine-tolerant regular register with bounded timestamps. In: Izumi, T., Kuznetsov, P. (eds.) SSS 2018. LNCS, vol. 11201, pp. 398–403. Springer, Cham (2018). https://doi.org/10.1007/978-3-030-03232-6_28
14. Bonomi, S., Pozzo, A.D., Potop-Butucaru, M., Tixeuil, S.: Approximate agreement under mobile Byzantine faults. Theor. Comput. Sci. **758**, 17–29 (2019)
15. Cachin, C., Kursawe, K., Petzold, F., Shoup, V.: Secure and efficient asynchronous broadcast protocols. IACR Cryptol. ePrint Arch. 2001, 6 (2001). http://eprint.iacr.org/2001/006
16. Correia, M., Neves, N.F., Veríssimo, P.: From consensus to atomic broadcast: time-free Byzantine-resistant protocols without signatures. Comput. J. **49**(1), 82–96 (2006)

17. Dijkstra, E.W.: Self-stabilizing systems in spite of distributed control. Commun. ACM **17**(11), 643–644 (1974)
18. Dolev, S.: Self-Stabilization. MIT Press, Cambridge (2000)
19. Dolev, S., Georgiou, C., Marcoullis, I., Schiller, E.M.: Practically-self-stabilizing virtual synchrony. J. Comput. Syst. Sci. **96**, 50–73 (2018)
20. Dolev, S., Georgiou, C., Marcoullis, I., Schiller, E.M.: Self-stabilizing Byzantine tolerant replicated state machine based on failure detectors. In: Dinur, I., Dolev, S., Lodha, S. (eds.) CSCML 2018. LNCS, vol. 10879, pp. 84–100. Springer, Cham (2018). https://doi.org/10.1007/978-3-319-94147-9_7
21. Dolev, S., Liba, O., Schiller, E.M.: Self-stabilizing Byzantine resilient topology discovery and message delivery. In: Gramoli, V., Guerraoui, R. (eds.) NETYS 2013. LNCS, vol. 7853, pp. 42–57. Springer, Heidelberg (2013). https://doi.org/10.1007/978-3-642-40148-0_4
22. Dolev, S., Welch, J.L.: Self-stabilizing clock synchronization in the presence of Byzantine faults. In: 14th Principles of Distributed Computing PODC'95, p. 256. ACM (1995)
23. Duan, S., Reiter, M.K., Zhang, H.: BEAT: asynchronous BFT made practical. In: Lie, D., Mannan, M., Backes, M., Wang, X. (eds.) Proceedings of the 2018 ACM SIGSAC Conference on Computer and Communications Security, CCS 2018, Toronto, ON, Canada, 15–19 October 2018, pp. 2028–2041. ACM (2018). https://doi.org/10.1145/3243734.3243812
24. Dubois, S., Potop-Butucaru, M., Tixeuil, S.: Dynamic FTSS in asynchronous systems: the case of unison. Theor. Comput. Sci. **412**(29), 3418–3439 (2011)
25. Dubois, S., Potop-Butucaru, M., Nesterenko, M., Tixeuil, S.: Self-stabilizing Byzantine asynchronous unison. J. Parallel Distrib. Comput. **72**(7), 917–923 (2012)
26. Fischer, M.J., Lynch, N.A.: A lower bound for the time to assure interactive consistency. Inf. Process. Lett. **14**(4), 183–186 (1982)
27. Fischer, M.J., Lynch, N.A., Paterson, M.: Impossibility of distributed consensus with one faulty process. J. ACM **32**(2), 374–382 (1985)
28. Georgiou, C., Gustafsson, R., Lindhé, A., Schiller, E.M.: Self-stabilization overhead: a case study on coded atomic storage. In: Atig, M.F., Schwarzmann, A.A. (eds.) NETYS 2019. LNCS, vol. 11704, pp. 131–147. Springer, Cham (2019). https://doi.org/10.1007/978-3-030-31277-0_9
29. Georgiou, C., Lundström, O., Schiller, E.M.: Self-stabilizing snapshot objects for asynchronous failure-prone networked systems. In: Atig, M.F., Schwarzmann, A.A. (eds.) NETYS 2019. LNCS, vol. 11704, pp. 113–130. Springer, Cham (2019). https://doi.org/10.1007/978-3-030-31277-0_8
30. Dolev, S., Petig, T., Schiller, E.M.: Brief announcement: robust and private distributed shared atomic memory in message passing networks. In: Distributed Computing, PODC'15, pp. 311–313. ACM (2015)
31. Dolev, S., Petig, T., Schiller, E.M.: Self-stabilizing and private distributed shared atomic memory in seldomly fair message passing networks. CoRR abs/1806.03498 (2018)
32. Georgiou, C., Marcoullis, I., Raynal, M., Schiller, E.M.: Loosely-self-stabilizing Byzantine-tolerant binary consensus for signature-free message-passing systems. CoRR abs/2103.14649 (2021)
33. Khanchandani, P., Lenzen, C.: Self-stabilizing Byzantine clock synchronization with optimal precision. Theory Comput. Syst. **63**(2), 261–305 (2019). https://doi.org/10.1007/s00224-017-9840-3
34. Lamport, L., Shostak, R.E., Pease, M.C.: The Byzantine generals problem. ACM Trans. Program. Lang. Syst. **4**(3), 382–401 (1982)

35. Lenzen, C., Rybicki, J.: Self-stabilising Byzantine clock synchronisation is almost as easy as consensus. J. ACM **66**(5), 32:1–32:56 (2019)
36. Lundström, O., Raynal, M., Schiller, E.M.: Self-stabilizing set-constrained delivery broadcast. In: 40th Distributed Computing Systems, ICDCS'20, pp. 617–627. IEEE (2020)
37. Lundström, O., Raynal, M., M. Schiller, E.: Self-stabilizing uniform reliable broadcast. In: Georgiou, C., Majumdar, R. (eds.) NETYS 2020. LNCS, vol. 12129, pp. 296–313. Springer, Cham (2021). https://doi.org/10.1007/978-3-030-67087-0_19
38. Lundström, O., Raynal, M., Schiller, E.M.: Self-stabilizing indulgent zero-degrading binary consensus. In: Distributed Computing and Networking ICDCN'21, pp. 106–115. ACM (2021)
39. Lundström, O., Raynal, M., Schiller, E.M.: Self-stabilizing multivalued consensus in asynchronous crash-prone systems. CoRR abs/2104.03129 (2021). https://arxiv.org/abs/2104.03129
40. Miller, A., Xia, Y., Croman, K., Shi, E., Song, D.: The honey badger of BFT protocols. In: Computer and Communications Security, CCS '16, pp. 31–42. ACM (2016)
41. Mostéfaoui, A., Hamouma, M., Raynal, M.: Signature-free asynchronous Byzantine consensus with $t < n/3$ and $O(n^2)$ messages. In: ACM Symposium on Principles of Distributed Computing, PODC'14, pp. 2–9. ACM (2014)
42. Pease, M.C., Shostak, R.E., Lamport, L.: Reaching agreement in the presence of faults. J. ACM **27**(2), 228–234 (1980)
43. Raynal, M.: Fault-Tolerant Message-Passing Distributed Systems - An Algorithmic Approach. Springer, Heidelberg (2018). https://doi.org/10.1007/978-3-319-94141-7
44. Salem, I., Schiller, E.M.: Practically-self-stabilizing vector clocks in the absence of execution fairness. In: Podelski, A., Taïani, F. (eds.) NETYS 2018. LNCS, vol. 11028, pp. 318–333. Springer, Cham (2019). https://doi.org/10.1007/978-3-030-05529-5_21
45. Sudo, Y., Nakamura, J., Yamauchi, Y., Ooshita, F., Kakugawa, H., Masuzawa, T.: Loosely-stabilizing leader election in a population protocol model. Theor. Comput. Sci. **444**, 100–112 (2012)
46. Toueg, S.: Randomized Byzantine agreements. In: Third Annual ACM Symposium on Principles of Distributed Computing PODC'84, pp. 163–178. ACM (1984)

Leader Election in Arbitrarily Connected Networks with Process Crashes and Weak Channel Reliability

Carlos López[1], Sergio Rajsbaum[1], Michel Raynal[2,3], and Karla Vargas[1(✉)]

[1] Instituto de Matemáticas, UNAM, Mexico City, Mexico
karla.vargas@ciencias.unam.mx
[2] Univ Rennes IRISA, 35042 Rennes, France
[3] Department of Computing, Polytechnic University, Kowloon, Hong Kong

Abstract. A channel from a process p to a process q satisfies the *ADD property* if there are constants K and D, unknown to the processes, such that in any sequence of K consecutive messages sent by p to q, at least one of them is delivered to q at most D time units after it has been sent. This paper studies implementations of an eventual leader, namely, an Ω failure detector, in an arbitrarily connected network of eventual ADD channels, where processes may fail by crashing. It first presents an algorithm that assumes that processes initially know n, the total number of processes, sending messages of size $O(\log n)$. Then, it presents a second algorithm that does not assume the processes know n. Eventually the size of the messages sent by this algorithm is also $O(\log n)$. These are the first implementations of leader election in the ADD model. In this model, only eventually perfect failure detectors were considered, sending messages of size $O(n \log n)$.

Keywords: ADD channel · Arbitrarily connected networks · Distributed algorithm · Eventual leader · Fault-tolerance · Process crash · Synchrony · System model · Unknown membership · Weak channel

1 Introduction

1.1 Leader Election

This is a classical problem encountered in distributed computing. Each process p_i has a local variable $leader_i$, and it is required that all the local variables $leader_i$ forever contain the same identity, which is the identity of one of the processes. A classical way to elect a leader consists in selecting the process with the smallest identity[1]. If processes may crash, the system is fully asynchronous, and the elected leader must

---

[1] A survey on election algorithms in failure-free message-passing systems appears in Chap. 4 of [19]. The aim is to elect a leader as soon as possible, and with as few messages as possible, and it can be done on a ring with $1.271\, n \log(n) + O(n)$ messages [11,17].

Partially supported by UNAM-PAPIIT grant IN106520. Full version of this paper is on *ArXiv* [15].

be a process that does not crash, leader election cannot be solved [21]. Not only the system must no longer be fully asynchronous, but the leader election problem must be weakened to the *eventual leader election problem*. This problem is denoted Ω in the failure detector parlance [2,3]. Notice that the algorithm must elect a new leader each time the previously elected leader crashes.

1.2 Related Work

Many algorithms for electing an eventual leader in crash-prone partially synchronous systems have been proposed. Surveys of such algorithms are presented in [20, Chapter 17] when communication is through a shared memory, and in [21, Chapter 18] when communication is through reliable message-passing.

In [1] there are proposed different levels of communication reliability and is its showed that in systems with only some timely channels and a complete network it is necessary that correct processes send messages forever even with just at most one process crash. An algorithm for implementing Ω in networks with unknown membership is presented in [13]. This algorithm works in a complete network and every process needs to communicate its name to every neighbor using a broadcast protocol.

In [7] it is presented an implementation of Ω for the case of the crash-recovery model in which processes can crash and then recover infinitely many times and channels can lose messages arbitrarily. The case of dynamic systems is addressed in [14], and the case where the underlying synchrony assumptions may change with time is addressed in [9]. Stabilizing leader election in crash-prone synchronous systems is investigated in [5].

The ADD Distributed Computing Model. This model was introduced in [22], as a realistic partially synchronous model of channels that can lose and reorder messages Each channel guarantees that some subset of the messages sent on it will be delivered in a timely manner and such messages are not too sparsely distributed in time. More precisely, for each channel there exist two constants K and D, not known to the processes (and not necessarily the same for all channels), such that for every K consecutive messages sent in one direction, at least one is delivered within D time units after it has been sent.

Even though ADD channels seem so weak, it is possible to implement an *eventually perfect failure detector*, $\Diamond P$, in a fully connected network of ADD channels, where asynchronous processes may fail by crashing [22]. Later on, it was shown that it is also possible to implement $\Diamond P$ in an arbitrarily connected network of ADD channels [16]. Recall that $\Diamond P$ is a classic failure detector, relatively powerful (more than sufficient to solve consensus), stronger than Ω [2] and yet realistically implementable [3].

The algorithm in [16] works for arbitrary connected networks of ADD channels, and sends messages of bounded size, improving on the previous unbounded size algorithm presented in [12]. However, the size of messages is exponential in n, the number of processes. More recently, an implementation of $\Diamond P$ using messages of size $O(n \log n)$, in an arbitrarily connected network of ADD channels was presented in [23].

1.3 Contribution

This paper shows that it is possible to implement Ω in an arbitrarily connected network where asynchronous processes may fail by crashing in a weaker model than the one presented in [23]. It first presents an implementation of Ω with messages of size $O(\log n)$, reducing the message size with respect to [23].

Most of the previous works related to Ω concentrated on communication-efficient algorithms in fully connected networks when considering the number of messages. They were considering neither the size of the messages nor arbitrarily connected networks.

The proposed algorithm works under very weak assumptions, requiring only that a directed spanning tree from the leader exists, composed of channels that eventually satisfy the ADD property. This algorithm requires that processes know n, the number of processes. Then, the paper shows how to extend the ideas to design an algorithm for the case where n is unknown in arbitrarily connected networks. Initially a process knows only its set of incident channels. Interestingly enough, eventually the size of the messages used by this algorithm is also $O(\log n)$.

We put particular attention to the size of the messages because it plays an important role in the time it takes for the processes to agree on the same leader, yet we show that our algorithms elect a leader in essentially optimal time. When designing ADD-based algorithms, it is challenging to transmit a large message by splitting it into smaller messages, due to the uncertainty created by the fact that, while the constants K and D do exist, a process knows neither them nor the time from which the channels forever satisfy them. This type of difficulty is also encountered in the design of leader election algorithms under weak eventual synchrony assumptions, e.g., [1,8,23]. Also in self-stabilizing problems, where ideas similar to our hopbound technique have been used [6], as well as in [23]. We found it even more challenging to work under the assumption that some edges might not satisfy any property at all; our algorithm works under the assumption that only edges on an (unknown to the processes) spanning tree are guaranteed to comply with the ADD property.

2 Model of Computation

Process Model. The system consists of a finite set of n processes $\Pi = \{p_1, p_2, ..., p_n\}$. Every process p_i has an identity, and without loss of generality we consider that the identity of p_i is its index i. As there is no ambiguity, we use indifferently p_i or i to denote the same process.

Every process p_i has also a read-only local clock $clock_i()$, which is assumed to generate ticks at a constant rate[2]. Local clocks need not to be synchronized to exhibit the same time value at the same time, local clocks are used only to implement timers. To simplify the presentation, it is assumed that local computations have zero duration.

Any number of processes may fail by crashing. A process is *correct* if it does not crash, otherwise, it is *faulty*.

---

[2] However, the algorithm presented below can be adapted in the case where local clocks can suffer bounded drifts, these drifts being known by the processes.

Virtual Global Clock. For notational simplicity, we assume the existence of an external reference clock which remains always unknown to the processes. The range of its ticks is the set of natural numbers. It allows to associate consistent dates with events generated by the algorithm.

Communication Network. It is represented by a directed graph $G = (\Pi, E)$, where an edge $(p_i, p_j) \in E$ means that there is a unidirectional channel that allows the process p_i to send messages to p_j. A bidirectional channel can be represented by two unidirectional channels, possibly with different timing assumptions. process p_i has a set of input channels and a set of output channels.

The graph connectivity requirement on the communication graph G depends on the problem to be solved. It will be stated in the Sect. 4 and 5 devoted to the proofs of the proposed algorithms.

Basic Channel Property. It is assumed that no directed channel creates, corrupts, or duplicates messages.

The ADD Property. A directed channel (p_i, p_j) satisfies the *ADD property* if there are two constants K and D (unknown to the processes[3]) such that

– for every K consecutive messages sent by p_i to p_j, at least one is delivered to p_j within D time units after it has been sent. The other messages from p_i to p_j can be lost or experience arbitrary delays.

Each directed channel can have its own pair (K, D). To simplify the presentation, and without loss of generality, we assume that the pair (K, D) is the same for all the channels.

The ◊ADD Property. The eventual ADD property, states that the ADD property is satisfied only after an unknown but finite period of time. Hence this weakened property allows the system to experience an initial anarchy period during which the behavior of the channels is arbitrary.

The Span-Tree Assumption. We consider that there is a time τ after which there is a directed spanning tree (i) that includes all the correct processes and only them, (ii) its root is the correct process with the smallest identity, and (iii) its channels satisfy the ◊ADD property. This behavioral assumption is called *Span-Tree* in the following.

Eventual Leader Election. Assuming a read-only local variable $leader_i$ at each process p_i, the leader failure detector Ω satisfies the following properties [2, 18]:

– *Validity:* For any process p_i, each read of $leader_i$ by p_i returns a process name.
– *Eventual leadership:* There is a finite (but unknown) time after which the local variables $leader_i$ of all the correct processes contain forever the same process name, which is the name of one of them.

---

[3] Always unknown, as the global time, also never known by the processes.

3 Eventual Leader Election with Known Membership

This section presents Algorithm 1 that implements Ω, assuming each process knows n. Parameter T denotes an arbitrary duration. Its value affects the efficiency of the algorithm, but not its correctness[4].

3.1 Local Variables at a Process p_i

Each process p_i manages the following local variables.

- $in\_neighbors_i$ (resp., $out\_neighbors_i$) is a (constant) set containing the identities of the processes p_j such that there is channel from p_j to p_i (resp., there is channel from p_i to p_j).
- $leader_i$ contains the identity of the elected leader.
- $timeout_i[1..n, 1..n]$ is a matrix of timeout values and $timer_i[1..n, 1..n]$ is a matrix of timers, such that the pair $\langle timer_i[j, n - k], timeout_i[j, n - k]\rangle$ is used by p_i to monitor the elementary paths from p_j to it whose length is k.
- $hopbound_i[1..n]$ is an array of non-negative integers; $hopbound_i[i]$ is initialized to n, while each other entry $hopbound_i[j]$ is initialized to 0. Then, when $j \neq i$, $hopbound_i[j] = n - k \neq 0$ means that, if p_j is currently considered as leader by p_i, the information carried by the last message ALIVE$(j, n - 1)$ sent by p_j to its out-neighbors (which forwarded ALIVE$(j, n - 2)$ to their out-neighbors, etc.) went through a path[5] of k different processes before being received by p_i. The code executed by p_i when it receives a message ALIVE$(j, -)$ is detailed in Sect. 3.2.
  The identifier $hopbound$ stands for "upper bound on the number of forwarding" that –due to the last message ALIVE$(j, -)$ received by p_i– the message ALIVE$(j, -)$ sent by p_i has to undergo to be received by all processes. It is similar to a *time-to-live* value.
- $penalty_i[1..n, 1..n]$ is a matrix of integers such that p_i increases $penalty_i[j, n - k]$ each time the $timer_i[j, n - k]$ expires. It is a penalization counter monitored by p_i with respect to the elementary paths of length k starting at p_j and ending at p_i.
- $not\_expired_i$ is an auxiliary local variable.

3.2 General Principle of the Algorithm

As many other leader election algorithms, Algorithm 1 elects the process that has the smallest identity among the set of correct processes by keeping as a candidate to be the leader the smallest identifier received as it is explained in the following sections. It is made up of three main sections: the one that generates and forwards the ALIVE() messages, the one that receives ALIVE() messages and the one that handles the timer expiration. Every section is described in detail below.

---

[4] If T is too big, the failure detection of a process currently considered as a leader can be delayed. On the contrary, a too small value of T can entail false suspicions of the current eventual leader p_j until the corresponding timer $timer_i[j]$ has been increased to an appropriate timeout value.

[5] In the graph theory, such a cycle-free path is called an *elementary* path.

Generating and Forwarding Messages. (Lines 6–9) Every T time units of clock clock$_i()$, a process p_i sends the message ALIVE($leader_i, hopbound_i[leader_i] - 1$).

A message ALIVE($*, n - 1$) is called *generating* message. A message ALIVE($*, n - k$) such that $1 < k < n - 1$, is called *forwarding* message (in this case it is the forwarding of the last message ALIVE($leader_i, hopbound$) previously received by p_i). Moreover, the value $n - k$ is called *hopbound value*. When a process p_i starts the algorithm, it proposes itself as candidate to be leader.

A message is sent if predicate $hopbound_i[leader_i] > 1$ of line 7 is true, The message sent is then ALIVE($leader_i, hopbound_i[leader_i] - 1$).

The message forwarding is motivated by the fact that, if $hopbound_i[leader_i] > 1$, maybe processes have not yet received a message ALIVE($leader_i, -$) whose sending was initiated by $leader_i$ and then forwarded along paths of processes (each process having decreased the carried hopbound value) has not reached all the processes. In this case, p_i must participate in the forwarding. To this end, it sends the message ALIVE($leader_i, hopbound_i[leader_i] - 1$) to each of its out-neighbors (line 8).

Let us observe that during the anarchy period during which, due to the values of the timeouts and the current asynchrony, channel behavior and process failure pattern, several generating messages ALIVE($*, n - 1$) can be sent by distinct processes (which compete to become leader) and forwarded by the other processes with decreasing hopbound values. But, when there are no more process crashes and there are enough directed channels satisfying the ADD property, there is a finite time from which a single process (namely, the correct process p_ℓ with the smallest identity) sends messages ALIVE($\ell, n - 1$) and no other process p_j sends the generating message ALIVE($j, n - 1$).

Message Reception. (Lines 10–17) When a process p_i such that $leader_i \neq i$ receives a message ALIVE($\ell, n - k$), it learns that (a) p_ℓ is candidate to be leader, and (b) there is a path with k hops from p_j to itself.

If $\ell \leq leader_i$, p_i considers ℓ as its current leader (line 11). Hence, if $\ell < leader_i$, p_ℓ becomes its new leader, otherwise it discards the message. This is due to the fact that p_i currently considers $leader_i$ as leader, and the eventual leader must be the correct process with the smallest identity.

Then, as the message ALIVE($\ell, n - k$) indirectly comes from $leader_i = \ell$ (which generated ALIVE($\ell, n - 1$)) through a path made up of k different processes, p_i increases the associated timeout value if the timer $timer_i[leader_i, hb]$ expired before it received the message ALIVE(ℓ, hb) (line 13). Moreover, whether $timer_i[leader_i, hb]$ expired or not, p_i resets $timer_i[leader_i, hb]$ (line 14) and starts a new monitoring session with respect to its current leader and the cycle-free paths of length hb from $leader_i$ to it.

The role of the timer $timer_i[\ell, hb]$ is to allow p_i to monitor p_ℓ with respect to the forwarding of the messages ALIVE(ℓ, hb) it receives such that $hb = n - k$ (i.e., with respect to the messages received from p_j along paths of length k).

Finally, p_i updates $hopbound_i[leader_i]$. To update it, p_i computes the value of $not\_expired_i$ (line 15) which is a bag (or multiset) of cycle-free path lengths x whose timers $timer_i[leader_i, n - x]$ is still running. To this end, the idea then is to select the less penalized path (hence the "smallest non-negative value" at line 16). But, it is possible that there are different cycle-free paths of lengths $x1$ and $x2$ such that we have $penalty_i[leader_i, n - x1] = penalty_i[leader_i, n - x2]$. In this case, in a conservative way, $\max(n - x1, n - x2)$ is selected to update the local variable $hopbound_i[leader_i]$.

---

initialization   —-Code for p_i—-

(1) $leader_i \leftarrow i$; $hopbound_i[i] \leftarrow n$; set $timer_i[i,n]$ to $+\infty$;
(2) **for each** $j \in \{1, \cdots, n\} \setminus \{i\}$ **and each** $x \in \{1, \cdots, n\}$ **do**
(3) 　　$timeout_i[j,x] \leftarrow 1$; set $timer_i[j,x]$ to $timeout_i[j,x]$;
(4) 　　set $penalty_i[j,x]$ to -1; $hopbound_i[j] \leftarrow 0$
(5) **end for**.

(6) **every** T **time units of** $clock_i()$ **do**
(7) 　**if** $(hopbound_i[leader_i] > 1)$
(8) 　　**then for each** $j \in out\_neighbors_i$
　　　　　do send ALIVE$(leader_i, hopbound_i[leader_i] - 1)$ to p_j **end for**
(9) 　**end if**.

(10) **when** ALIVE(ℓ, hb) **such that** $\ell \neq i$ **is received**
　　% from a process in $in\_neighbors_i$
(11) 　**if** $(\ell \leq leader_i)$ **then**
(12) 　　$leader_i \leftarrow \ell$;
(13) 　　**if** $([timer_i[leader_i, hb]$ expired)
　　　　　then $timeout_i[leader_i, hb] \leftarrow timeout_i[leader_i, hb] \times 2$ **end if**;
(14) 　　set $timer_i[leader_i, hb]$ to $timeout_i[leader_i, hb]$;
(15) 　　$not\_expired_i \leftarrow \{x \mid timer_i[leader_i, x]$ not expired $\}$;
(16) 　　$hopbound_i[leader_i] \leftarrow$
　　　　　$\max\{x \in not\_expired$ with smallest non-negative $penalty_i[leader_i, x]\}$
(17) 　**end if**.

(18) **when** $timer_i[leader_i, hb]$ **expires and** $(leader_i \neq i)$ **do**
(19) 　$penalty_i[leader_i, hb] \leftarrow penalty_i[leader_i, hb] + 1$;
(20) 　**if** $\Big(\wedge_{1 \leq x \leq n} ([timer_i[leader_i, x]$ expired$) \Big)$ **then**
(21) 　　　$leader_i \leftarrow i$
(22) 　**else**　　same as lines 15-16
(23) 　**end if**.

Algorithm 1: Eventual leader election in the $\Diamond$ADD model with known membership

Timer Expiration. (Lines 18–23) Given a process p_i, when the timer currently monitoring its current leader through a path of length $k = n - hb$ expires (line 18), it increases its $penalty_i[leader_i, n-k]$ entry (line 19).

The entry $penalty_i[j, n-k]$ is used by p_i to cope with the negative effects of the channels which are on cycle-free paths of length k from p_j to p_i and do not satisfy the ADD property. More precisely we have the following. If, while p_i considers p_j is its current leader (we have then $leader_i = j$), and $timer_i[j, n-k]$ expires, p_i increases $penalty_i[j, n-k]$. The values in the vector $penalty_i[j, 1..n]$ are then used at lines 15–16 (and line 22) to update $hopbound_i[leader_i]$ which (if p_j is the eventually elected leader) will contain the length of an cycle-free path from p_j to p_i made up of $\Diamond$ADD channels (i.e., a path on which $timer_i[j, n-k]$ will no longer expire).

Then, if for all the hopbound values, the timers currently monitoring the current leader have expired (line 20), p_i becomes candidate to be leader (line 21).

If one (or more) timer monitoring its current leader has not expired, p_i recomputes the path associated with the less penalized hopbound value in order to continue monitoring $leader_i$ (line 22).

4 Proof of Algorithm 1

This section shows that Algorithm 1 elects an eventual leader while assuming the Span-Tree behavioral assumption.

We have to prove that the algorithm satisfies *Validity* and *Eventual Leader Election*. For *Validity*, let us observe that the local variables $leader_i$ of all the processes always contain a process identity. Hence, we must only prove *Eventual Leader Election*, i.e. we must only show that the variables $leader_i$ of all the correct processes eventually converge to the same process identity, which is the identity of one of them.

Due to space limitation, the proof of the lemmas are in the full version of the paper that can be found in [15].

Lemma 1. *Let p_i and p_j be two correct processes connected by a $\Diamond$ADD channel, from p_i to p_j. There is a time after which any two consecutive messages received by p_j on this channel are separated by at most $\Delta = (K-1) \times T + D$ time units.*

Given any run r of Algorithm 1, let correct(r) denote the set of processes that are correct in this run and crashed(r) denote the set of processes that are faulty in this run.

The following lemma shows that there is a time after which there are no ALIVE($j, n - k$) messages with $p_j \in$ crashed(r), i.e. eventually all correct processes stop sending the ALIVE messages from a failed process which proves that once a leader fails, eventually all processes elect a new leader.

Lemma 2. *Let $p_i \in$ crashed(r) and $1 \le a < n - 1$. Given a run r there is a time after which there are no messages ALIVE($i, n - a$).*

Theorem 1. *Given a run r satisfying the Span-Tree property, there is a finite time after which the variables $leader_i$ of all the correct processes contain the smallest identity $\ell \in$ correct(r). Moreover, after p_ℓ has been elected, there is a finite time after which the only messages sent by processes are ALIVE($\ell, -$) messages.*

Theorem 2. *The size of a message is $O(\log n)$.*

Proof. The proof follows directly from the fact that a message carries a process identity which belongs to the set $\{1, \cdots, n\}$ and a hopbound number $hopbound$ such that $2 \le hopbound \le n - 1$. Since an integer bounded with n can be represented with exactly $\log n$ bits and we have two integers bounded with n we have that the size of every message is $O(\log n)$.

4.1 Time Complexity

Given a run r, let ℓ denote the smallest identity such that $\ell \in$ correct(r). Let t^a be the time given by Lemma 2, i.e. a time from which no message from crashed processes is till in transit (they have been received or are lost). Let $t^a \le t^r$ be the time after which:

1. All failures already happened.
2. All $\Diamond$ ADD channels satisfy their constants K and D.

After t^r, let Δ be the constant given by Lemma 1.

Lemma 3. *Let p_i be a correct process such that for every $t > t^r$, $hopbound_i[\ell] = n - k$. Then, for every correct process p_j such there is a $\Diamond$ ADD channel from p_i to p_j, $timeout_j[\ell, n - (k+1)] \leq C + 2^{log(\lceil \Delta \rceil)}$ with $timeout_j[\ell, n - (k+1)] = C$ before t^r.*

Lemma 3 states that after t^r, a timeout value is increased a finite number of times. Let t^c be the time after which all timeouts have reached their maximum, namely, no timeout is increased again. The following claims refer to the communication graph after t^c.

Lemma 4. *For every correct process p_i such that there is a minimum length path of $\Diamond$ ADD channels of length k from p_ℓ to p_i, $leader_i = \ell$ at time $t^c + (k \times \Delta)$.*

Let $\mathcal{D}$ be the diameter of the underlying spanning-tree of $\Diamond$ ADD channels.

Theorem 3. *For every correct process p_i it takes $O(\mathcal{D} \cdot \Delta)$ time to have $leader_i = \ell$.*

Proof. The proof is direct from Lemma 4.

4.2   Simulation Experiments

This section presents simulation experiments related to the performance predicted by Theorem 3 of Algorithm 1. Only a few experiments are presented, a more detailed experimental study is beyond the scope of this conference version. Our experiments show that a leader is elected in time proportional to the diameter of the network, in two network topologies: a ring and a random regular graph of degree 3.

Considering the constants K and D satisfied by an $\Diamond$ ADD once it stabilizes, Lemma 1 shows that for a given T (the frequency with which the messages are sent), then $\Delta = (K - 1) \times T + D$ is an upper bound on the time of the consecutive reception of two messages by a process. According to Theorem 3, the time to elect a leader is proportional to the diameter of the network, where the K, D and T determine the slope of the function.

For the (time and memory) efficiency of the experiments we assume some simplifying assumptions, which seem sufficient to a preliminary illustration of the results:

- All the channels are $\Diamond$ ADD to avoid the need of a penalization array.
- All the messages are delivered within time at most D or not delivered at all. This is sufficient to illustrate the convergence time to a leader. Additional experimental work is needed to determine the damage done by messages that are delivered very late.
- We selected $K = 4$, $D = 12$ and $T = 1, 5, 10$.

Convergence Experiments. The experiments of the ring in Fig. 1 and Fig. 2, are when the probability of a message being lost is 1%, and 99% respectively. The case of a random graph of degree 3 up to 50,000 nodes is in Fig. 3 when the probability of a message being lost is 1%. These experiments verify that indeed the convergence time is proportional to the diameter. The constants appear to be smaller than Δ, the one predicted by Theorem 3.

Simulation Details. We performed our simulation results in a 48 multicore machine with 256 GB of memory, using a program based on the Discrete Event Simulator *Simpy*, a framework for Python. We used the *Networkx* package to model graph composed of ADD channels. For the ring simulations, experiments were performed for each n from 10 up to 400 nodes, and taking the average of 10 executions, for each value of n. For the random regular networks, the degree selected was 3, and experiments starting with n starting in 100, up to 10, 000, taking the average of 5 executions. The n was incremented by 100 to reach 10,000 and from then on until 50,000 we incremented n by 10,000 each time. A performance impediment was indeed the large amount of memory used.

The convergence time curves we obtained for the ring experiment are functions of the form $f(x) = c \cdot x$, where x represents the diameter of the network, and the constant c is, roughly, between 2.5 and 4.5 as T goes from 1 to 10. While for the random regular networks, we again got a constant that doubled in size, roughly, as T goes from 1 to 10. This behavior seems to be better than the one predicted by Theorem 3, which says that the constant c should have grown 10 times.

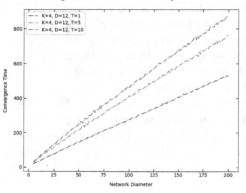

Fig. 1. A ring with drop rate of 1%

Re-Election Convergence Simulation. If an elected leader fails, we would like to know in how much time a new leader is elected.

Note that the ◊ ADD channels can arbitrarily delay the delivery of some messages. This condition has a great impact in the time it takes to Algorithm 1 to change a failed leader. For the following simulations again we assume that all the messages are delivered within time at most D or not delivered at all. But note that in a realistic scenario, we can ease the impact of the arbitrarily delayed messages by

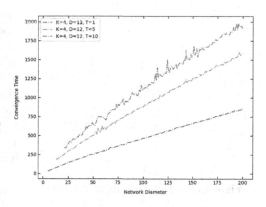

Fig. 2. A ring with drop rate of 99%

adding a timestamp to every message and keeping track for every neighbor of this timestamp. If the timestamp of the recently received message is smaller than the current one, just ignore the message. This timestamp does not have a bound, but if we use an integer and increase it by one every second that a message is sent, this integer can hold on

up for a century without overflowing [6]. By adding an integer to the message, we keep
messages of size $O(log\ n)$.

For the simulation of Fig. 4 we
selected $K = 4$, $D = 12$, $T = 1$ and the probability of a message
being lost is 1%. We performed this
simulation on a ring. The algorithm
starts at time t_0 and continues its
execution till the average time in
which a leader is elected (the curve
represented in orange). In this time,
the candidate to be the leader fails
and then a timer from an external
observer is started in every process.
This timer is used to know the average time needed for each process to
discard the failed leader (curve rep-

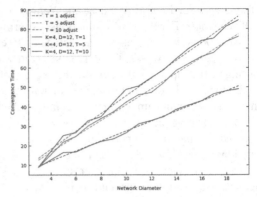

Fig. 3. A 3-regular random graph with drop rate of 1%

resented in purple) and then converge to a new leader (curve represented in blue). This
experiment verify that indeed the convergence time after the current leader fails is proportional to the diameter since $\Delta = (K - 1) \times T + D = 3 + 12 = 15$.

5   Eventual Leader Election with Unknown Membership

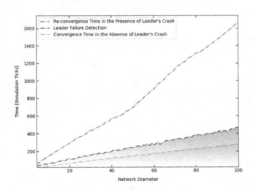

Fig. 4. Convergence time for re-election

Here, while n exists and has a fixed
value, it is no longer assumed that processes know it. Consequently, the processes have an "Unknown Membership"
of how many and which are the processes
in the network. Nevertheless, for conve-
nience, the proposed algorithm still uses
the array notation for storing the values
of timers, timeouts, hopbounds, etc. (in
an implementation dynamic data structures –e.g., lists– should be used).

Algorithm 2 solves eventual leader
election in the $\Diamond$ ADD model with
unknown membership, which means
that, initially, a process knows nothing about the network, it knows only its input/output
channels.

Our goal is to maintain the $O(\log n)$ bound on the size of the messages even in
this model. It seems that it is not easy to come up with a minor modification of the first
algorithm. For instance, a classic way of ensuring that forwarding the ALIVE message is
cycle-free is to include the path information in the message along which the forwarding

---

[6] An unsigned integer can be encoded with 32 bits, so its maximum value can be 4294967296.
A year has 31536000 seconds.

occurred, as done in the paper [16]. This would result in message sizes of exponential size, while assuming a slightly different model, we show how to eventually stay with $O(\log n)$ messages.

Furthermore, since we want the complexity to be $O(\log n)$ eventually, we need to design a mechanism that works as a *broadcast* in which once a process p_i knows a new process name from p_j, the later does not need to send to p_i the same information but only the leader information. The proposed mechanism in this paper is not the same as the proposed in [23] since we are preventing processes to send all the known names but eventually, only the leader information.

Since no process has knowledge about the number of participating processes, this number must be learned dynamically as the names of processes arrives. In order to the leader to reach every process in the network, there must be a path of $\lozenge$ ADD channels from every correct process to the leader. It follows that an algorithm for eventual leader election in networks with unknown membership cannot be a straightforward extension of the previous algorithm. More precisely, instead of the unidirectional channels and *Span-Tree* assumptions, Algorithm 2 assumes that (i) all the channels are bidirectional $\lozenge$ ADD channels, and (ii) the communication network restricted to the correct processes remains always connected (namely, there is always a path –including correct processes only– connecting any two correct processes).

In Algorithm 1, every process p_i uses n to initialize its local variable $hopbound_i[i]$ (which thereafter is never modified). In the unknown membership model, $hopbound_i[i]$ is used differently, namely it represents the number of processes known by p_i so far. So its initial value is 1. Then, using a technique presented in [23], $hopbound_i[i]$ is updated as processes know about each other: every time a process p_i discovers a new process identity it increases $hopbound_i[i]$.

5.1   General Principle of the Algorithm

Initially each process p_i only knows itself and how many channels are connected to it. So the first thing p_i needs to do is communicate its identity to its neighbors. Once its neighbors know about it, p_i no longer sends its identity. The same is done with other names that p_i learns. For that, p_i keeps a *pending set* for every channel connected to it that tracks the information it needs to send to its neighbors. So initially, p_i adds the pair (new, i) to every pending set.

During a finite amount of time, it is necessary to send an ALIVE() message to every neighbor without any constraint because the set of process names needs to be communicated to other processes. That is, information about a leader might be empty and the message only contains the corresponding pending set.

When process p_i receives an ALIVE() message from p_j, this message can contain information about the leader and the corresponding pending set that p_j saves for p_i. First, p_i processes the information contained in the pending set and then processes the information about the leader.

How p_i Learns New Process Names. If p_i finds a pair with a name labeled as new and is not aware of it, it stores the new name in the set $known_i$, increases its hopbound value, and adds to every pending set (except to the one belonging to p_j) this information labeled as new. In any case, p_i needs to communicate p_j that it already knows that information, so p_i adds this information to the pending set of p_j but labeled as an acknowledgment.

When p_j receives $name$ labeled as an acknowledgment from p_i, i.e. $(\text{ack}, name)$, it stops sending the pair $(\text{new}, name)$ to it, so it deletes that pair from p_i's pending set. Eventually, it receives a pending set from p_j not including $(\text{new}, name)$, so p_i deletes $(\text{ack}, name)$ from p_j's pending set.

How p_i Processes the Leader Information. As in Algorithm 1, every process keeps as leader a process with minimum id. Since it is assumed that all the channels are $\Diamond$ ADD, there is no need to keep a timer for every hopbound value or a penalty array. In this case, process p_i keeps the greatest $n - k$, i.e. hopbound value that it receives from the process it considers to be the leader. If this value (or a greater one) does not arrive on time, p_i proposes itself as the leader. In case a smaller hopbound value of the leader arrives, it is only taken if its timer expired.

5.2   Local Variables at Each Process p_i

Each process p_i manages the following local variables.

- $leader_i$ contains the identity of the candidate leader.
- $hopbound_i[1..)$ is an array of natural numbers; $hopbound_i[i]$ is initialized to 1.
- $timeout_i[\cdot]$ and $timer_i[\cdot]$ have the same meaning as in Algorithm 1. So, when p_i knows p_j, the pair $\langle timer_i[j], timeout_i[j]\rangle$ is used by p_i to monitor the sending of messages by p_j (which is not necessarily a neighbor of p_i).
- $known_i$ is a set containing the processes currently known by p_i. At the beginning, p_i only knows itself.
- $out\_neighbors_i$ is a set containing the names of the channels connecting p_i to its neighbor processes. The first time p_i receives through channel m a message sent by a process p_j, p_j and m become synonyms
- $pending_i[1, ..., k]$ is a new array in which, when p_i knows p_j, $pending_i[j]$ contains the pairs of the form $(label, id)$ that are pending to be send through channel connecting p_i and p_j. There are two possible labels, denoted new and ack.

5.3   Detailed Behavior of a Process p_i

The code of Algorithm 2 addresses two complementary issues: the management of the initially unknown membership, and the leader election.

Initialization. (Lines 1–4) Initially, each process p_i knows only itself and how many input/output channels it has. Moreover, it does not know the name of the processes connected to these channels (if any) and how many neighbors it has (the number of channels is higher or equal to the number of neighbors). So when the algorithm begins, it proposes itself as the leader and in the pending sets of every channel adds its pair (new, i) for neighbors to know it.

```
(1)   initialization                                    —Code for p_i—
(2)   leader_i ← i; hopbound_i[i] ← 1;
(3)   known_i ← {i}; out_neighbors_i initialized to the channels of p_i;
(4)   for each m ∈ out_neighbors_i do pending_i[m] ← {(new, i)} end for.

(5)   every T time units of clock_i() do
(6)       for each channel m ∈ out_neighbors_i (let p_j be the associated neighbor) do
(7)           if (hopbound_i[leader_i] > 1)
(8)               then send ALIVE(leader_i, hopbound_i[leader_i] − 1, pending_i[j]) to p_j
(9)               else  send ALIVE(⊥, ⊥, pending_i[j]) to p_j
(10)          end if
(11)      end for.

(12)  when ALIVE(ℓ, hb, pending) is received from p_j through channel m
      % from then on: p_j and m are synonyms from an addressing point of view
(13)      set_i ← ∅;
(14)      for each (label, k) ∈ pending do
(15)          if (label = new)
(16)              then set_i ← set_i ∪ {k};
(17)                  if (k ∉ known_i)
(18)                      then known_i ← known_i ∪ {k}; hopbound_i[i] ← hopbound_i[i] + 1;
(19)                          add an entry in timeout_i, timer, hopbound_i;
(20)                          add (new, k) to every pending[p] with p ≠ m
(21)                      else if ((new, k) ∈ pending_i[m])
(22)                          then pending_i[m] ← pending_i[m] \ (new, k) end if;
(23)                          pending_i[m] ← pending_i[m] ∪ (ack, k)
(24)                      end if
(25)              else pending_i[m] ← pending_i[m] \ (new, k)
(26)          end if
(27)      end for;
(28)      for each (ack, k) ⊂ pending_i[m] such that k ∉ set_i do
(29)          pending_i[m] ← pending_i[m] \ {(ack, k)} end for;
(30)      if (ℓ ≤ leader_i and ℓ ≠ i)
(31)          then leader_i ← ℓ;
(32)              if (hb ≥ hopbound_i[leader_i]) ∨ (timer_i[leader_i] expired)
(33)                  then hopbound_i[leader_i] ← hb;
(34)                      if ([timer_i[leader_i] expired)
(35)                          then timeout_i[leader_i, hb] ← timeout_i[leader_i, hb] × 2 end if;
(36)                      set timer_i[leader_i] to timeout_i[leader_i]
(37)              end if
(38)      end if.

(39)  when (timer_i[leader_i] expires) do leader_i ← i.
```

Algorithm 2: Eventual leader election in the ◇ ADD model with unknown membership

Sending a Message. (Lines 5–11) Every T units of time, p_i sends a message through every channel m. In some cases the leader information is empty because of the condition of line 7. But in any case, it must send a message that includes information about the network that is included in the set $pending_i[j]$.

Receiving a Message. (Lines 12–38) When p_i receives a message (line 12) from process p_j (through channel m), at the beginning it knows from which channel it came and eventually knows from whom is from. When the message is received, the information included in *pending* (lines 14–29) is processed, and then the leader information is processed (lines 30–38).

Processing New Information. (Lines 14–29) The input parameter set *pending* includes pairs of the form $(label, id)$, where $label \in \{$new, ack$\}$ and id is the name of some process. When p_i processes the pairs that it received from p_j there can be two kind of pairs. The first is a pair with label new (line 15), which means that p_j is sending new information (at least for p_j) to p_i. When this information is actually new for p_i (line 17) then, it stores this new name, increases its hopbound entry and adds to every pending set(but not the one from which it received the information) this new information (line 20).

In case that p_i already knows the information labeled as new for p_j (line 21), then p_i needs to check if it is included in the pending set to p_j this information as new too. If that is the case, then it deletes from $pending[m]$ this pair (line 22). In any case, p_i adds to the pending set the pair (ack, k) for sending through the channel from where this message was received (line 23).

If p_i receives the pair (ack, k) (line 25), then it deletes the pair (new, k) from the set $pending_i[m]$, because the process that sent this pair, already knows k.

Processing the Leader Related Information. (Lines 30–38) If the leader related information is not empty, p_i processes it. As in the first algorithm, if the identity of the proposed leader is smaller than the current one, then it is set as p_i's new leader (line 31). Then, it processes the hopbound. If the recently arrived hopbound is greater than the one currently stored, then the recently arrived is set as the new hopbound (line 33). If the timer for the expected leader expired, it needs more time to arrive to p_i, so the timeout is increased (line 35) and the timer is set to timeout (line 36).

Deleting Pairs. (Lines 21, 25 and 28) If some process p_i wants to send some information k to p_j, it adds to the pending set of p_j the pair (new, k). When p_j receives this pair, it looks if this is already in its set, in that case, it deletes the pair from p_i's pending set (line 21). Then, p_j adds an (ack, k) to the pending set of p_i. As soon as p_i receives this pair from p_j, it deletes from p_j's pending set the pair (new, k) (line 25). So when p_j receives a pending set from p_i without the pair (new, k), it means that p_i already received the acknowledgment message, so p_j deletes (ack, k) from p_i's pending set (line 28).

Timer Expiration. (Line 39) When the timer for the expected leader expires, p_i proposes itself as the leader.

Notice that, when compared to Algorithm 1, Algorithm 2 does not use the local arrays $penalty_i[1..n, 1..n]$ employed to monitor the paths made of non-ADD channels.

6 Underlying Behavioral Assumption and Proof of Algorithm 2

In the following we consider that there is a time τ after which no more failures occur, and the network is such that (i) all the channels are bidirectional $\Diamond ADD$ channels, and (ii) the communication network restricted to the correct processes remains always connected. Assuming this, this section shows that Algorithm 2 eventually elects a leader despite initially unknown membership. All the proofs of this algorithm are in full version of this paper that can be found on [15].

7 Conclusion

The $\Diamond ADD$ model has been studied in the past as a realistic, particularly weak communication model. A channel from a process p to a process q satisfies the ADD property if there are two integers K and D (which are unknown to the processes) and a finite time τ (also unknown to the processes) such that, after τ, in any sequence of K consecutive messages sent by p to q at least one message is delivered by q at most D time units after it has been sent. Assuming first that the correct processes are connected by a spanning tree made up of $\Diamond$ ADD channels, this article has presented an algorithm that elects an eventual leader, using messages of only size $O(\log n)$. Previous algorithms in the $\Diamond ADD$ model implemented an eventually perfect failure detector, with messages of size $O(n \log n)$. In addition to this, the article has presented a second eventual leader election algorithm in which no process initially knows the number of processes. This algorithm sends larger messages, to be able to estimate n, but only for a finite amount of time, after which the size of the messages is again $O(\log n)$. We conjecture that it is necessary, that the process identities are repeatedly communicated to the potential leader. Although we proved that our algorithms elect a leader in time proportional to the diameter of the graph, many interesting question related to performance remain open.

References

1. Aguilera, M., Delporte-Gallet, C., Fauconnier, H., Toueg, S.: Communication-efficient leader election and consensus with limited link synchrony. In: 23th ACM Symposium on Principles of Distributed Computing (PODC'04), pp. 328–337. ACM press (1996)
2. Chandra, T.D., Hadzilacos, V., Toueg, S.: The weakest failure detector for solving consensus. J. ACM **43**(4), 685–722 (1996)
3. Chandra, T.D., Toueg, S.: Unreliable failure detectors for reliable distributed systems. J. ACM **43**(2), 225–267 (1996)
4. Chandy, K.M., Lamport, L.: Distributed snapshots: determining global states of distributed systems. ACM Trans. Comput. Syst. **3**(1), 63–75 (1985)
5. Delporte-Gallet, C., Devismes, S., Fauconnier, H.: Stabilizing leader election in partial synchronous systems with crash failures. J. Parallel Distrib. Comput. **70**(1), 45–58 (2010)
6. Altisen, K., Devismes, S., Dubois, S., Petit, F.: Introduction to Distributed Self-Stabilizing Algorithms. Distributed Computing Theory. Morgan and Claypool Publishers, San Rafael (2019)
7. Fernández-Campusano, C., Larrea, M., Cortiñas, R., Raynal, M.: A distributed leader election algorithm in crash-recovery and omissive systems. Inf. Process. Lett. **118**, 100–104 (2017)

8. Fernández, A., Jimenez, E., Raynal, M., Trédan, G.: A timing assumption and two t-resilient protocols for implementing an eventual leader service in asynchronous shared-memory systems. Algorithmica **56**(4), 550–576 (2010). https://doi.org/10.1007/s00453-008-9190-2
9. Fernández, A., Raynal, M.: From an asynchronous intermittent rotating star to an eventual leader. IEEE Trans. Parallel Distrib. Syst. **21**(9), 1290–1303 (2010)
10. Fetzer, Ch.: Fail-awareness in timed asynchronous systems. In: 15th ACM Symposium on Principles of Distributed Computing (PODC'96), pp. 314–341. ACM press (1996)
11. Higham, L., Przytycka, T.: A simple efficient algorithm for maximum finding on rings. Inf. Process. Lett. **58**, 319–324 (1996)
12. Hutle, M.: An efficient failure detector for sparsely connected networks. In: IASTED International Conference on Parallel and distributed Computing and Networks, pp. 369–374 (2004)
13. Jimenez, E., Arevalo, S., Fernandez, A.: Implementing unreliable failure detectors with unknown membership. Inf. Process. Lett. **100**(2), 60–63 (2006)
14. Larrea, M., Soraluze, I., Cortiñas, R., Raynal, M.: Specifying and implementing an eventual leader service for dynamic systems. Int. J. Web Grid Serv. **8**(3), 204–224 (2012)
15. López, C., Rajsbaum, S., Raynal, M., Vargas, K.: Leader election in arbitrarily connected networks with process crashes and weak channel reliability. Arxiv https://arxiv.org/abs/2105.02972 (2021)
16. Kumar, S., Welch, J.L.: Implementing $\Diamond P$ with bounded messages on a network of ADD channels. Parallel Process. Lett. **29**(1) 1950002, 12p (2019)
17. Mendivil, J.-R., Arrieta, I., Raynal, M.: Leader election: from Higham-Przytycka's algorithm to a gracefully degrading algorithm. In: 6th International Conference on Complex, Intelligent, and Software Intensive Systems (CISIS'12), pp. 225–232. IEEE Computer Press (2012)
18. Raynal, M.: A short introduction to failure detectors for asynchronous distributed systems. ACM SIGACT News **36**(1), 53–70 (2005)
19. Raynal, M.: Distributed Algorithms for Message-Passing Systems, p. 510. Springer, Heidelberg (2013). https://doi.org/10.1007/978-3-642-38123-2. ISBN 978-3-642-38122-5
20. Raynal, M.: Concurrent Programming: Algorithms, Principles And Foundations, p. 515. Springer, Heidelberg (2013). https://doi.org/10.1007/978-3-642-32027-9. ISBN 978-3-642-32026-2
21. Raynal, M.: Fault-Tolerant Message-passing Distributed Systems: An Algorithmic Approach, p. 492. Springer, Heidelberg (2018). https://doi.org/10.1007/978-3-319-94141-7. ISBN 978-3-319-94140-0
22. Sastry, S., Pike, S.M.: Eventually perfect failure detectors using ADD channels. In: Stojmenovic, I., Thulasiram, R.K., Yang, L.T., Jia, W., Guo, M., de Mello, R.F. (eds.) ISPA 2007. LNCS, vol. 4742, pp. 483–496. Springer, Heidelberg (2007). https://doi.org/10.1007/978-3-540-74742-0_44
23. Vargas, K., Rajsbaum, S., Raynal, M.: An eventually perfect failure detector for networks of arbitrary topology connected with ADD channels using time-to-live values. Parallel Process. Lett. **30**(2), 2050006, 23 p (2020). A preliminary version appeared in the 49th IEEE/IFIP International Conference on Dependable Systems and Networks (DSN'2019), pp. 264–275

Fault-Tolerant Termination Detection with Safra's Algorithm

Georgios Karlos[1]([✉]), Wan Fokkink[1], and Per Fuchs[2]

[1] Vrije Universiteit Amsterdam, Amsterdam, The Netherlands
{g.karlos,w.j.fokkink}@vu.nl
[2] Technische Universität München, München, Germany
per.fuchs@cs.tum.edu

Abstract. Safra's distributed termination detection algorithm employs a logical token ring structure within a distributed network; only passive nodes forward the token, and a counter in the token keeps track of the number of sent minus the number of received messages. We adapt this classic algorithm to make it fault-tolerant. The counter is split into counters per node, to discard counts from crashed nodes. If a node crashes, the token ring is restored locally and a backup token is sent. Nodes inform each other of detected crashes via the token. Our algorithm imposes no additional message overhead, tolerates any number of crashes as well as simultaneous crashes, and copes with crashes in a decentralized fashion. Experiments with an implementation of our algorithm were performed on top of two fault-tolerant distributed algorithms.

1 Introduction

Termination detection is a fundamental problem in distributed systems which was introduced independently in [9] and [12]. Termination can be announced when all nodes in the network have become passive and no messages are in transit. Distributed termination detection is applied in e.g. workpools, routing, diffusing computations, self-stabilization, and checking stable system properties such as deadlock and garbage in memory. Many (mostly failure-sensitive) termination detection algorithms have been proposed in the literature, see [17,18].

In Safra's algorithm [7,10] a token repeatedly visits all nodes in the network via a predetermined logical ring structure; a node passes on the token when it is passive. Each node keeps track of the number of outgoing minus incoming messages, and these counts are accumulated in the token. Nodes that receive a message are colored black, as the count in the token may be unreliable, if the message overtook the token in the ring. The black color is transferred to the token at its next visit. If the token returns to the initiator without a black color and with counter 0, the initiator can announce termination.

Safra's algorithm imposes only little message overhead when nodes remain active over a long period of time, unlike termination detection algorithms for which every message needs to be acknowledged (e.g. [9]). Additionally, it does not require idle messages to be sent out when nodes become passive and does not

© Springer Nature Switzerland AG 2021
K. Echihabi and R. Meyer (Eds.): NETYS 2021, LNCS 12754, pp. 71–87, 2021.
https://doi.org/10.1007/978-3-030-91014-3_5

run into underflow issues, as opposed to weight-throwing schemes ([19,22]). In [6] an optimized version of Safra's algorithm was proposed, that does not always color receiving nodes black and detects termination within a single round trip of the token after actual termination has occurred.

We propose a fault-tolerant algorithm based on an improved version of Safra's algorithm [6]. A node crash is handled locally by its predecessor in the ring; a new token is issued, as the old token may have been lost in the crash. A numbering scheme in the token makes sure only a single token is being passed on; if the old token was not lost in the crash, the new token will be dismissed. Only the message exchange between alive nodes is counted. For this purpose the counter at the nodes and in the token is split into N counters, with N the number of nodes in the network. Nodes have a failure detector and inform each other of crashes through the token, so that they uniformly count the message exchange with the same set of nodes. A node reporting a new crash makes sure the token completes another round trip, to avoid inconsistent message counts in the token.

Next to the aforementioned strong points of Safra's algorithm, our fault-tolerant variant has some additional advantages compared to existing fault-tolerant termination detection algorithms, which will be discussed in Sect. 2. (Following [11,21], we call the distributed algorithm for which termination is checked basic and the termination detection algorithm the control algorithm.) First, the basic algorithm can be decentralized, meaning there can be multiple initiator nodes. Second, if the initiator of our termination detection algorithm crashes before sending out the first token, this role is automatically taken over by its predecessor. Thus our algorithm can cope with any number of node crashes, and it is robust against simultaneous node crashes. Third, only one additional message is required for each crash, and no relatively expensive schemes like leader election or taking a global snapshot are employed. The price to pay is that, in the absence of stable storage, the bit complexity of a message is $\Theta(N)$, compared to $\Theta(1)$ for the failure-sensitive version of Safra's algorithm. Considering current network technologies, with Gbits/second throughput and microseconds latency, this token size incurs a tolerable overhead in network load, especially since the token is only forwarded by idle nodes.

We tested our algorithm in a multi-threaded emulation environment and performed experiments on two fault-tolerant distributed algorithms from the literature. Compared with the failure-sensitive version of Safra's algorithm, our algorithm exhibits a satisfactory performance, in the sense that it imposes no additional message overhead. Of course it does impose some overhead, by adding extra concurrency at each node and additional synchronization. However, even with a large number of failures, profiling of our experiments shows that the execution time of the basic algorithm remains the dominant factor for the overall performance. The two basic distributed algorithms employed in our experiments are quite different in nature, which suggests this conclusion holds more generally, admittedly only up to the network sizes we analyzed.

Developing our algorithm was a delicate matter. Still, owing to a correctness proof (omitted here) in combination with many test runs with an implementation, we can confidently claim that our algorithm correctly detects termination.

2   Related Work

We discuss some existing fault-tolerant termination detection algorithms, mainly from a functional point of view. Only [16] reports on performance results based on an actual implementation. Generally a complete network topology and a perfect failure detector are required, as such assumptions are essential for developing a fault-tolerant termination detection algorithm, see [20].

Lai and Wu [15] presented a fault-tolerant variant of the Dijkstra-Scholten algorithm [9] for centralized basic algorithms, meaning there is a single initiator node. Active nodes are in the tree, rooted in the initiator, which announces termination when the tree has disappeared. In the event of a crash, all alive nodes communicate with the designated root node, causing a sequential bottleneck.

Lifflander et al. [16] proposed a series of algorithms based on [9] that avoid the bottleneck of [15]. These algorithms are resistant to single-node failures but are only probabilistically tolerant to multi-node failures and incur additional control messages even in crash-free executions. In case of a crash the tree is reconstructed locally. If two nodes fail concurrently, the algorithms may not be able to recover. The algorithm then detects that this is the case. Failure of the root node cannot be handled. Performance results are reported based on an experimental setup, consisting of three mock-up parallel algorithm implementations. Results show acceptable processing time overhead. Message overhead results are not reported.

Tseng [22] developed a fault-tolerant variant of weight-throwing [19] for centralized basic algorithms. Nodes donate part of their weight to the basic messages they send. The receiver claims that weight on receipt. A node that becomes passive returns its weight to the leader, who announces termination when it is passive and reclaimed its original weight. The number of control messages increases linearly with the number of basic messages. The algorithm is vulnerable to underflow of weight values and control messages require space to represent floats at high precision. A global snapshot is taken when a new crashed node is detected. When the leader crashes, an election scheme is employed.

In Venkatesan's algorithm [23] a leader node is in charge of announcing termination. If the leader crashes, an election is held. The local stacks at the nodes must be continuously replicated by the leader and its backup Upon learning of a crashed node, the leader simulates the state of every node in the system to determine whether it has terminated.

Hursey and Graham [14] developed a termination detection scheme for their fault-tolerant ring-based MPI application. Their algorithm relies on a leader election scheme and fault-tolerant primitives provided by MPI.

Mittal et al. [20] introduced a general framework for transforming any failure-sensitive termination detection algorithm into a fault-tolerant variant that can cope with any number of node crashes. The basic idea is to restart termination detection after each node crash. When applied to existing failure-sensitive algorithms, the resulting fault-tolerant algorithms have a significant overhead in control messages, even when no nodes become passive or crash.

3 System Model

We assume a fully asynchronous message-passing system with no shared memory or global clock. Messages may arrive in any order and delays are unbounded but finite. The N nodes are *logically* organized in a ring and are assigned unique, totally-ordered IDs. Failures are permanent; once a node crashes, it halts and never recovers. Like most fault-tolerant distributed algorithms in the literature, we require that the underlying physical network provides reliable bidirectional communication channels between each pair of nodes.

Nodes are either *active* or *passive*. An active node can send/receive basic messages, perform internal events, or become passive when it *terminates locally*. A passive node cannot send basic messages or perform internal events and only becomes active upon receipt of a basic message. Termination may be announced while basic messages from crashed nodes are in transit. It is therefore required that passive nodes never become active by the receipt of a basic message sent by a node they know has crashed. An execution of the basic algorithm has terminated if all alive nodes are passive and for all basic messages in transit, the destination node either has crashed or knows that the sender has crashed. The termination detection problem consists of two parts: *Liveness*: if the system has terminated, this is eventually detected by an alive node; and *Safety*: when termination is detected, the system terminated at some point in the past.

A *perfect* failure detector [4] is required to solve termination detection in the presence of failures [13,20]. Such a failure detector, which never falsely suspects that a node crashed, and eventually detects each node crash, can be built if there is a known upper bound on network latency.

4 Safra's Algorithm

Safra's (failure-sensitive) termination detection algorithm [7,10] generalizes the Dijkstra-Feijen-van Gasteren algorithm [8] from synchronous to asynchronous message passing networks. We give a detailed description of Safra's algorithm, including improvements from [6]. This will serve as a basis for the description of our fault-tolerant version later on.

Safra's algorithm is centralized, with node 0 as initiator. The basic algorithm however is allowed to be decentralized and does not need to be ring-based. A token t circulates the ring, starting at the initiator of the control algorithm when it becomes passive for the first time, and being forwarded by the other nodes once they are passive. The field $count_t$ in t represents the number of basic messages in transit during the round trip of t. Each node i records in $count_i$ the number of basic messages it sent minus the number of basic messages it received, since the last time it forwarded the token. Each time t is received by a node i, $count_i$ is added to $count_t$, and $count_i$ is reset to zero. Upon return of t to the initiator, after it has become passive, termination is detected if $count_t$ is zero.

The token can during its round trip underestimate the number of basic messages in transit, if the receipt of a message is accounted for in the token before the send of this message. To recognize this, colors black and white are used. Initially all nodes are white, and when the initiator sends out a fresh token, the token is white. When a node i receives a basic message m, it may be that the send of m was not yet recorded in $count_t$. Therefore upon receipt of m, i marks itself black. When t visits a black node, t becomes black and the node white; from then on t remains black for the rest of the round trip. When the initiator has received back t, has become passive, and has added the value of $count_0$ to $count_t$, it decides whether termination can be detected. If t is black or $count_t$ is not zero, the initiator sends out a fresh white token again. Otherwise, it can safely announce that the execution of the basic algorithm has terminated.

Two enhancements of Safra's algorithm to reduce detection delay were given in [6]. One inefficiency is that a basic message always blackens its receiver. Actually an inconsistent snapshot can only exist when the receipt of a message is recorded before its send. This can happen when a basic message overtakes the token, meaning that it is sent after the sender was visited by the token, but reaches the receiver before it is visited by the token. A second inefficiency of the original algorithm is that termination is only detected at the initiator. Another enhancement allows detection to occur at any node. When multiple nodes can detect termination, there is a second situation in which an inconsistent snapshot can occur. When both the sender and the receiver of a message are ahead of the token, but the receiver will be visited by the token before the sender, it is possible for the receiver to detect termination before the sender is visited. (This case is omitted in [6], which may result in erroneous detection.)

To deal with the aforementioned scenarios, a sequence number seq_i is introduced at every node i, starting at zero. When a node forwards the token, it increases its sequence number by one, so that nodes in the visited region have a higher sequence number from those in the unvisited region. A node piggybacks its sequence number seq_m to every basic message m it sends. Using the sequence number, an offending message can be detected if it has a higher sequence number than the receiver, or they both have the same sequence number but the sender has a higher ID than the receiver. Since multiple nodes can detect termination, an offending message should not only blacken the receiver but also all subsequent nodes in the ring up to (but not including) the sender. At all these nodes, the token represents an inconsistent snapshot. So none should detect termination.

The field $black_t$ in the token is now a node ID, expressing that all nodes the token visits from now up to (but not including) $black_t$ are black. When the token is sent by the initiator for the first time, $black_t = N - 1$, so that all nodes from 1 up to $N - 1$ are initially considered black. Hence termination can only be detected after the token has visited all nodes at least once. Likewise, $black_i$ at a node i represents that all nodes that the token visits from i up to $black_i$ are black. Initially $black_i = i$ at all nodes i, meaning that i considers all nodes white. If a node i receives a basic message m of which the send may not have been accounted for in the token, then $black_i$ is set to the furthest node from i among $black_i$ and the sender of m. The function $furthest_i(j, k)$ computes whether node j or k is furthest away from i in the ring. It is defined by:

k if $i \leq j \leq k$ or $k < i \leq j$ or $j \leq k < i$; and j otherwise.

If the token reaches a node i, it must wait until i is passive. Then i adds the value of $count_i$ to the value of $count_t$. If i is white, meaning that $black_t = black_i = i$, it can determine termination in the same way the initiator does in Safra's algorithm: check whether the value of $count_t$ is zero. If i is black or detects no termination, it forwards the token to its successor. Before doing so, it sets $black_t$ to $furthest_i(black_t, black_i)$ if this value is not i, or else to $(i+1) \bmod N$. The latter means the successor of i in the ring will consider the token white. Finally, i sets $count_i$ to zero and $black_i$ to i and increases seq_i by one.

Algorithms 1–4 present the pseudocode of the four procedures available at each node i for the improved version of Safra's algorithm: initialization, sending/receiving a basic message m to/from a node j (SBM/RBM) , and receiving a token (RT). Subscript i of a procedure name represents the node where the procedure is performed. Action $send(m, j)$ denotes that message m is sent to node j, and the Boolean field $passive_i$ is $true$ only when node i is passive. Procedures are executed without interruption, except that while waiting to become passive, in line 1 of RT, a node is allowed to perform SBM and RBM calls.

Algorithm 1: Initialization$_i$

---

1 $count_i \leftarrow 0$; $black_i \leftarrow i$; $seq_i \leftarrow 0$;
2 **if** $i = 0$ **then**
3 $\quad$ $wait(passive_0)$;
4 $\quad$ $count_t \leftarrow count_0$; $black_t \leftarrow N - 1$; $send(t, 1)$; $count_0 \leftarrow 0$; $seq_0 \leftarrow 1$;

Algorithm 2: SendBasicMessage$_i (m, j)$

---

1 $seq_m \leftarrow seq_i$; $send(m, j)$; $count_i \leftarrow count_i + 1$;

Algorithm 3: ReceiveBasicMessage$_i (m, j)$

---

1 **if** $seq_m = seq_i + 1 \vee (j > i \wedge seq_m = seq_i)$ **then**
2 $\quad$ $black_i \leftarrow furthest_i(black_i, j)$;
3 $count_i \leftarrow count_i - 1$;

Algorithm 4: ReceiveToken$_i$

---

1 $wait(passive_i)$;
2 $count_t \leftarrow count_t + count_i$; $black_i \leftarrow furthest_i(black_i, black_t)$;
3 **if** $count_t = 0 \wedge black_i = i$ **then**
4 $\quad$ Announce;
5 $black_t \leftarrow furthest_i(black_i, (i+1) \bmod N)$;
6 $send(t, (i+1) \bmod N)$;
7 $count_i \leftarrow 0$; $black_i \leftarrow i$; $seq_i \leftarrow seq_i + 1$;

5   Fault-Tolerant Version

From now on we assume nodes may spontaneously and permanently crash. It is customary to assume for fault-tolerant distributed algorithms that there is a bidirectional channel between each pair of distinct nodes (see e.g. [21]), because else a node failure may result in disconnected subnetworks. Actually, it suffices if at any time a channel can be established between any two alive nodes.

Mittal et al. [20] showed that a perfect failure detector is required to solve termination detection in the presence of failures. In a fully asynchronous setting, such a detector cannot be built. A practical compromise is to assume an upper bound on the network latency. Each node sends out heartbeat messages at regular time intervals. When a node i has not received a heartbeat from another node j within some time interval, then i permanently considers j as crashed.

Each node i stores the identities of crashed nodes in one of the sets CRASHED_i and REPORT_i. The latter contains the identities that i has not yet reported to the other alive nodes by means of the token; this will be explained below.

Since counts of messages to and from crashed nodes need to be discarded, the token contains N counters, one per node; moreover, each node needs to count its message exchange with each other node separately and from the start of the execution run (instead of since the last token visit). So we split the field $count_i$ for each node i into a sequence $[count_i^0, \ldots, count_i^{N-1}]$. For each node j, the field $count_i^j$ stores the number of basic messages i has sent to j minus the number of basic messages i has received from j. (The fields $count_i^i$ are redundant as they always carry the value zero.) If (the failure detector of) i detects that a node j has crashed, then i permanently disregards the value of $count_i^j$. Likewise, to separately keep track of the counters at the different nodes in the token, the field $count_t$ is split into a sequence $[count_t^0, \ldots, count_t^{N-1}]$. If these counters were lumped together into a single counter $count_t$, and say a node i sent a basic message to a node j which then crashed, there might be no way of telling whether or not j received this message and updated $count_j^i$ and $count_t$.

If a node i learns from its failure detector that some other node j crashed, it must share this information with the other alive nodes via the token. Else there would be the risk that although i from now on disregards $count_i^j$, some other alive node k may still take into account $count_k^j$, which could lead to a premature termination detection at k. For this purpose the token contains a set CRASHED_t. When i forwards the token with $j \in \text{CRASHED}_t$, it adds j to CRASHED_i, to avoid that it announces the same crashed node multiple times.

Each node i keeps track of its successor $next_i$ in the ring; initially $(i+1) \bmod N$. Each time i detects $next_i$ has crashed, the value of this field is changed into i's nearest alive successor. We must ensure that the token is not lost; this could happen if the token was traveling to or being handled by $next_i$ at the moment it crashed. Therefore, after having determined its new successor $next_i'$, i forwards the token once again, to $next_i'$. For this purpose i stores the last token it forwarded. These local variables are updated as soon another token (with a higher sequence number) arrives.

In case $next_i$ forwarded the token before crashing, $next_i'$ will receive the same token twice. Therefore the token has a sequence number seq_t, which is increased by one at each consecutive round trip of the token. In the first round $seq_t = 1$. Each node i keeps track of the highest sequence number it has passed on so far in seq_i (initially $seq_i = 0$), and ignores incoming tokens with $seq_t \leq seq_i$. The last node in the ring, initially $N - 1$, increases the sequence number every time it forwards the token. If it crashes, this task is taken over by its predecessor. A node i can determine whether it is the last node by checking if $next_i < i$.

As in the failure-sensitive variant of Safra's algorithm, $black_t$ and $black_i$ express which nodes are considered black, and when the token is sent by the initiator for the first time, $black_t = N - 1$ to guarantee it visits all nodes at least once. If a node receives an offending basic message, it colors all nodes in the ring between itself and the sender black. If the failure detector of a node i reports a crashed node and, at the next token visit, i does not detect termination, then i colors all other nodes black, as they must all be visited by the token.

The pseudocode of the procedures at each node is given in Algorithms 5–10. Again, the procedures should be executed without interruption, except that while waiting to become passive, in line 2 of procedure ReceiveToken, a node may perform SendBasicMessage, ReceiveBasicMessage and FailureDetector calls.

In the initialization phase, nodes provide their local variables with initial values; node 0 holds the token. At sending/receiving a basic message to/from a noncrashed process, the sender/receiver updates the corresponding counter. Basic messages received from a crashed node in REPORT_i may still be accounted for by i in the control algorithm, to allow for termination detection at the next token visit to i. If the receipt of a message is accounted for in the token before its send, the receiver colors the nodes up to the sender black.

Algorithm 5: Initialization $_i$

1 **for** $j = 0$ **to** $N - 1$ **do**
2    $count_i^j \leftarrow 0$; $count_t^j \leftarrow 0$;
3 $black_i \leftarrow i$; $seq_i \leftarrow 0$; $next_i \leftarrow (i + 1) \bmod N$;
4 $\text{CRASHED}_i \leftarrow \emptyset$; $\text{CRASHED}_t \leftarrow \emptyset$; $\text{REPORT}_i \leftarrow \emptyset$;
5 **if** $i = 0$ **then**
6    $black_t \leftarrow N - 1$; $seq_t \leftarrow 1$; ReceiveToken $_0$;
7 **else**
8    $black_t \leftarrow i$;

Algorithm 6: SendBasicMessage $_i (m, j)$

1 **if** $j \notin \text{CRASHED}_i \cup \text{REPORT}_i \cup \text{CRASHED}_t$ **then**
2    $seq_m \leftarrow seq_i$; $send(m, j)$; $count_i^j \leftarrow count_i^j + 1$;

Algorithm 7: ReceiveBasicMessage $_i (m, j)$

1 **if** $j \notin \text{CRASHED}_i$ **then**
2   **if** $seq_m = seq_i + 1 \vee (j > i \wedge seq_m = seq_i)$ **then**
3       $black_i \leftarrow furthest_i(black_i, j)$;
4    $count_i^j \leftarrow count_i^j - 1$;

Algorithm 8: ReceiveToken$_i$

---

1 **if** $seq_t = seq_i + 1$ **then**
2      $wait\,(passive_i);$ $black_i \leftarrow furthest_i(black_i, black_t);$
3      $\text{CRASHED}_t \leftarrow \text{CRASHED}_t \setminus \text{CRASHED}_i;$
4      $\text{CRASHED}_i \leftarrow \text{CRASHED}_i \cup \text{CRASHED}_t;$
5      $\text{REPORT}_i \leftarrow \text{REPORT}_i \setminus \text{CRASHED}_t;$
6   **if** $black_i = i \vee \text{REPORT}_i = \emptyset$ **then**
7          $count_t^i \leftarrow 0;$
8   **for all** $j \in \{0, \ldots, N-1\} \setminus \text{CRASHED}_i$ **do**
9              $count_t^i \leftarrow count_t^i + count_i^j;$
10   **if** $black_i = i$ **then**
11         $sum_i \leftarrow 0;$
12   **for all** $j \in \{0, \ldots, N-1\} \setminus \text{CRASHED}_i$ **do**
13             $sum_i \leftarrow sum_i + count_t^j;$
14   **if** $sum_i = 0$ **then**
15   Announce;
16   **if** $next_i \in \text{CRASHED}_t$ **then**
17   NewSuccessor$_i$;
18   **if** $next_i < i$ **then**
19         $seq_t \leftarrow seq_t + 1;$
20   **if** $\text{REPORT}_i \neq \emptyset$ **then**
21         $\text{CRASHED}_t \leftarrow \text{CRASHED}_t \cup \text{REPORT}_i;$ $black_t \leftarrow i;$
22         $\text{CRASHED}_i \leftarrow \text{CRASHED}_i \cup \text{REPORT}_i;$ $\text{REPORT}_i \leftarrow \emptyset;$
23   **else**
24         $black_t \leftarrow furthest_i(black_i, next_i);$
25     $send(t, next_i);$ $black_i \leftarrow i;$ $seq_i \leftarrow seq_i + 1;$

Procedure ReceiveToken$_i$ (RT$_i$) is executed when a token arrives at node i. It only proceeds if i did not receive an instance of this token before (line 1). It then waits until it becomes passive, because in the meantime the values of $count_i^j$, $black_i$ and REPORT$_i$ may still change.

Once passive, $black_i$ is set to the furthest of $black_i$ and $black_t$ (line 2). Then, the set CRASHED$_t$ is relieved of the nodes that i reported through the token before (line 3). The remaining nodes in CRASHED$_t$ are copied to CRASHED$_i$, because they will be reported when i forwards t (line 4). REPORT$_i$ is relieved of nodes in CRASHED$_t$ (line 5). The values $count_i^j$ for nodes $j \notin$ CRASHED$_i$ are accumulated in $count_t^i$ (lines 7–9); but only if i is white or REPORT$_i$ is empty (line 6), because then it may be employed in termination detection at i (in lines 10–15) or at other nodes, respectively. If i is white (line 10), the values $count_t^j$ for nodes $j \notin$ CRASHED$_i$ are accumulated in sum_i (lines 11–13); if this sum is 0, i announces termination (lines 14–15). If no termination is detected, i checks whether its successor is in CRASHED$_t$; if so, NewSuccessor$_i$ is called to select another successor (lines 16–17). Next, i checks whether it is the last node in the ring, and if so increases the sequence number of t by 1 (lines 18–19). If REPORT$_i$ is nonempty (line 20), then it is added to CRASHED$_t$, so that t will report these crashed nodes to all alive nodes; $black_t$ is set to i, to ensure that the token visits all nodes up to i again before termination can be detected, as all alive nodes must first achieve a consistent view on the set of crashed nodes (line 21). Next all nodes in REPORT $_i$ are moved to CRASHED $_i$ (line 22). If REPORT $_i$ is empty, then $black_t$ is set to $furthest_i(black_i, next_i)$ (lines 23–24). Finally, i forwards t to $next_i$, colors itself white, and increases seq_i by one (line 25).

FailureDetector $_i$ (FD $_i$) is invoked if i's failure detector reports that a node j crashed ($crashed(j)$ in line 1). If i was not yet aware of this crash (line 2), then j is added to REPORT $_i$ (line 3), so that this crash will be reported to other nodes via the token. If j is the successor of i in the ring, NewSuccessor $_i$ is invoked to compute a new successor of i (lines 4–5). A backup token (possibly updated compared to the original token) is sent to the new successor (line 11), if i received the token at least once (first disjunct in line 6); the second disjunct in line 6 ensures a backup token is sent when the initiator crashes before ever becoming passive. REPORT $_i$ is added to CRASHED $_t$ (line 7); nodes in REPORT $_i$ are not transposed to CRASHED $_i$ yet, because the backup token may be discarded in favor of the original token. By $black_t \leftarrow i$ (in line 8) it is guaranteed that the backup token visits all nodes up to i again before termination can be detected, as all alive nodes must take into account the crash of j. If no alive node has an identity greater than i, then seq_t is increased by one (lines 9–10).

Algorithm 9: FailureDetector $_i$

---

1 $crashed(j)$;
2 **if** $j \notin$ CRASHED $_i \cup$ REPORT $_i$ **then**
3   REPORT $_i \leftarrow$ REPORT $_i \cup \{j\}$;
4   **if** $j = next_i$ **then**
5   NewSuccessor $_i$;
6   **if** $seq_i > 0 \vee next_i < i$ **then**
7   CRASHED $_t \leftarrow$ CRASHED $_t \cup$ REPORT $_i$;
8              $black_t \leftarrow i$;
9   **if** $next_i < i$ **then**
10                 $seq_t \leftarrow seq_i + 1$;
11             $send(t, next_i)$;

Algorithm 10: NewSuccessor $_i$

---

1 $next_i \leftarrow (next_i + 1) \bmod N$;
2 **while** $next_i \in$ CRASHED $_i \cup$ REPORT $_i$ **do**
3      $next_i \leftarrow (next_i + 1) \bmod N$;
4 **if** $next_i = i$ **then**
5      $wait(passive_i)$;
6   Announce;
7 **if** $black_i \neq i$ **then**
8      $black_i \leftarrow furthest_i(black_i, next_i)$;

NewSuccessor $_i$ (NS $_i$) computes i's new successor after $next_i$ crashed. First, $next_i$ is changed into $(next_i + 1) \bmod N$ (line 1). Then, it is repeatedly checked whether the new value of $next_i$ is a crashed node (line 2), and if so its value is increased by one, modulo N (line 3). After the value of $next_i$ has stabilized, i checks whether it is the only remaining alive node in the network (line 4), and if so, waits until it has become passive to announce termination (lines 5–6). Else, if $black_i \neq i$, then $black_i$ is set to $furthest_i(black_i, next_i)$ (lines 7–8).

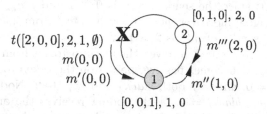

(a) Node 0 sends two messages and forwards the token before crashing

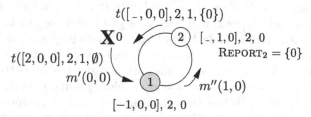

(b) After the crash is detected, the two messages are discarded.

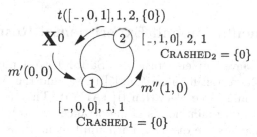

(c) Termination is detected in the next round

Fig. 1. Example run on a faulty network of three nodes

Example 1. We consider one possible run of our fault-tolerant algorithm on a ring of three nodes in Fig. 1. Initially all nodes are active, all counters carry the value 0, and $black_i = i$ and $seq_i = 0$ for $i = 0, 1, 2$. Node 0 sends basic messages m and m' to node 1, node 1 sends basic message m'' to node 2, and node 2 sends basic message m''' to node 1 (all with their node ID and sequence number 0 attached); $count_0^1$ is set to 2, and $count_1^2$ and $count_2^1$ are set to 1. Nodes 0 and 2 now become passive. Node 0 sends the token to node 1 (with $count_t^0 = 2$, $count_t^1 = count_t^2 = 0$, $black_t = 2$, $seq_t = 1$ and $\text{CRASHED}_t = \emptyset$), and crashes. This leads to Fig. 1a where the cross at node 0 represents that it has crashed, the sequences of $count$ values at alive nodes are placed between square brackets, and empty CRASHED and REPORT sets at nodes have been omitted.

In Fig. 1b, node 2 detects node 0 crashed and sets REPORT_2 to $\{0\}$; from now on node 2 ignores $count_2^0$. Since $next_2 = 0$, node 2 makes node 1 its new successor. Since $1 < 2$, node 2 sends a backup token to node 1 (with $count_t^1 = count_t^2 = 0$, $black_t = 2$, $seq_t = 1$ and $\text{CRASHED}_t = \{0\}$). Node 1 receives m from node 0 and sets $count_1^0$ to -1; since the sender of m is $0 < 1$ and $seq_m = 0 = seq_1$, $black_1$ remains unchanged; moreover, node 1 receives m''' from node 2 and sets $count_1^2$ to 0; since the sender of m''' is $2 > 1$ and $seq_{m'''} = 0 = seq_1$, $black_1$ is set to 2. Then, in Fig. 1c, node 1 receives the backup token from node 2 and sets CRASHED_1 to $\{0\}$. It becomes passive, passes on the token to node 2 with $count_t^1$ set to $count_1^2 = 0$, and sets both $black_1$ and seq_1 to 1. Node 1 does not detect termination since $black_t = 2$. Next, node 1 receives the original token from node 0, which is dismissed. Node 2 receives the token and sets CRASHED_2 to $\{0\}$ and REPORT_2 to $\emptyset$. It does not detect termination because it sets $count_t^2$ to $count_2^1 = 1$, and $count_t^1 + count_t^2 = 0 + 1 > 0$. It passes on the token to node 1 with $black_t = 1$ and $seq_t = 2$, and sets $black_2$ to 2 and seq_2 to 1.

When the token arrives, node 1 sets CRASHED_t to $\emptyset$, computes $count_t^1 = 1$, passes on the token to node 2 with $black_t = 2$, and sets $black_1$ to 1 and seq_1 to 2. In the meantime node 2 receives m'' from node 1 and sets $count_2^1$ to 0; since the sender of m'' is $1 < 2$ and $seq_{m''} = 0 < seq_2$, $black_2$ remains unchanged. Node 2 becomes passive again. When the token arrives, node 2 computes $count_t^1 + count_t^2 = 0 + 0 = 0$. Since also $black_t = 2$, it announces termination. Finally node 1 ignores message m' from node 0, because $0 \in \text{CRASHED}_1$.

6   Implementation and Experimental Results

We applied a Java implementation of SafraFT to a fault-tolerant version of the Chandy-Misra routing algorithm [5] (CM) and the Afek-Kutten-Yung self-stabilizing spanning tree algorithm [1] (AKY)[1]. They form a good test bench because detecting termination is of importance for both algorithms while their messaging behaviors are distinct. Their implementations are on top of the Ibis distributed programming platform [2]. Experiments were conducted on the DAS-4 supercomputer [3]. Multiple network nodes were run on each DAS-4 compute node to achieve decently sized networks, up to 2000 network nodes. When more network nodes are placed on a single compute node, profiling shows this starts to influence the outcome of experiments. For this reason the experiments with the CM and AKY algorithms were limited to 2000 network nodes. Before each run a certain percentage of nodes, up to 90%, was randomly selected to crash after performing a certain number of events. As an aside, the experiment unveiled a delicate implementation issue. Updates of token variables in the ReceiveToken procedure must be atomic because otherwise incorrect behavior may occur if a node receives a backup token while handling the token.

---

[1] https://github.com/PerFuchs/safra-termination-detection-fault-tolerant.

Moreover, we abstractly emulated activity of a basic algorithm and network behavior under randomized execution scenarios[2]. The emulation experiments were performed on a single compute node of DAS-4. We used networks of 16, 48 and 144 nodes and two probability distributions (uniform and Gaussian) for the randomized choices. We emulated a decentralized basic algorithm, with half of the nodes initially active. For each version, network size and probability distribution we performed a test with no nodes crashing and, for SafraFT only, a test for each 20% interval ($[1, 20]$, $[21, 40]$, etc.) of crashing nodes. We repeated each test 1000 times, for a total of 42,000 runs (two probability distributions, three network sizes, and six intervals for SafraFT; one interval $[0, 0]$ for SafraFS).

Emulation results in Fig. 2 (Emu, left) confirm that SafraFT imposes no additional control message overhead, in the absence of crashes, compared to SafraFS. Both variants tend to require the same number of token steps to detect termination after it has occurred (T_{post}), incurring on average half a round of extra token steps (R_{post}). They also require the same number of token steps before termination (T_{pre}). This is to be expected since in the absence of crashes, the operation of SafraFT is almost identical to that of SafraFS. These results are stable across the two probability distributions used in the emulator.

Performance results of emulations have to be taken with a grain of salt. In practice workloads do not always follow a smooth probability distribution, thread-scheduling policies as well as the hardware platform may to introduce biases, and basic algorithms may exhibit behavior to deal with actual node crashes. Still these synthetic results give some indication of the performance overhead SafraFT may impose, and importantly the large number of emulations helped to further increase confidence in the correctness of this algorithm.

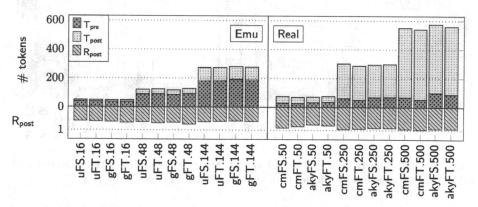

Fig. 2. Top: Tokens sent on crash-free networks. Uniform, Gaussian distributions denoted by **u**, **g**. **Bottom**: Detection delay, in token rounds after termination.

The CM/AKY results in Fig. 2 (Real, right) also show that SafraFT incurs no extra control message overhead on crash-free runs. Compared to the emulation results, there is a small increase in R_{post}. For emulation, T_{pre} is considerably larger than for CM/AKY. This difference can be attributed to the generally

---

[2] https://github.com/gkarlos/FTSEmu.

higher workload of the emulated basic algorithm compared to CM/AKY. For instance, the initiator of the CM algorithm does not take part in the computation after the initial broadcast, and thus is mostly passive. Moreover, in CM, nodes send estimate messages to their neighbors but not to their parent. This leaves more nodes passive compared to the randomized activities of the emulator.

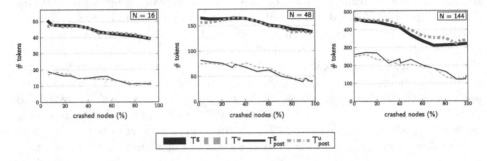

Fig. 3. Tokens sent by SafraFT on faulty networks in the emulator. Uniform, Gaussian distributions denoted by **u**, **g**.

Figure 3 shows the effect of crashes on emulation runs of SafraFT. The total number of token steps (T) and T_{post} decrease roughly in a linear fashion as failures increase, because when nodes crash, fewer nodes remain to forward the token, and our emulations of basic algorithms do not react to crashes.

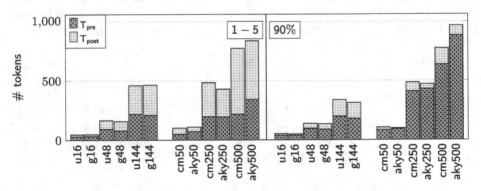

Fig. 4. Tokens sent by SafraFT with 1–5 and 90% crashed nodes in the emulator and CM/AKY experiments. Uniform, Gaussian distributions denoted by **u**, **g**.

Such a decrease is not to be expected for real-world distributed algorithms. A node crash may cause active nodes to activate other nodes, or the workload of the crashed node may be reassigned to other alive nodes, extending the trip of the token. A significant increase of T_{pre} is indeed observed for CM/AKY in Fig. 4 if many nodes (90%) crash, compared to if 1–5 nodes crash. By contrast, crashes have little effect on emulated basic algorithms, for reasons discussed before. T_{pre} remains roughly the same for small networks and actually shows a decrease on the largest one, due to the fact that the overall activity produced by the emulator on 16 and 48 nodes is relatively small compared to that of 144 nodes.

Table 1 shows the sum of processing times (pt), detection overhead (ov), and time to detect termination after it happened (tt), for SafraFS and SafraFT, when applied to CM and AKY. For $N \leq 1000$, pt-FS and pt-FT grow roughly linear to N. An analysis of processing times shows that the main factor in the higher time consumption of SafraFT, compared to SafraFS, is the growth in token size when N increases. The outlier $N = 2000$, for SafraFS but especially SafraFT, turned out to be caused by our experimental setup. Each physical compute node may simultaneously host up to 100 network nodes, depending on N. Each of these instances uses multiple threads. Altogether there are at least four times as many threads as cores on each machine. This leads to threads being preempted by the operating system, which happens more often for threads that try to send large messages (and in SafraFT, the token size grows linearly with N).

Table 1. CM/AKY results of SafraFS/FT on crash-free networks. Times in s.

N	Crash-Free CM			Crash-Free AKY		
	pt-FS/FT	ov-FS/FT	tt-FS/FT	pt-FS/FT	ov-FS/FT	tt-FS/FT
50	0.02/**0.03**	4.8/7.4%	0.01/**0.01**	0.04/**0.05**	4.1/ 5.7%	0.01/**0.01**
250	0.12/**0.24**	3.7/7.1%	0.03/**0.10**	0.15/**0.26**	6.1/10.5%	0.02/**0.08**
500	0.26/**0.64**	2.4/5.7%	0.06/**0.34**	0.30/**0.52**	6.4/11.1%	0.05/**0.18**
1000	0.59/**1.15**	2.2/4.3%	0.12/**0.54**	0.68/**1.46**	4.9/10.6%	0.11/**0.70**
2000	1.43/**2.66**	1.2/2.2%	0.23/**1.05**	1.63/**6.03**	3.4/12.7%	0.30/**3.85**

A relatively large part of the overall processing time is spent on detecting termination because CM and AKY complete their tasks relatively quickly. For basic algorithms that take a long time to complete, the time taken for termination detection can be expected to be negligible. The processing time overhead of termination detection (between 2.2% and 12.7% in all runs for SafraFT) would reduce significantly for long-running jobs, owing to the fact that Safra's algorithm tends to impose only little control message overhead, unlike termination detection algorithms in which every basic message is acknowledged (e.g. [9]). Remarkably, for CM the overhead of SafraFT decreases when N grows, while for AKY it increases. The reason is that the number of times nodes become passive grows significantly slower, in terms of N, for CM than for AKY.

Table 2. CM/AKY results of SafraFT on faulty networks. Times in s.

N	Faulty CM				Faulty AKY			
	pt-1-5	pt-90	tt-1-5	tt-90	pt-1-5	pt-90	tt-1-5	tt-90
50	0.06	0.11	0.02	0.01	0.07	0.11	0.01	0.002
250	0.29	0.52	0.10	0.04	0.29	0.60	0.07	0.01
500	0.71	1.18	0.30	0.08	0.65	1.48	0.19	0.03
1000	1.59	2.43	0.70	0.11	1.91	4.40	0.42	0.10
2000	5.11	6.88	2.22	0.31	4.25	9.93	1.70	0.17

Table 2 shows the sum of the processing times at all nodes and the time to detect termination after it happened for SafraFT, applied to CM and AKY, when 1 to 5 nodes (1–5) and when 90% of the nodes (90%) crash. When more nodes crash, the processing time increases, due to backup tokens, while the time to detect termination decreases, because there are fewer alive nodes.

7   Conclusion

We presented a fault-tolerant algorithm for distributed termination detection based on an improved version of Safra's algorithm. In our fault-tolerant variant message counters are maintained per node, so that counts to and from crashed nodes can be discarded. If a node crashes, the ring structure is restored locally and a backup token is sent. Strong points are: little message overhead when nodes remain active for a long time; robust against any number of and simultaneous node crashes; only one additional message per crash; the basic algorithm can be decentralized; no leader election scheme; no underflow issues. Compared to other algorithms, our algorithm generates far fewer, but larger control messages. For overall performance, fewer messages tend to be better, since more messages mean more processing at each node, as well as at the network stack.

Experiments indicate our algorithm imposes no significant extra overhead in control messages compared to its failure-sensitive counterpart. Despite the $O(N)$ bit complexity of the token, the available throughput and low latency of current network technologies, as well as the low message complexity of our algorithm, may render our approach feasible for large networks. This needs to be validated in experiments with real-life distributed networks under realistic and diverse workloads on many machines.

Testing the behavior of fault-tolerant distributed algorithms on very large networks turned out to be challenging. Emulating basic algorithms by means of unrestricted randomization results in executions that refuse to terminate on large networks and do not faithfully mimic all aspects of real-life distributed algorithms. Moreover, in experiments on top of two actual algorithms, allocating multiple network nodes on a single compute node influences the results. These challenges may partly explain why [16] is the only related paper we are aware of to report experimental results, for networks of up to 2048 nodes.

Next to performing realistic experiments for larger networks, future work is to develop a version of our fault-tolerant algorithm in the presence of stable storage. In that case the memory overhead of splitting the counter in the token can be avoided, at the cost of storing message counts in stable storage.

Acknowledgement. Ceriel Jacobs provided valuable feedback on the design and implementation of our algorithm.

References

1. Afek, Y., Kutten, S., Yung, M.: The local detection paradigm and its applications to self-stabilization. Theoret. Comput. Sci. **186**(1–2), 199–229 (1997)

2. Bal, H.E., et al.: Real-world distributed computer with Ibis. IEEE Comput. **43**(8), 54–62 (2010)
3. Bal, H.E., et al.: A medium-scale distributed system for computer science research: infrastructure for the long term. IEEE Comput. **49**(5), 54–63 (2016)
4. Chandra, T.D., Toueg, S.: Unreliable failure detectors for reliable distributed systems. J. ACM **43**(2), 225–267 (1996)
5. Chandy, K.M., Misra, J.: Distributed computation on graphs: shortest path algorithms. Commun. ACM **25**(11), 833–837 (1982)
6. Demirbas, M., Arora, A.: An optimal termination detection algorithm for rings, Technical report, The Ohio State University (2000)
7. Dijkstra, E.W.: Shmuel Safra's version of termination detection. EWD Manuscript 998, The University of Texas at Austin (1987)
8. Dijkstra, E.W., Feijen, W.H.J., van Gasteren, A.J.M.: Derivation of a termination detection algorithm for distributed computations. Inf. Process. Lett. **16**, 217–219 (1983)
9. Dijkstra, E.W., Scholten, C.S.: Termination detection for diffusing computations. Inf. Process. Lett. **11**(1), 1–4 (1980)
10. Feijen, W.H.J., van Gasteren, A.J.M.: Shmuel Safra's termination detection algorithm. In: Feijen, W.H.J., van Gasteren, A.J.M. (eds.) On a Method of Multiprogramming. Monographs in Computer Science, pp. 313–332. Springer, Heidelberg (1999). https://doi.org/10.1007/978-1-4757-3126-2_29
11. Fokkink, W.J.: Distributed Algorithms: An Intuitive Approach, 2nd edn. MIT Press, Cambridge (2018)
12. Francez, N.: Distributed termination. ACM Trans. Program. Lang. Syst. **2**, 42–55 (1980)
13. Helary, J.-M., Hurfin, M., Mostefaoui, A., Raynal, M., Tronel, F.: Computing global functions in asynchronous distributed systems with perfect failure detectors. IEEE Trans. Parallel Distrib. Syst. **11**(9), 897–907 (2000)
14. Hursey, J., Graham, R.L.: Building a fault tolerant MPI application: a ring communication example. In: Proceedings of IPDPS Workshop on High Performance Computing, pp. 1549–1556. IEEE (2011)
15. Lai, T.-H., Wu, L.-F.: An (N-1)-resilient algorithm for distributed termination detection. IEEE Trans. Parallel Distrib. Syst. **6**(1), 63–78 (1995)
16. Lifflander, J., Miller, P., Kale, L.: Adoption protocols for fanout-optimal fault-tolerant termination detection. In: Proceedings of PPoPP, pp. 13–22. ACM (2013)
17. Matocha, J., Camp, T.: A taxonomy of distributed termination detection algorithms. J. Syst. Softw. **43**(3), 207–221 (1998)
18. Mattern, F.: Algorithms for distributed termination detection. Distrib. Comput. **2**(3), 161–175 (1987). https://doi.org/10.1007/BF01782776
19. Mattern, F.: Global quiescence detection based on credit distribution and recovery. Inf. Process. Lett. **30**(4), 195–200 (1989)
20. Mittal, N., Freiling, F., Venkatesan, S., Penso, L.: On termination detection in crash-prone distributed systems with failure detectors. J. Parallel Distrib. Comput. **68**(6), 855–875 (2008)
21. Tel, G.: Introduction to Distributed Algorithms, 2nd edn. Cambridge University Press, Cambridge (2000)
22. Tseng, T.C.: Detecting termination by weight-throwing in a faulty distributed system. J. Parallel Distrib. Comput. **25**(1), 7–15 (1995)
23. Venkatesan, S.: Reliable protocols for distributed termination detection. IEEE Trans. Reliab. **38**(1), 103–110 (1989)

Weak Amnesiac Flooding of Multiple Messages

Zahra Bayramzadeh[1], Ajay D. Kshemkalyani[2], Anisur Rahaman Molla[3], and Gokarna Sharma[1(✉)]

[1] Kent State University, Kent, OH 44242, USA
zbayramz@kent.edu, sharma@cs.kent.edu
[2] University of Illinois at Chicago, Chicago, IL 60607, USA
ajay@uic.edu
[3] Indian Statistical Institute, Kolkata 700108, India
molla@isical.ac.in

Abstract. Flooding is a fundamental concept in distributed computing. In flooding, typically, a node forwards a message to its neighbors for the first time when it receives a message. Later if the node receives the same message again, it simply ignores the message and does not forward it. The nodes store a "message record" to ensure that the same message is not forwarded again.

Hussak and Trehan [STACS'20] introduced *amnesiac flooding* where nodes do not require to keep the message record. They established a surprising result that the amnesic flooding of a single ($k = 1$) message starting from some source node always terminates in bipartite graphs in e rounds and in non-bipartite graphs in $e < j \leq e + D + 1$ rounds, where e is the eccentricity of the source node and D is the diameter of the graph. Recently, Hussak and Trehan [arXiv'20] introduced *dynamic amnesiac flooding* initiated in possibly multiple rounds with possibly multiple ($k > 1$) messages from possibly multiple source nodes. They showed that the *partial-send* case where a node only sends a message to neighbours from which it did not receive *any* message in the previous round and the *ranked full-send* case where a node sends some highest ranked message to all neighbors from which it did not receive *that* message in the previous round, both terminate. However, they showed that the *unranked full-send* case, where a node sends some random message (not necessarily the highest ranked message) to all the neighbors from which it did not receive *that* message in the previous round, does not terminate.

In this paper, we show that the unranked full-send case also terminates, provided that diameter D is known to graph nodes. We further show that the termination time is $D \cdot (2k - 1)$ rounds in bipartite graphs and $(2D + 1) \cdot (2k - 1)$ rounds in non-bipartite graphs.

1 Introduction

Flooding is one of the fundamental and most useful primitives in distributed computing. In flooding, the task is to disseminate message(s) from source nodes

K. Echihabi and R. Meyer (Eds.): NETYS 2021, LNCS 12754, pp. 88–94, 2021.
https://doi.org/10.1007/978-3-030-91014-3_6

to all the nodes of the network. Suppose a distinguished source node has a message θ initially. The goal is to disseminate θ to all the nodes of the network. In a synchronous, round-based distributed network, flooding is typically performed as follows: In the first round, the distinguished source node sends θ to all its neighbors. From the next round onwards, when a node receives θ for the first time, it sends a copy of θ to its neighbors (except the neighbors from which it receives θ). If it receives θ again, it doesn't do anything. This essentially requires each node in the network to maintain a "message record" of θ to indicate whether that node has seen θ in some previous round. If a node receives θ and it has a record that it has seen θ before, then it does not forward θ. This ensures that the node never floods θ twice. It is well-known that this *classic* flooding process always terminates and the number of rounds until termination is $D + 1$, the diameter of the network. The message record is of size at least 1 bit for a message.

Moving from single message flooding to multiple message flooding, the flooding approach for a single message has to be applied to each of the messages separately. Therefore, each node has to have the message record of at least 1 bit per message, i.e., $\Omega(k)$ bits for $k > 1$ messages, which may be a problem for resource-constrained devices [23,24].

Hussak and Trehan [11] asked an interesting question for the single message flooding starting from a distinguished source node: What will happen if nodes do not keep the record of the message θ? Will the flooding process still terminate? Not keeping a record means that message travels on its own without depending on a message record. Not having a message record simplifies client-server application design as well as makes it scalable due to the fact that servers do not need to keep track of session information [25]. It will also provide fault tolerance even when network nodes crash.

Intuitively, if the nodes do not keep any record, they may forward the message again and again when received in subsequent rounds. Thus, the absence of a message record raises the possibility that θ may be circulated infinitely. Hussak and Trehan [11] formally studied flooding without the message record, calling it *amnesiac flooding*, and showed that the single message ($k = 1$) flooding that starts from a distinguished source node terminates in bipartite graphs in e rounds and in non-bipartite graphs in $e < j \le e + D + 1$ rounds, where e is eccentricity of the source node. Using two rounds to initiate flooding with the second round dependant on the first, termination time was improved to $e + 1$ rounds in any (non-bipartite) network by Turau [24], reducing the $e+D+1$ rounds of [11] by D rounds. However, the dependency on the first two rounds makes the result from Turau [24] not truly amnesiac compared to Hussak and Trehan [11]. Interestingly, the result of Turau [24] matches the termination time of classic flooding, since $e \le D$, and the termination time of classic flooding is $D+1$ rounds. In the recent followup work, Hussak and Trehan [12] showed that the same termination time of e rounds in bipartite graphs and $e \le j \le e + D + 1$ rounds in non-bipartite graphs can be achieved for a single message θ starting from multiple source nodes concurrently. Essentially, Hussak and Trehan [12] showed that the proofs of [11] for the single source case carry over with simple modifications mainly to the definitions for the multiple source case. Turau [25] gave an alternative detailed proof.

Recently in [12], Hussak and Trehan considered *dynamic* amnesiac flooding of multiple $k > 1$ messages, where the messages may be initiated in possibly different rounds (i.e., not necessarily in the same first round) by different source nodes in the graph. Dynamic flooding arises in different real-world applications. One prominent example is disaster monitoring [25] where a distributed system of sensors is deployed to monitor a disaster event. As soon as sensors detect an event which may happen at different times for different sensors, they start flooding this information in the network. Furthermore, one source node may initiate multiple (different) messages (the source nodes may not be all different, i.e., $1 \le k' < k$ source nodes for k messages). They considered the following three cases (problems) of dynamic amnesiac flooding in the synchronous message passing setting where each node receives messages from neighbors, performs internal computation, and sends messages to neighbors in synchronized rounds:

- *partial-send*: a node only sends a message to its neighbors from which it did not receive *any* message in the previous round.
- *ranked full-send*: a node sends some highest ranked message to all neighbors from which it did not receive *that* message in the previous round.
- *unranked full-send*: a node sends some random message (not necessarily the higest ranked message) to all neighbors from which it did not receive *that* message in the previous round.

Hussak and Trehan [12] showed that both the partial-send and ranked full-send problems terminate, but the unranked full-send problem does not terminate.

In this paper, we establish that the unranked full-send problem also terminates, provided that diameter D is known to network nodes. We further prove the termination time for the unranked full-send problem in both bipartite and non-bipartite graphs.

Overview of the Model and Results. Let the communication network be modeled as an undirected and unweighted but connected graph $G = (V, E)$, where V is the network nodes and $E \subseteq V \times V$ is the edges of G. Every node is assumed to have a unique identifier (e.g., its IP address). The nodes are allowed to communicate through the edges of the graph G. We consider a *synchronous message passing*[1] network, where computation proceeds in synchronous rounds with a node performing the following three tasks in each round: (i) receive messages from its neighbors, (ii) perform local computation, and (iii) send messages to its neighbors. No message is lost in transit. The messages are assumed to have unique IDs (which may not necessarily be consecutive and the smallest message ID may not be 1). A message θ is called globally i-th ranked if and only if the ID of θ is i-th largest among the IDs of all the messages in the set. The (global)

---

[1] In the asynchronous message passing framework, it was shown by Hussak and Trehan [11] that amnesiac flooding does not terminate.

rank of the messages is not known to graph nodes (i.e., the unranked problem), otherwise it becomes the ranked problem which terminates.

We prove the following theorem for the unranked full-send problem.

Theorem 1. (unranked full-send). *Given a set $\{\theta^1, \ldots, \theta^k\}$ of $k > 1$ messages positioned on $1 \le k' \le k$ nodes of a network G initiated at possibly different rounds, the unranked full-send problem terminates in bipartite graphs in $D \cdot (2k - 1)$ rounds and in non-bipartite graphs in $(2D + 1) \cdot (2k - 1)$ rounds[2] with each node storing $O(\log(\max\{k, D\}))$ bits, provided that the diameter D is known to the graph nodes.*

Theorem 1 is interesting and important since it was shown in Hussak and Trehan [12] that the unranked full-send problem does not terminate.

Comparison to Amnesiac and Classic Flooding. We first compare our result to amnesiac flooding and then to classic (non-amnesiac) flooding. Nodes do not need to store any information in the amnesiac flooding definition of Hussak and Trehan [11]. However, the assumption of graph nodes knowing D in our algorithm is a stronger condition than the amnesiac flooding definition of [11]. This is because knowing D requires each graph node to keep $\lceil \log D \rceil$ bits record in memory. Therefore, the storage requirement for any algorithm knowing D is at least $\Omega(\log D)$ bits. The total storage $O(\log(\max\{k, D\}))$ bits at each node in our algorithm is due to the fact that it also uses a *wait* variable which needs $O(\log k)$ bits. Therefore, our algorithm provides a trade-off between two parameters k and D regarding memory; $O(\log D)$ bits when $k = O(D)$ and $O(\log k)$ bits otherwise. Nodes need to store record of each message in classic (non-amnesiac) flooding, i.e., at least $\Omega(k)$ bits memory to flood k different messages. Therefore, the memory requirement in our algorithm is a significant reduction on the memory requirement at graph nodes compared to classic flooding when $k > \Omega(\log D)$.

The above comparison to amnesiac and classic flooding shows that our algorithm provides a '*weak*' variant of amnesiac flooding, that is, it reduces storage requirement of classic flooding but does not completely remove it as in amnesiac flooding [11,24]. An interesting direction for future research is whether a weaker assumption than D is enough to make the unranked full-send problem terminate. Finally, we prove the termination time of our algorithm using the single message termination time of [11]. One interesting property of our algorithm is that if a better termination time is available for the single message flooding, then the termination time improves proportionally.

Techniques. Suppose all messages are initiated in the beginning of round 1. Knowing D, the proposed algorithm asks messages to start their flooding process in the interval of $(2D + 1)$ rounds, i.e., at rounds $1, (2D + 1) + 1, 2 \cdot (2D + 1) + 1, \ldots, (k-1) \cdot (2D+1)+1$. Suppose the source nodes of $k > 1$ messages $\theta^1, \ldots, \theta^k$

---

[2] If eccentricity $e_1, e_2, \ldots, e_{k'}$ of the k' source nodes is known instead of D, then the bounds translate to $e_{\max} \cdot (2k - 1)$ in bipartite graphs and $(2e_{\max} + 1) \cdot (2k - 1)$ in non-bipartite graphs with memory $O(\log(\max\{k, e_{\max}\}))$ bits, where $e_{\max} := \max_{1 \le l \le k'} e_l$.

know the rank (ID) of all the messages, say, $1, \ldots, k$, with message θ_i having rank i. Let us call this rank order as *global rank*. Knowing the global rank, θ^i can immediately decide how long to wait before starting the flooding process. Since it is known that a single message θ^i finishes flooding in $(2D + 1)$ rounds [11] ($e \leq D$), all k messages finish flooding by $k \cdot (2D + 1)$ rounds. That is, the ranked full-send problem terminates in $k \cdot (2D + 1)$ rounds.

The challenge to overcome is when the source nodes do not know the global rank of the messages (the unranked problem). We devise an algorithm that takes into account local ranks of the messages (i.e., the positions in the ranks of the messages at a node) in deciding the wait time for the messages. Except the globally lowest ranked message, the wait time assigned at round 1 may not be equal to its wait time knowing its global rank. The algorithm asks locally lowest ranked messages to start amnesiac flooding at round $(\kappa - 1) \cdot (2D + 1) + 1$, $\kappa \geq 1$ following the single message algorithm of Hussak and Trehan [11]. If the message that starts flooding at round $(\kappa - 1) \cdot (2D + 1) + 1$, $\kappa \geq 1$, is globally κ ranked, we show that it terminates by round $\kappa \cdot (2D + 1)$; otherwise during the round between $(\kappa - 1) \cdot (2D + 1) + 2$ and $\kappa \cdot (2D + 1) + 1$ (inclusive), it finds that its global rank is higher than κ and starts waiting increasing its wait time proportional to its local rank at that time. We will also show that the wait time update stops at round $(\kappa' - 1) \cdot (2D) + 1$ for the globally κ' ranked message. This altogether guarantees that the algorithm terminates in $k \cdot (2D + 1)$ rounds for $k > 1$ messages.

Finally, we show that this approach extends to the case of messages initiated at different rounds with termination time at most $(2k - 1) \cdot (2D + 1)$. For bipartite graphs, the only change is replacing $(2D + 1)$ with D so that the bound becomes $(2k - 1) \cdot D$.

Related Work. Hussak and Trehan [11] were the first to consider amnesiac flooding. They showed that amnesiac flooding of a single message θ starting from a distinguished source node in the beginning of round 1 terminates in e rounds in bipartite graphs and in $e + D + 1$ rounds in non-bipartite graphs, $e \leq D$. They showed in [12] that this result also holds even when a single message θ starts flooding in the beginning of round 1 from multiple source nodes. In the asynchronous setting, they showed that amnesiac flooding does not terminate even for a single message starting from a source node. Recently, Hussak and Trehan [12] introduced dynamic amnesiac flooding initiated in multiple rounds by possibly multiple source nodes with possibly multiple messages. They showed that the partial-send and ranked full-send problems terminate but the unranked full-send problem does not terminate. In this paper, we show that the unranked full-send problem also terminates, provided that D is known.

Turau [24] improved the result of Hussak and Trehan [11] such that the amnesiac flooding terminates in $e + 1$ rounds, even in non-bipartite graphs. This result is interesting since this termination time matches the classic flooding termination time of $D + 1$, since $e \leq D$. This result also applies to the single message starting flooding from multiple source nodes in the beginning of round 1. However, the assumption behind this result – the second round depending on the first – makes this result not truly amnesiac. Turau [24] also proved that the problem of selecting κ source nodes with minimal termination time is NP-hard. Particularly, Turau showed that unless NP = P there is no approximation algorithm for amnesiac flooding with approximation ratio $3/2 - \epsilon$. For asynchronous systems, Turau proved that deterministic amnesiac flooding is only possible if a large enough part of the message can be updated by each node. Very recently, Turau [25] provided an alternative detailed proof for the single message flooding starting from multiple source nodes in the beginning of round 1. Specifically, Turau showed that, for every non-bipartite graph G and every set V' of source nodes that start flooding simultaneously, there exists a bipartite graph $G(V')$ such that the execution of amnesiac flooding on both graphs G and $G(V')$ is strongly correlated and termination times coincide. This led to bounds that are independent of the diameter as well as it allowed to determine source nodes for which amnesiac flooding terminates in minimal time. Turau also gave tight lower and upper bounds for the time complexity in special cases of $|V'| = 1$ and $|V'| > 1$. In fact, the case of $|V'| > 1$ was reduced to the case of $|V'| = 1$.

Flooding is a fundamental concept used in solving a diverse set of fundamental problems in distributed computing, e.g., leader election [14,15], spanning tree construction [2,13,16,17,21], shortest paths computation [9,10,20], aggregation [5], routing [18], etc. Flooding of multiple messages is a must in many distributed applications, e.g., k-information dissemination or gossiping [1,3–5,7,16,17,19,22].

Amnesiac flooding uses the most recent edges from which the message is received to a node to decide which neighboring edges of that node are used to flood the message from that node. This concept finds applications and uses in social networks [6], broadcasting [8], and client-server application design [25]. More details in [11,12,23–25].

References

1. Ahmadi, M., Kuhn, F., Kutten, S., Molla, A.R., Pandurangan, G.: The communication cost of information spreading in dynamic networks. In: Proceedings of 39th IEEE International Conference on Distributed Computing Systems (ICDCS), pp. 368–378 (2019)
2. Attiya, H., Welch, J.L.: Distributed Computing - Fundamentals, Simulations, and Advanced Topics. Wiley Series on Parallel and Distributed Computing, 2nd edn. Wiley, Hoboken (2004)
3. Borokhovich, M., Avin, C., Lotker, Z.: Tight bounds for algebraic gossip on graphs. In: IEEE International Symposium on Information Theory, ISIT 2010, Austin, Texas, USA, 13–18 June 2010, Proceedings, pp. 1758–1762 (2010)
4. Boyd, S.P., Ghosh, A., Prabhakar, B., Shah, D.: Randomized gossip algorithms. IEEE Trans. Inf. Theory 52(6), 2508–2530 (2006)
5. Chen, J., Pandurangan, G.: Almost-optimal gossip-based aggregate computation. SIAM J. Comput. 41(3), 455–483 (2012)

6. Doerr, B., Fouz, M., Friedrich, T.: Social networks spread rumors in sublogarithmic time. In: Proceedings of the Forty-Third Annual ACM Symposium on Theory of Computing, STOC '11, pp. 21–30. Association for Computing Machinery, New York (2011). https://doi.org/10.1145/1993636.1993640

7. Dutta, C., Pandurangan, G., Rajaraman, R., Sun, Z., Viola, E.: On the complexity of information spreading in dynamic networks. In: SODA, pp. 717–736 (2013)

8. Elsässer, R., Sauerwald, T.: The power of memory in randomized broadcasting. In: Proceedings of the Nineteenth Annual ACM-SIAM Symposium on Discrete Algorithms (SODA), pp. 218–227 (2008)

9. Ghaffari, M., Li, J.: Improved distributed algorithms for exact shortest paths. In: STOC, pp. 431–444 (2018)

10. Holzer, S., Wattenhofer, R.: Optimal distributed all pairs shortest paths and applications. In: PODC, pp. 355–364 (2012)

11. Hussak, W., Trehan, A.: On the termination of flooding. In: STACS, pp. 17:1–17:13 (2020)

12. Hussak, W., Trehan, A.: Terminating cases of flooding. CoRR abs/2009.05776 (2020). https://arxiv.org/abs/2009.05776

13. Kshemkalyani, A.D., Singhal, M.: Distributed Computing: Principles, Algorithms, and Systems. Cambridge University Press, Cambridge (2011)

14. Kutten, S., Pandurangan, G., Peleg, D., Robinson, P., Trehan, A.: On the complexity of universal leader election. J. ACM **62**(1), 7:1–7:27 (2015). https://doi.org/10.1145/2699440

15. Kutten, S., Pandurangan, G., Peleg, D., Robinson, P., Trehan, A.: Sublinear bounds for randomized leader election. Theor. Comput. Sci. **561**, 134–143 (2015). https://doi.org/10.1016/j.tcs.2014.02.009

16. Lynch, N.A.: Distributed Algorithms. Morgan Kaufmann Publishers Inc., San Francisco (1996)

17. Peleg, D.: Distributed Computing: A Locality-Sensitive Approach. Society for Industrial and Applied Mathematics, USA (2000)

18. Rahman, A., Olesinski, W., Gburzynski, P.: Controlled flooding in wireless ad-hoc networks. In: International Workshop on Wireless Ad-Hoc Networks, pp. 73–78 (2004)

19. Sarma, A.D., Molla, A.R., Pandurangan, G.: Distributed computation in dynamic networks via random walks. Theor. Comput. Sci. **581**, 45–66 (2015)

20. Tanenbaum, A.S., Wetherall, D.J.: Computer Networks, 5th edn. Prentice Hall Press, Hoboken (2010)

21. Tel, G.: Introduction to Distributed Algorithms, 2nd edn. Cambridge University Press, Cambridge (2001)

22. Topkis, D.M.: Concurrent broadcast for information dissemination. IEEE Trans. Softw. Eng. **11**(10), 1107–1112 (1985)

23. Turau, V.: Analysis of amnesiac flooding. CoRR abs/2002.10752 (2020). https://arxiv.org/abs/2002.10752

24. Turau, V.: Stateless information dissemination algorithms. In: Richa, A.W., Scheideler, C. (eds.) SIROCCO 2020. LNCS, vol. 12156, pp. 183–199. Springer, Cham (2020). https://doi.org/10.1007/978-3-030-54921-3_11

25. Turau, V., et al.: Amnesiac flooding: synchronous stateless information dissemination. In: Bureš, T. (ed.) SOFSEM 2021. LNCS, vol. 12607, pp. 59–73. Springer, Cham (2021). https://doi.org/10.1007/978-3-030-67731-2_5

Optimal Exclusive Perpetual Grid Exploration by Luminous Myopic Robots Without Common Chirality

Arthur Rauch[1], Quentin Bramas[1]([✉])[ID], Stéphane Devismes[2],
Pascal Lafourcade[3][ID], and Anissa Lamani[1][ID]

[1] ICUBE, CNRS, University of Strasbourg, Strasbourg, France
bramas@unistra.fr
[2] VERIMAG, Université Grenoble Alpes, Grenoble, France
[3] CNRS UMR 6158, LIMOS, University Clermont Auvergne,
Clermont-Ferrand, France

Abstract. We consider swarms of luminous myopic robots that run in synchronous Look-Compute-Move cycles. These robots evolve in a finite grid and are disoriented, *i.e.*, they have neither global compass nor a common chirality. In this context, we propose optimal solutions for the perpetual exploration of a finite grid. Precisely, we investigate optimality in terms of the visibility range, number of robots, number of colors. In more detail, under the optimal visibility range one, we give an algorithm which is optimal w.r.t. the number of robots: it uses three robots and three colors. Under visibility two, we design an algorithm that uses five robots and only one color, *i.e.*, robots are oblivious.

Keywords: Luminous myopic robots · Perpetual exploration · Finite grid · Exclusiveness

1 Introduction

We consider swarms of *luminous robots* [16], *i.e.*, autonomous robots endowed with visibility sensors, motion actuators, and lights of different colors. Those robots operate in *synchronous* Look-Compute-Move cycles, where they first sense the environment (Look), then choose a destination and update their light color (Compute), and finally move to the chosen destination (Move).

Our goal is to investigate the computational power of such robot swarms. Hence, we consider luminous robots with limited capabilities. First, they are *myopic*, *i.e.*, they are only able to sense their surroundings within a bounded visibility range. Furthermore, they are fully disoriented since they have neither a *global compass* nor a *common chirality*. Finally, except from their lights, robots have neither persistent memories nor communication capabilities.

This study was partially supported by the French ANR projects ANR-16-CE40-0023 (DESCARTES) and ANR-16 CE25-0009-03 (ESTATE).

© Springer Nature Switzerland AG 2021
K. Echihabi and R. Meyer (Eds.): NETYS 2021, LNCS 12754, pp. 95–110, 2021.
https://doi.org/10.1007/978-3-030-91014-3_7

We are interested in coordinating such weak robots to solve a infinite global task called the *perpetual exploration*. Given a space which is partitioned into locations, it requires each of these locations to be visited infinitely often by at least one robot. Here, we conveniently discretize the space by a finite graph, where nodes represent locations and edges represent the possibility for a robot to move from one location to another.

In this paper, we look for optimal *exclusive* solutions to the perpetual exploration of a *finite grid*, both in terms of visibility range and number of robots. Exclusiveness [1] requires any two robots to never simultaneously occupy the same position nor traverse the same edge.

Related Work. The exploration problem is a problem that has been extensively investigated. Various topologies have been considered, *e.g.*, lines [13], rings [2,7, 10,14,15], trees [12], torus [9], finite [3,8] and infinite grids [4,5]. In particular, it is shown in [4] that, without a common chirality, exploring an infinite grid with oblivious[1] synchronous robots is impossible under visibility range one, whatever be the number of robots. This latter result is established by proving that, under these settings, robots are not able to move from an arbitrary distance. Hence, it also applies to grid of unbounded size.

In finite graphs, many papers [7–10,12–14] consider the terminating version of the exploration (henceforth called *terminating* exploration), which requires that all robots eventually stop moving after all nodes have been visited. The perpetual exploration requires each location to be visited infinitely often by all or a part of robots. Perpetual exploration of finite graphs has been, for example, considered in [2,3,6].

A large part of the literature deals with "*non-myopic*" robots, *i.e.*, robots with an unbounded visibility range allowing them to sense the whole graph and the positions of all the robots; see [2,3,8–10,12–14]. In such a context, robots are always assumed to be anonymous and oblivious. Exploration algorithms satisfying exclusiveness are proposed in both finite [2,3,6] and infinite graphs [4,5].

Chirality is usually assumed in the 2D Euclidean plan; see for example [11]. However, recently, few works dedicated to discrete environment, *e.g.*, in (infinite) graphs [4], investigated the exploration problem assuming robots which have a common chirality. Chirality is important when dealing with optimal solutions. For example, with visibility range one and few colors ($O(1)$), five (resp. six) synchronous robots are necessary and sufficient to explore an infinite grid with (resp. without) the common chirality assumption [4,5].

A recent work [6] studies the exploration problem in finite grid, with myopic, synchronous, and luminous robot (like our model here), yet assuming robots agree on a common chirality. In a nutshell, it is shown in [6] that two robots with three colors and a common chirality are necessary and sufficient to solve the problem under visibility range one. Moreover, under visibility range two and assuming a common chirality, three robots are necessary and sufficient when robots have only one color.

---

[1] Oblivious robots have no state and cannot remember past actions.

Table 1. Summary of our results.

Visibility	# Robots	# Colors	Algorithm
1	2	Finite	Impossible (Theorem 1)
1	3	3	Vone_3^3
2	5	1	Vtwo_1^5

Contribution. To the best of our knowledge, the present work is the first study of the (perpetual) exclusive exploration with myopic (luminous) robots in finite grids with robots without chirality.

Our contribution is threefold. We prove that, under any finite visibility range, the perpetual exploration is not solvable using only two robots, whatever be the finite number of available colors. Then, we present a perpetual exploration algorithm that is optimal in terms of visibility range (1) and number of robots (3). Moreover, this algorithm only requires 3 colors per robots. Finally, we propose an algorithm that requires five oblivious robots, *i.e.*, each of those five robots needs only one color (the optimal), yet assuming visibility range two. Nevertheless, following results in [4], visibility range two is the smallest range where a solution with oblivious robots is possible. Table 1 summarizes our contributions.

Roadmap. Section 2 is devoted to the computational model and definitions. In Sect. 3, we prove our impossibility result. We present our algorithm in Sects. 4 and 5. We make concluding remarks in Sect. 6.

2 Model

We consider a set of $n > 0$ robots located on a *finite grid* made of $\mathcal{L} \geq n$ lines and $\mathcal{C} \geq n$ columns,[2] *i.e.*, robots evolve in an undirected graph $G(V, E)$ where $V = \{(i, j) : i \in [0, \mathcal{C} - 1], j \in [0, \mathcal{L} - 1]\}$ and $E = \{\{(i, j), (k, l)\} : (i, j) \in V \wedge (k, l) \in V \wedge |i - k| + |j - l| = 1\}$. The size of the grid is then $\mathcal{L} \times \mathcal{C}$. Grid coordinates are used for the analysis only, *i.e.*, robots cannot access them.

We assume a discrete time where, at each *round*, the robots synchronously perform a *Look-Compute-Move* cycle. In the *Look* phase, a robot gets a snapshot of the subgraph induced by the nodes within distance $\Phi \in \mathbb{N}^*$ from its position. Φ is called the *visibility range* of the robots. The snapshot is not oriented in any way as the robots do not agree on a common North. However, it is implicitly ego-centered since the robot that performs a Look phase is located at the center of the subgraph in the obtained snapshot. Then, each robot *computes* a destination (either Up, Left, Down, Right or Idle) based only on the snapshot it received. Finally, it *moves* towards its computed destination.

We forbid any two robots to occupy the same node simultaneously. A node is *occupied* when a robot is located at this node, otherwise it is *empty*. Robots

---

[2] The requirement on the numbers of lines and columns is only made for the sake of simplicity.

may have *lights* with different colors that can be seen by robots within distance Φ from them. We denote by Cl the set of all possible colors.

The *state* of a node is either the color of the light of the robot located at this node, if it is occupied, or $\perp$ otherwise. In the Look phase, the snapshot includes the state of the nodes (within distance Φ). During the compute phase, a robot may decide to change the color of its light.

In all our algorithms, we also prevent any two robots from traversing the same edge simultaneously. Since we already forbid them to occupy the same position simultaneously, this means that we additionally prevent robots from swapping their position. Algorithms verifying this property are said to be *exclusive*. However, to be as general as possible, we do not make this additional assumption in our impossibility results.

Configurations. A *configuration* C in a grid $G(V, E)$ is a set of pairs (p, c), where $p \in V$ is an occupied node and $c \in Cl$ is the color of the robot located at p. A node p is empty if and only if $\forall c, (p, c) \notin C$. We sometimes just write the set of occupied nodes when the colors are clear from the context.

Views. We denote by G_r the *globally oriented view* centered at the robot r, i.e., the subset of the configuration containing the states of the nodes at distance at most Φ from r, translated so that the coordinates of r is $(0, 0)$. We use this globally oriented view in our analysis to describe the movements of the robots (see, for example, Fig. 1): when we say "the robot moves Up", it is according to the globally oriented view. However, since robots do not agree on a common North, they have no access to the globally oriented view. When a robot looks at its surroundings, it instead obtains a snapshot. To model this, we assume that the *local view* acquired by a robot r in the Look phase is the result of an arbitrary *indistinguishable transformation* on G_r. The set $\mathcal{IT}$ of indistinguishable transformations contains:

1. the rotations of angle 0 (to have the identity), $\pi/2$, π and $3\pi/2$, centered at r,
2. the mirroring (robots cannot distinguish between clockwise and counterclockwise), and
3. any combination of rotation and mirroring.

Here, we assume that robots are *self-inconsistent*, meaning that different transformations may be applied at different rounds.

It is important to note that when a robot r computes a destination d, it is relative to its local view $f(G_r)$, which is the globally oriented view transformed by some $f \in \mathcal{IT}$. So, the actual movement of the robot in the *globally oriented view* is $f^{-1}(d)$. For example, if $d = Up$ but the robot sees the grid upside-down (f is the π-rotation), then the robot moves $Down = f^{-1}(Up)$. In a configuration C, $V_C(i, j)$ denotes the globally oriented view of a robot located at (i, j).

A robot is said to be *isolated* when the only robot in its view is itself.

Algorithm. An algorithm A is a tuple $(Cl, Init, T)$ where Cl is the set of possible colors, $Init$ is a mapping from any considered grid to a non-empty set of initial configurations in that grid, and T is the transition function $Views \rightarrow \{Idle, Up, Left, Down, Right\} \times Cl$, where $Views$ is the set of local views. When the robots are in Configuration C, a configuration C' obtained after one round satisfies: for all $((i, j), c) \in C'$, there exists a robot in C with color $c' \in Cl$ and a transformation $f \in \mathcal{IT}$ such that one of the following conditions holds:

- $((i, j), c') \in C$ and $f^{-1}(T(f(V_C(i, j)))) = (Idle, c)$,
- $((i-1, j), c') \in C$ and $f^{-1}(T(f(V_C(i-1, j)))) = (Right, c)$,
- $((i+1, j), c') \in C$ and $f^{-1}(T(f(V_C(i+1, j)))) = (Left, c)$,
- $((i, j-1), c') \in C$ and $f^{-1}(T(f(V_C(i, j-1)))) = (Up, c)$, or
- $((i, j+1), c') \in C$ and $f^{-1}(T(f(V_C(i, j+1)))) = (Down, c)$.

We denote by $C \mapsto C'$ the fact that C' can be reached in one round from C (*n.b.*, $\mapsto$ is then a binary relation over configurations). An execution of Algorithm A in a grid G is then a sequence $(C_i)_{i \in \mathbb{N}}$ of configurations such that $C_0 \in Init(G)$ and $\forall i \geq 0$, $C_i \mapsto C_{i+1}$.

Perpetual Finite Grid Exploration. An execution $(C_i)_{i \in \mathbb{N}}$ in a grid $G = (V, E)$ achieves the *Perpetual Finite Grid Exploration* (PFGE) if for every node $u \in V$ and for every time t, there exists a time $t' \geq t$ such that u is occupied in $C_{t'}$.

An algorithm A that uses n robots solves the *Perpetual Finite Grid Exploration* (PFGE) problem if for every finite grid $G = (V, E)$ with at least n lines and n columns and every initial configuration $C_0 \in Init(G)$, we have every execution of A in G starting from C_0 that achieves the PFGE.

An Algorithm as a Set of Rules. We write an algorithm as a set of rules, where a *rule* is a triplet $(V, d, c) \in Views \times \{Idle, Up, Left, Down, Right\} \times Cl$.

We say that an algorithm $(Cl, Init, T)$ includes the rule (V, d, c), if $T(V) = (d, c)$. By extension, the same rule applies to indistinguishable views, *i.e.*, $\forall f \in \mathcal{IT}, T(f(V)) = (f(d), c)$. Consequently, we forbid an algorithm to contain two rules (V, d, c) and (V', d', c') such that $V' = f(V)$ for some $f \in \mathcal{IT}$. Hence, an algorithm corresponds to a set of rules if each destination is the result of applying one of its rules.

As an illustrative example, consider the rule R_1 given in Fig. 1. This rule is defined for robots having a visibility range of two. This rule means that, when a blue robot B sees two robots with color R, one on top and one in diagonal, then the blue robot is dictated to move Up. By extension the same rule applied if the view is rotated by π, but in that case, the destination would be Down.

In the same figure, Rule R_2 is a rule where the three black nodes represent a part of the outer boundary of the grid, that we call *a wall* in the remaining of the paper. In our algorithms, we often define similar rules that apply regardless of the presence of a wall in some part of the view. Thus, to avoid defining several time rules with very similar views, we propose a notation to represent several rules in just one picture. For example, Rule R_3 in Fig. 1 has three nodes hatched with vertical lines, which means that the rule applies regardless of the presence of a wall located at those nodes. In practice, every rule that contains such vertical (resp. horizontal) hatched lines, represents a set of rules obtained by replacing each of those lines either by walls, or by empty nodes. For example, Rule R_3 in Fig. 1 is a concise representation of Rules R_1 and R_2.

Notice also that, due to the absence of orientation and chirality, a rule (V, d, c) may be ambiguous, meaning that there exists $f \in \mathcal{IT}$ such that $T(V) \neq f^{-1}(T(f(V)))$. In the figures, we illustrate such ambiguities by depicting the possible destinations with several arrows. For example, Fig. 2 shows an ambiguous rule where the robot has a symmetric view. Hence depending on the transformation f chosen by the adversary, the robot moves either left or right when executing this rule.

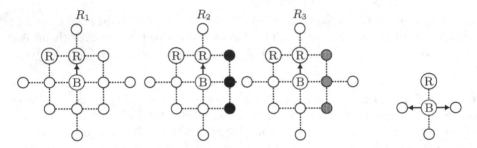

Fig. 1. Examples of rules.   **Fig. 2.** Example of an ambiguous rule.

Algorithms Having Locally-Defined Initial Configurations. In a given grid, the set of possible initial configurations of an algorithm can be reduced to a singleton. In such a case, the scalability and flexibility of the algorithm is weak. To be more general, we propose algorithms with *locally-defined* sets of initial configurations. Configurations in a locally-defined set of initial configurations are defined by one and the same pattern which fixes the colors and relative positions of the robots. Hence, for a given grid, every two possible initial configurations are equal up to a translation applied to all robot positions and the set of all possible initial configurations is closed by such translations.

3   Impossibility Results

It has been shown in [6] that the PFGE problem is not solvable using only one robot for any finite visibility range. We now extend this result by proving that the PFGE problem is also not solvable using two robots if they have a visibility range of one. Hence, throughout this section, **we assume two robots under visibility range one.**

First, we observe that in large enough grids, if robots travel a long distance without seeing any wall, or seeing one and the same wall without reaching its corner, then they must execute a periodic sequence of movements. Indeed, the maximum number of distinct relative positions and colors two robots endowed with $|Cl|$ colors can have is the number of 2-combination with repetitions $B = \binom{|Cl|}{2} = \frac{|Cl|(|Cl|+1)}{2}$. Thus, if robots travel a distance at least B without seeing a wall, or seeing one and the same wall without reaching its corner, then they are actually executing a periodic sequence of movements. Of course, the value of B depends on the algorithm, yet it is always finite. Notice also that $|Cl| > 1$, since it has been shown in [6] that two oblivious robots with visibility range 1 are not sufficient to solve the PFGE problem. Hence, $B \geq 3$.

The above observations are important to prove our impossibility results. First, we use them to show that once robots move far away from the wall, their movements are restricted. In more detail, they can only move in straight line; see Lemmas 1 and 2.

Lemma 1. *Let* A *be an algorithm solving the PFGE problem with two robots. If there exists an execution of* A *containing a configuration* C *where the two robots are at distance at least $2B$ from any wall and, from* C, *the robots perform a periodic sequence of movements with no ambiguous rules, then the robots move in a straight line until reaching a wall.*

Sketch of proof: When a robot executes unambiguous rules, it can only move from or towards the other robots, hence remains on the same line. Indeed, any view containing another robot has an axis of symmetry passing through the other robot (recall that we assume visibility range 1), and the destination of an unambiguous rule must be on the axis as well.   □

Lemma 2. *Let* A *be an algorithm solving the PFGE problem with two robots. If there exists an execution of* A *containing a configuration* C *where robots are at distance at least $2B$ from any wall and, from* C, *robots perform a periodic sequence of movements, then this sequence does not include any ambiguous rule.*

Sketch of proof: Every time robots execute an ambiguous rule, robots are making a turn, and the adversary can decide on which side the robots are turning. If the periodic sequence of movements contains an ambiguous move, the robots will make at least one turn per period, hence the adversary can make the robots remain in the same square grid of size B (the period of the sequence is at most B). While doing so, the robots do not see any wall, and do not explore the whole grid.   □

Due to the limited visibility range, the two robots cannot be to far from each other, as stated in the following three lemmas.

Lemma 3. *Robots are always at distance at most 6 for each other.*

Lemma 4. *No exploration algorithm can reach a configuration where the two robots are at distance at least 3, one robot sees no wall, and the other sees zero or one wall.*

Lemma 5. *If both robots see no wall, then they should be at distance one from each other.*

The next Lemma states that, if two robots are on the same line, this line must be an axis of symmetry of their views and they cannot break this symmetry without executing an ambiguous rule (due to the lack of chirality agreement). Hence, the adversary can decide on which side of the axis it will keep the robots.

Lemma 6. *Let A be an algorithm solving the PFGE problem using two robots. Let C be a configuration where the two robots are on the same line L. Let R be a set of nodes delimiting a rectangle for which L is an axis of symmetry. Let $R_1 \subset R$ such that the union of R_1 and the symmetric of R_1, with respect to L, is equal to R. Then, from C, a configuration where a robot is located at a node in R_1 is reachable.*

We now prove our main lemma, which states that the robots cannot move further than a distance of $4B$ from all walls. To achieve this, we need two additional definitions. A *corner box* is the set of nodes forming a square of size $2B$ including a corner of the grid. We say robots are in a T-configuration when they are adjacent, only one is adjacent to a wall, and they are both at distance at most $3B$ from another wall.

Lemma 7. *If A solves the PFGE problem with two robots, then, if at a given time $t > B$, a robot is in a corner box or if robots are in a T-configuration, then there exists an execution after C such that a robot ends up a time $t' > t$ either in a corner box or in a T-configuration, and between time t and t' the robots remain at distance at most $4B$ from a wall.*

Proof. We consider a grid of size greater than $4B$, otherwise the lemma is proven regardless of what the robots are doing (a robot is infinitely often in a corner box and any wall at distance $4B$).

Then, assume a robots is in a corner box in a configuration C (the case where robots are in a T-configuration is treated in the last paragraph of this proof) at a given time $t > B$. To explore the grid, the robots must leave the corner box. Indeed, if a robot stays forever in a corner box, then both robots remain as distance at most $2B + 6$ (by Lemma 3) from that corner and, since, $B \geq 3$, $2B + 6 < 4B$ meaning that some node are only finitely often visited. We denote by W_1 and W_2 the two walls adjacent to the corner contained in the corner box where a robot was located in C; see Fig. 3. Without the loss of generality, we assume that at a given time t_0, the last robot, say r, leaving the corner box of size $2B$ is at distance $2B + 1$ from W_1, and so at distance at most $2B$ from W_2.

Claim 1: *After leaving the corner box from a given side, either (i) the robots move until reaching the wall opposite to W_1, in a T-configuration, while remaining at distance $2B$ from Wall W_2, or (ii) end up in a line L parallel and at distance at most $4B$ to W_1, while remaining at distance at most $2B+1$ from Wall W_2.*

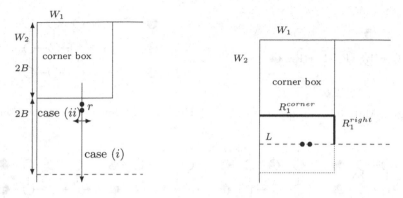

Fig. 3. The different cases in Lemma 7.

From the previous claim, we saw that two cases can occur; see Fig. 3. In the first case, the lemma is proven.

In the second case, robots end up in a line L parallel to W_1 at time a given $t_1 \geq t_0$, while remaining at distance at most $2B + 1 < 4B$ from wall W_2. We consider the set of nodes $R_1 = R_1^{corner} \cup R_1^{right}$ where R_1^{corner} is the segment of nodes at distance $2B$ from the wall W_1 and with distance to W_2 in the interval $[0, 2B+1]$, and R_1^{right} is the segment of nodes at distance $2B+1$ from W_2 and at distance from W_1 in the interval $[2B, d_1]$, where d_1 is the distance of the robots to W_1; see Fig. 3 (from the previous Claim, $d_1 \leq 4B$). The union of R_1 with its symmetric with respect to L delimits a rectangle (dotted line in the figure) so that, using Lemma 6, there exists an execution such that a robot reaches R_1.

If a robot reaches R_1^{corner}, then a robot reaches a corner box and the lemma is proven. If a robot reaches R_1^{right}, then the robots have traveled a distance at least B without seeing a wall, hence are executing a periodic sequence of movements. The sequence cannot contain an ambiguous rule (by Lemma 2) because the robots are at distance at least $2B$ from any wall, so they are moving in a straight line (by Lemma 1), and they end up in the wall opposite to W_2 and reach a T-configuration, while remaining at distance at most $4B$ from W_1.

We now consider the case where robots are in a T-configuration in configuration C. Then, they are on a line L perpendicular to a wall, say W_2. Using a similar argument, we know that either the robot enter the closest corner box, or move in a straight line to the opposite wall until they reach a T-configuration.

$\square$

We can now prove our impossibility result.

Theorem 1. *The PFGE is not solvable with a only two robots with visibility range 1, for any bounded number of colors.*

Proof. Assume that algorithm A solves the PFGE problem and consider a grid of size $9B \times 9B$. Since the robots should explore the entire grid, a robot is eventually in a corner box. Using Lemma 7 repeatedly, we can construct a execution from

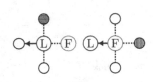 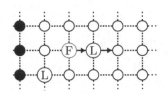

Fig. 4. Move in a straight line.

Fig. 5. Beginning of the exploration of a line.

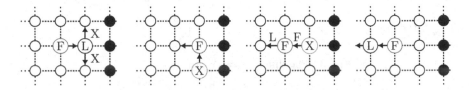

Fig. 6. Sequence of configurations during a turn around.

there where the two robots remain at distance at most $4B$ from any wall. Hence, nodes at distance more than $4B$ from all the walls are not visited anymore, a contradiction. □

4   Visibility Range One: Vone_3^3

In this section, we present an algorithm, denoted by Vone_3^3, which assumes visibility range one (the optimal) and uses three robots endowed with three colors. By Theorem 1, Vone_3^3 is optimal in terms of number of robots. We encourage the reader to take a look at the online animation illustrating the behavior of Vone_3^3 [17] while reading the following explanation.

The algorithm defines three roles for the robots using the colors: L (*leader*), F (*follower*), X (*landmark*). The roles are not fixed, robots will alternate between several roles along the execution. Moreover, in few particular situations, roles will not exactly correspond to their default meanings.

Initially, the three robots are aligned, two of them have color L while the third one has color F; moreover the two robots with color L are adjacent. In the following, we denote this pattern by LLR. Since initial configurations are locally-defined, the possible initial configurations are then all those containing the pattern LLR.

Since we assume the synchronous model and we consider the perpetual exploration, the execution is necessarily eventually periodic. So, from an initial configuration, the goal is to lead robots to a configuration C_p from which they will start to perform periodic journeys around the grid. We first explain how periodic journeys are built. Then, we will see how robots can easily reach a configuration of the journey starting from any initial configuration.

The main idea of the algorithm is to make the leader and the follower move and explore a given line while the landmark robot remains idle to keep track of

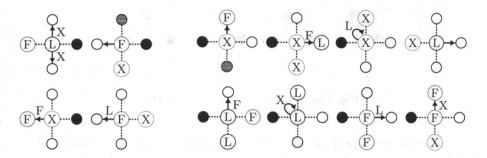

Fig. 7. Turn around.   **Fig. 8.** Up and turn.

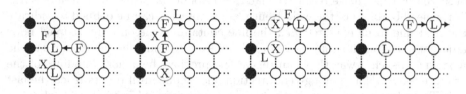

Fig. 9. Up and turn. This sequence occurs when the robots are not in a corner. The case when the robots are in a corner is presented in Fig. 11.

the next line to explore. Every time a line is explored, the three robots, including the landmark robot, move "up" by one row (assuming, for illustration purpose, that robots are visiting a line from left to right). Then, once the robots reach a corner, they change their direction and repeat the same process.

It is easy to make move the leader and the follower on the same direction to explore a line: the leader moves away from the follower while the follower, as suggested by its name, follows the leader. The rules executed by those two robots to move along a straight line are presented in Fig. 4.

During the line exploration by the leader and the follower, the landmark robot is left beside a wall on a line adjacent to the line traversed by the two other robots; refer to Fig. 5. When the leader and the follower reach the other wall, the idea is to make them move back and cross the same line again since they do not have any sense of direction. For this purpose, they need to swap their respective positions. This is done as follows: the first robot that detects the wall is the leader, in this case, it moves to an adjacent empty node (except for the last line, there is a symmetry and so the scheduler chooses which direction to take) and changes its color to X. In the next round, the follower reaches the wall and observes the landmark, *i.e.*, the previous leader. Since the follower sees only one other robot, it detects that they are moving back to traverse the same line in the other direction. So, the follower moves back to its previous position followed by the landmark. Moreover, the follower becomes the leader while the landmark becomes a follower. Finally, they both start moving straight on the opposite direction. The rules executed during the moving back process are those

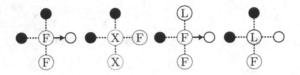

Fig. 10. Last corner preparation - rules.

given in Fig. 7 plus the first rule of Fig. 9 (this latter will also be used when switching to an upper line).

When the leader and follower reach again the wall again, the leader can observe this time that a robot (the landmark robot) is located on a different line in the neighborhood of the wall. Hence, an orientation can be defined to indicate the next line to be explored *i.e.*, the line containing the only unoccupied node adjacent to the one hosting the leader. The idea is to make the robots move to the next line in such a way they can repeat the previous behavior. The lines of the grid are then explored in a given direction one by one until robots reach the last line. The rules that are executed to make a line change, when the landmark robot is reached, are presented in Fig. 8. Figure 9 shows the sequence of configurations occurring during a line change.

Given an orientation of the grid, assume without the loss of generality that the robots are exploring the grid line by line in a given direction. As the grid is finite, eventually the robots reach the last line with respect to the current exploring direction. When this happens, the robots change the exploring direction by a clockwise angle of π. The robots then exhibit the same behavior as previously: they explore the lines of the grid with respect to the new orientation. Note that this change of direction is initiated by the first robot to join the last line (the leader) as it is located at a corner. The change of direction is done through several rules that are presented in Fig. 10, while the sequence of configurations composing this process are presented in Fig. 11.

Assume initially the robots are all adjacent to a wall (remember that they are aligned and their colors are respectively F, L and L). Then, we have defined few rules in order for the robot to reach, after one round, a configuration of the periodic journey. After that, robots behave exactly as previously explained. Starting from any other initial configuration, the goal is to move straight toward a wall. Once the leader robots see the wall, it moves to an unoccupied node and the reached configuration is exactly the same as the first one of an "up and turn" sequence. Hence after that, the periodic journeys start. The rules used by the robots to do this are shown in Fig. 13.

For the grids with 3 lines or 3 columns, a specific rule is needed as any "up and turn" sequence is considered to be done at a corner. The rule is shown in Fig. 14 and the sequence of movements when the grid has only 3 lines or 3 columns is shown in Fig. 15 (the specific rule is used in the fifth round of the sequence).

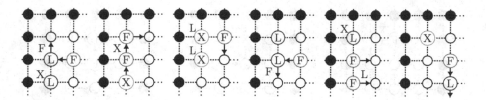

Fig. 11. Corner turn. After the sequence, the exploration continues as before, but everything is rotated by a clockwise angle of π.

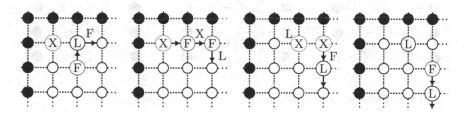

Fig. 12. First "up and turn" after the corner turn.

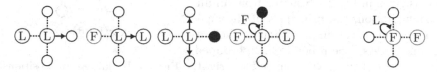

Fig. 13. Rules used by the robots to reach the wall starting from a configuration where they shape an LLF pattern.

Fig. 14. Rules used by robots to handle grid with only 3 columns.

Theorem 2. Vone_3^3 *solves the perpetual exploration problem with three robots, having three colors and visibility range one.*

Sketch of proof: Using our simulation tool, we were able to prove that our algorithm is correct for any grid $n \times m$, with $m, n \in \{3, 4\}$. Then, we have shown that when a group of robots is traveling along a row, adding a column does not change the relative position of the robots when they reach a wall. Similarly, adding a row does not change the relative position of the robots when they reach a corner. The sequence of movement performed in a corner does not depend on the size of the grid, so that, regardless of the size of the grid, the robots explore the entire grid in a perpetual manner.   □

5 Visibility Range Two : Vtwo_1^5

We now outline our second algorithm, Algorithm Vtwo_1^5, which requires five oblivious robots (*i.e.*, they all have the same fixed color) and visibility range of 2. Again, we encourage the reader to follow the explanation of the algorithm while looking at the animations available online [17].

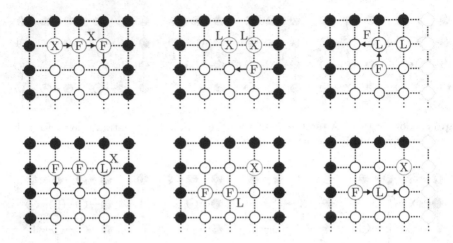

Fig. 15. Sequence of configurations when the grid has only 3 columns.

The initial relative position of robots in Vtwo_1^5 is given in Fig. 16. Starting from an initial configuration, the principles of the algorithm are similar, yet simpler, than for the previous one. Indeed, the robots remain grouped together, and they move from left to right,

Fig. 16. Initial relative positions.

without making any rotation when reaching a wall. Every time the group of robots reaches a wall, they perform a turn sequence to move back to the opposite wall, one row above or below, depending on the current orientation of the group (see Fig. 19 for a turn one row below). After moving straight (see Fig. 17) to the opposite wall, everything is mirrored, so they do the same. They move back an forth until they reach the top wall. After following the top wall (using a specific periodic sequence of movements, see Fig. 18), they make a special turn in order to move back and forth in the other direction.

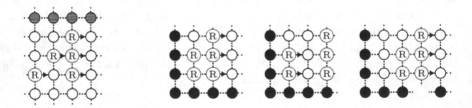

Fig. 17. Move in a straight line.   **Fig. 18.** Follow the wall.

The proof of the next theorem is similar to that of Theorem 2.

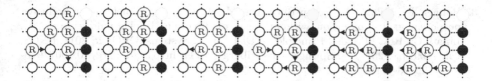

Fig. 19. Turn around.

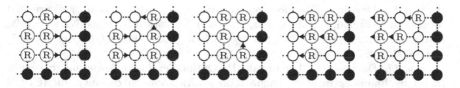

Fig. 20. Turn at a corner.

Theorem 3. Vtwo$_1^5$ *solves the perpetual exploration problem with five oblivious robots under visibility range two.*

6 Conclusion

We have investigated the perpetual exclusive exploration of a finite grid by a swarm of myopic luminous synchronous robots that have neither a common sense of direction nor a common chirality. In these settings, We have proposed optimal solutions with respect to either the number of robots, the visibility range, and the number of colors.

In more detail, we have first shown that if robots have only a visibility range one, then the problem is not solvable with two robots, regardless of the number of colors. Then, we have proposed Vone$_3^3$ which uses three robots and three colors. This algorithm is optimal both in terms of visibility range and number of robots.

Next, under visibility range two, we gave Algorithm Vone$_1^5$. This latter requires five oblivious robots, *i.e.*, five robots that use the minimal number of color (one). Following the impossibility result of [4], visibility range two is the smallest range admitting a solution in our settings.

The immediate open questions related to this work are about determining whether Vone$_3^3$ is also optimal with respect to the number of colors and whether Vone$_1^5$ is optimal with respect to the number of robots. Finally, it would be interesting to extend our study to other topologies such as torus-shaped networks.

References

1. Baldoni, R., Bonnet, F., Milani, A., Raynal, M.: Anonymous graph exploration without collision by mobile robots. Inf. Process. Lett. **109**(2), 98–103 (2008)
2. Blin, L., Milani, A., Potop-Butucaru, M., Tixeuil, S.: Exclusive perpetual ring exploration without chirality. In: Lynch, N.A., Shvartsman, A.A. (eds.) Distributed Computing, DISC 2010. LNCS, vol. 6343, pp. 312–327. Springer, Heidelberg (2010). https://doi.org/10.1007/978-3-642-15763-9_29

3. Bonnet, F., Milani, A., Potop-Butucaru, M., Tixeuil, S.: Asynchronous exclusive perpetual grid exploration without sense of direction. In: Fernàndez Anta, A., Lipari, G., Roy, M. (eds.) Principles of Distributed Systems, OPODIS 2011. LNCS, vol. 7109, pp. 251–265. Springer, Heidelberg (2011). https://doi.org/10.1007/978-3-642-25873-2_18

4. Bramas, Q., Devismes, S., Lafourcade, P.: Finding water on poleless using melomaniac myopic chameleon robots. In: FUN 2020, 10th International Conference on Fun with Algorithms. LiPICs, Favignana, Sicily, Italy, 28–30 September 2020).

5. Bramas, Q., Devismes, S., Lafourcade, P.: Infinite grid exploration by disoriented robots. In: Georgiou, C., Majumdar, R. (eds.) Networked Systems, NETYS 2020. LNCS, vol. 12129, pp. 129–145. Springer, Cham (2021). https://doi.org/10.1007/978-3-030-67087-0_9

6. Bramas, Q., Devismes, S., Lafourcade, P.: Optimal exclusive perpetual grid exploration by luminous myopic opaque robots with common chirality. In: ICDCN 2021: International Conference on Distributed Computing and Networking, Virtual Event, pp. 76–85. ACM, Nara, Japan, 5–8 January 2021

7. Datta, A.K., Lamani, A., Larmore, L.L., Petit, F.: Enabling ring exploration with myopic oblivious robots. In: 2015 IEEE International Parallel and Distributed Processing Symposium Workshop, IPDPS 2015, pp. 490–499. IEEE Computer Society, Hyderabad, India, 25–29 May 2015

8. Devismes, S., Lamani, A., Petit, F., Raymond, P., Tixeuil, S.: Terminating exploration of a grid by an optimal number of asynchronous oblivious robots. Comput. J. 64(1), 132–154 (2020)

9. Devismes, S., Lamani, A., Petit, F., Tixeuil, S.: Optimal torus exploration by oblivious robots. Computing 101(9), 1241–1264 (2019)

10. Devismes, S., Petit, F., Tixeuil, S.: Optimal probabilistic ring exploration by semi-synchronous oblivious robots. Theor. Comput. Sci. (TCS) 498, 10–27 (2013)

11. Dieudonné, Y., Petit, F., Villain, V.: Leader election problem versus pattern formation problem. In: Lynch, N.A., Shvartsman, A.A. (eds.) Distributed Computing, DISC 2010. LNCS, vol. 6343, pp. 267–281. Springer, Heidelberg (2010). https://doi.org/10.1007/978-3-642-15763-9_26

12. Flocchini, P., Ilcinkas, D., Pelc, A., Santoro, N.: Remembering without memory: tree exploration by asynchronous oblivious robots. Theor. Comput. Sci. 411(14–15), 1583–1598 (2010)

13. Flocchini, P., Ilcinkas, D., Pelc, A., Santoro, N.: How many oblivious robots can explore a line. Inf. Process. Lett. 111(20), 1027–1031 (2011)

14. Flocchini, P., Ilcinkas, D., Pelc, A., Santoro, N.: Computing without communicating: ring exploration by asynchronous oblivious robots. Algorithmica 65(3), 562–583 (2013)

15. Ooshita, F., Tixeuil, S.: Ring exploration with myopic luminous robots. In: Izumi, T., Kuznetsov, P. (eds.) Stabilization, Safety, and Security of Distributed Systems, SSS 2018. LNCS, vol. 11201, pp. 301–316. Springer, Cham (2018). https://doi.org/10.1007/978-3-030-03232-6_20

16. Peleg, D.: Distributed coordination algorithms for mobile robot swarms: new directions and challenges. In: Pal, A., Kshemkalyani, A.D., Kumar, R., Gupta, A. (eds.) Distributed Computing – IWDC 2005, IWDC 2005. LNCS, vol. 3741, pp. 1–12. Springer, Heidelberg (2005). https://doi.org/10.1007/11603771_1

17. Rauch, A., Bramas, Q., Devismes, S., Lafourcade, P., Lamani, A.: Optimal exclusive perpetual grid exploration by luminous myopic robots without chirality: the animations (2021). https://doi.org/10.5281/zenodo.4640462

AUCCCR: Agent Utility Centered Clustering for Cooperation Recommendation

Amaury Bouchra Pilet[✉], Davide Frey, and François Taïani

IRISA, Inria, Univ Rennes, CNRS, Rennes, France
amaury.bouchra-pilet@irisa.fr

Abstract. Providing recommendation to agents (e.g. people or organizations) regarding whom they should collaborate with in order to reach some objective is a recurring problem in a wide range of domains. It can be useful for instance in the context of collaborative machine learning, grouped purchases, and group holidays. This problem has been modeled by hedonic games, but this generic formulation cannot easily be used to provide efficient algorithmic solutions. In this work, we define a class of hedonist games that allows us to provide an algorithmic solution to a wide class of collaboration recommendation problems by means of a clustering algorithm. We evaluate our algorithm, theoretically and experimentally, and show that it performs better than other well-known clustering algorithms in this context.

1 Introduction

In this work, we aim to provide an algorithmic solution for people or organizations who wish to collaborate towards some task (buying, machine-learning...) but do not want to cooperate with members whose individual objectives are too different from their own. This problem can be modeled as a set of rational agents that may or may not form coalitions depending on the utility they might derive from such coalitions. More specifically, we consider a model, inspired by prescriptive decision theory [17], in which an agent obtains a utility that is positively correlated with the size of the group it belongs to (the larger the better), and negatively correlated with his/her distance from the group's barycenter (the closer the better). Such a model can capture various practical collaboration problems. In collaborative machine learning, for instance, learners with similar but different tasks may or may not collaborate with each other, depending on the effect of this collaboration on the efficiency of their learning process. In grouped purchases, potential buyers will search for other people with similar buying habits to save money by placing grouped orders, but will not benefit if the products they buy deviate too much from their preferred options. Similarly, when planing organized vacations, most people want to save money with group rates, but not at the cost of visiting too many places they are not interested in. We also consider the case where people want to be in large enough groups but not too large.

© Springer Nature Switzerland AG 2021
K. Echihabi and R. Meyer (Eds.): NETYS 2021, LNCS 12754, pp. 111–125, 2021.
https://doi.org/10.1007/978-3-030-91014-3_8

An algorithm that solves this problem should provide collaboration recommendations (e.g. with whom each agent should perform a grouped order) that agents find acceptable and from which they do not deviate. Since, in this context, the relative utility of different options for an agent depends on the choices of other agents, it is crucial to prevent situations where a few agents reject the recommendation, as this could lead to a complete collapse of the solution. The departure of a dissatisfied agent can decrease the group's value for other agents, which may in turn leave the group, etc.

In economic terms, this problem can be modeled as a *hedonic game* [7], but this formalization tends unfortunately to be too general to allow for effective algorithmic solutions. Existing algorithms that solve generic hedonic games, summarized in [4], need to constrain them by requiring the existence of some kind of equilibrium (Nash equilibrium or a similar definition), which is not guaranteed by the generic definition of hedonic games and may not exist in practice. In our work, we instead consider a more restricted class of games, and we provide a solution that also works in cases where no kind of equilibrium exists.

Hedonic games designed for specific problems have been proposed in the past, for example in [15], but the practical problems they consider do not apply to our context. In particular, these works tend to adopt a formalization that ensures the existence of some kind of equilibrium, while we want solutions for cases where no equilibrium exists.

Our insight is that the hedonic games corresponding to our coalition formation problem can be interpreted as a clustering problem. We want to identify groups of close and numerous agents, which is essentially a clustering task. However, commonly used clustering algorithms, while technically applicable, are not adapted to this particular task. To address this gap, this paper proposes a novel clustering algorithm, that is specifically designed to address the task we want to solve.

We begin this paper with a formal definition of the class of hedonic games we use to model collaboration. We then propose AUCCCR (pronounced "okr", IPA: [okʁ]), a clustering algorithm able to provide solutions to the considered problem. We present a theoretical and experimental evaluation of our algorithm on both synthetic and real data. From this evaluation, we conclude that our algorithm respects individual agents interests better than other clustering algorithms.

2   Our Approach

2.1   Problem Statement and Formalization

We consider a set of agents $a_0, ..., a_m \in A$ that seek to form groups so that every agent in a group effectively benefits from belonging to this group. We represent agents' preferences using an ℓ-dimensional real vector given by a projector function, $p : A \to \mathbb{R}^\ell$, which allows us to measure the similarity/distance between agents. In our model, the benefit from being in a group depends on two factors: the size of the group (the larger the better), and the similarity between an agent and the (average of) the group (the more similar the better). In practice, the chosen projector function is likely to associate each agent a vector of real values taken from a database. For example (what is used in our experiments) statistics on review for holiday destinations or annual spending in different types of goods.

More concretely, we define a *utility* function $U_p(a, g)$ that expresses the interest of an agent, a, in a group, g, using a projector, p, as the product of two factors: *(i)* the value of the group (function v), which grows with the group's size; and *(ii)* a decreasing function $(n(...))$ of the distance between an agent and the group's barycenter (as measured by a distance function $d(...)$). This is formally captured by the following formula:

$$U_p(a, g) : A \times \mathcal{P}(A) \to \mathbb{R}^+ = n\Big(d\big(p(a), bary_p(g)\big)\Big) \times v(\#g) \qquad (1)$$

Where $\mathcal{P}(A)$ is the power set of A, $\#g$ is the size of group g, $bary_p(g)$ is the barycenter of group g with projector p, and the functions d, v, and n are defined as follows: $d(x, y) : \mathbb{R}^l \times \mathbb{R}^l \to \mathbb{R}$ is the *distance* between x and y; $v(n)$: $\mathbb{N} \to [1, +\infty[$ is the *value* of a group of size n (increasing, $v(1) = 1$); and $n(d) : \mathbb{R}^+ \to [0, 1]$ is the *normalizer* for an agent at distance d from its group's barycenter (decreasing, $n(0) = 1$). In the following we will also use $g(a)$ to denote the group of agent a. These definitions associate a utility value of 1 with an agent that remains alone (since the interest of joining a group with a barycenter at distance 0 and consisting of 1 element (itself) is $n(0) \times v(1) = 1 \times 1$).

Given A, $p()$, $d()$, $n()$ and $v()$, the group formation problem we seek to solve consists in finding a partition of A (i.e. set of groups), that maximizes the sum of every agent's utility, while minimizing (or even eliminating) the benefit individual agents could gain by either changing group or remaining alone (to ensure the solution's stability). Table 1 summarizes these notations and defines the variables and parameters appearing in Algorithm 1.

2.2   Algorithm

The algorithm we propose (Algorithm 1) uses a greedy clustering procedure similar to k-means [11] at its core (lines 13–17), but extends it with additional search heuristics that take into account the specificity of the group formation problem (Sect. 2.1). We use a k-means++ initialization [3] to bootstrap the algorithm (lines 2–7), in which each group is initialized with one single random agent, whose probability of being chosen increases with its distance from already initialized groups (line 5).

Once initialized, the process uses two nested loops: an inner loop (lines 12–18) that optimizes group sizes, but keeps their barycenters fixed, and an outer loop (lines 9–21) that updates barycenters (line 19) whenever group sizes have stabilized. This is in contrast to k-means which only uses one optimization loop, and is due to the fact that we need to take group sizes into account.

At each iteration of the inner loop, agents compute their utility for each group (taking into account their potential move) and choose their best option assuming other agents do not move (line 16). An agent may choose to be on its own (represented by a choice of $\perp$ for its group). Since groups are initially

Table 1. Symbols used in algorithm

Ident	Description
$[\![a, b]\!]$	The range of integers from a to b (inclusive)
$\#A$	Size of set A
$d.invertkv()$	Key-value inverted version of dictionary d
$rand(A)$	Random (uniform) element of A
$rand(A, d)$	Random element of A with distribution d
$d(x, y)$	Distance between x and y
$bary(g)$	Barycenter of group g
$p(a)$	Projector function
A	Set of agents
k	Number of clusters
$dmin2$	k-means++ distribution
grp	Groups
$ngrp$	New groups (in building)
$size$	Groups' sizes
$nsize$	New groups' sizes (in building)
$bgrp$	Estimated best groups for each agent (inner loop)
val	Affectation's value (sum of utilities)
$nval$	New affectation's (just built) value (sum of utilities)

empty, the first iteration of the inner loop differs slightly from the following ones: in particular, we artificially force the variable storing the size of each group to $\#A$, the total number of agents (line 11). This makes all groups look equal in terms of size during this first iteration, and agents initially migrate to the group with the closest barycenter (line 16), irrespective of this group's actual size. Only in subsequent iterations of the inner loop are the group sizes of iteration i used in iteration $i + 1$ (using the variables $nsize$ and $size$, which are updated at lines 13 and 17). This dual-loop architecture is necessary because the size of the group influences the utility of agents, unlike in k-means.

The outer loop (lines 9–21) updates the variable $ngrp$ at line 19 with the new groups returned by the inner loop. This causes the next inner-loop iteration to use the new barycenters when computing group assignments on line 16. Note that this algorithmic structure causes barycenters to remain fixed throughout each iteration of the inner loop.

The termination conditions of the inner and outer loops (lines 18 and 21) determine when the algorithm stops and come in two variants (shown in red and blue in the pseudo-code). Conditions in ♣red♣ belong to the *base* variant, in which the loop termination conditions are based on the absence of changes in groups (or more precisely in the size of groups at line 18 for the inner loop, and in the composition of these groups at line 21 for the outer loop). This variant

---

Algorithm 1: Clustering algorithm

```
1 function CLUSTER(A, p, k) is
2     ra ← rand(A); ngrp[0] ← {ra};                                    // k-means++ init
3     foreach a ∈ A do  dmin2[a] ← d(p(a), p(ra))² ;
4     for 1 ⩽ i < k do
5         na ← rand(A, dmin2); ngrp[i] ← {na};
6         foreach a ∈ A do
7             ⌊ dmin2[a] ← min(d(p(a), p(na))², dmin2[a]);
8     ♠nval ← #A♠;
9     repeat                                                           // main loop
10        ♠val ← nval♠;
11        grp ← ngrp; ngrp.clear(); nsize ← [#A...#A];
12        repeat
13            size ← nsize; nsize ← [0...0];
14            foreach a ∈ A do
15                bgrp[a] ← argmax_{i∈⊥∪⟦0,k−1⟧} [n(d(p(a), bary(
16                    grp[i] ∪ a))) × v(size[i] + 1_{a∉grp[i]})];
17                ⌊ nsize[bgrp[a]] + +;
18        until ♣nsize = size♣ ♠ ∑ nsize ⩾ ∑ size♠;
19        ngrp ← bgrp.invertkv();
20        ♠nval ← ∑_{a∈A} n(d(p(a), bary(ngrp.group(a)))) × v(#ngrp.group(a))♠;
21    until ♣ngrp = grp♣ ♠nval ⩽ val♠;
22    return grp;
```

---

ensures that a converged state is a Nash equilibrium, but does not guarantee that the algorithm does converge (notably in the obvious case where there is no such equilibrium). By contrast, lines and conditions in ♠blue♠ belong to the variant that *guarantees convergence*, by using termination conditions that are based on the variation of numeric values: at line 18, when the number of nodes in non-singleton group stops decreasing, and at line 21, when the overall utility (sum of agents utilities) of the solution (measured by *nval*) stops improving.

Like k-means, Algorithm 1 requires a hyper-parameter k, the number of groups, to operate. To determine k, we use a greedy control loop (not shown) which increases k until no significant gain in terms of global utility (the sum of every agent's utility) is achieved. To reduce the randomness of this process (since our clustering algorithm is randomized), each k value is further tried *prc* times, and the control loop only terminates once no gain in utility has been achieved over a pre-configured number *mmt* of iterations (a momentum mechanism).

3 Theoretical Analysis

Here we prove the essential property of the clustering algorithm presented in Algorithm 1 in its base variant.

Proposition 1 (Nash equilibrium). *Let us consider a game where each agent's (player's) possible choices consist in choosing any of k different groups, or choosing to remain alone (choosing ⊥). In particular, let $U_p(a,g)$ represent the payoff of each agent, a, for choosing group g, and 1 represent the payoff for not choosing any group. Algorithm 1 (base) outputs, if it exists, a (weak) Nash equilibrium for this game.*

Proof (Nash equilibrium). Let us assume that the algorithm terminates: we want to prove that it returns a Nash equilibrium.

To this end, we consider the last iteration of the outer loop guarded by $ngrp = grp$ (lines 9–21) and, in this iteration, the last iteration of the inner loop guarded by $nsize = size$ (lines 12–18). During this last inner loop iteration, variable grp is identical to the output of the algorithm, as the last update of grp is before the inner loop on line 11. Moreover, at the end of each iteration of the inner loop, $nsize$ contains the sizes of the groups in $ngrp$. As a result, at the end of the last iteration of the two loops, variable $size$ corresponds to the sizes of the groups in grp since $ngrp = grp$ and $nsize = size$. This is also true each time line 16 is executed during the last iteration of the inner loop as $size$ is only updated at line 13. This means that at line 16, each agent selects its best group based on the current group assignment (grp) and on the current group sizes ($size$) by maximizing the utility we defined in Eq. 1. Moreover, this does not change the current group assignments since $grp = ngrp$, meaning that the assignment in grp indeed constitutes a Nash equilibrium.   □

Since the existence of a Nash equilibrium for this game is not guaranteed, it is possible that Algorithm 1's base variant does not terminate. Now, we prove that Algorithm 1's guaranteed convergence variant, for which we do not have an equilibrium proof, will always terminate.

Proposition 2 (Convergence). *Algorithm 1 (guaranteed convergence) always terminates.*

Proof (Convergence). Among the control structures used in Algorithm 1's guaranteed convergence variant, the two **repeat...until** loops are the only ones that do not trivially terminate (**for** loops terminate trivially). For the inner loop, we replaced $nsize = size$ by $nsize.sum() \geqslant size.sum()$ as termination condition. This implies that, during the loop, $size.sum()$'s values are a strictly decreasing natural (ℕ) sequence. Such a sequence cannot be infinite, thus, the inner loop terminates. For the outer loop, we replaced $ngrp = grp$ by $nval \leqslant val$ as termination condition. This implies that, during the loop, val's values are a strictly increasing real sequence. Moreover, those values are given by a formula taking as parameter an affectation of a finite number of agents in a finite number of groups. The possible inputs for this formula for a given execution of the algorithm (fixed parameters) are a finite set. This implies that the possible values for val (output of this formula) are also a finite set (for a given execution of the algorithm). Thus, val's values are a strictly increasing sequence of elements of a finite set (the order is given by the classical order for real number). Such a sequence cannot be infinite, thus, the outer loop terminates. We can now conclude that our algorithm will always terminate. □

We note that Algorithm 1's guaranteed convergence variant is not guaranteed to produce a Nash equilibrium. Such an equilibrium may simply not exist, but even if it exists, the algorithm is not guaranteed to find it. Due to the end condition of its outer loop, Algorithm 1's guaranteed convergence variant, while still based on agents' self-interest, is more centered on attaining general optimality (maximizing the sum of all agents utilities) than Algorithm 1's base variant.

An important difference between our algorithm and generic approaches to solve hedonic games presented in [4] lies in the inner loop of our algorithm, which allows groups to grow to a point that is better for every agent even in situations where individual rational decisions could not. Let us consider four agents, A, B, C and D. $n(x) = \frac{1}{1+x}$, $v(x) = x$, p's values in $\mathbb{R}^2$, $p(A) = [0,0]$, $p(B) = [0,2.5]$, $p(C) = [2.5,0]$ and $p(D) = [2.5,2.5]$ (a square). In this example, having all agents in a single group, $\mathcal{G}$, is the best solution. $\forall_{a \in \{A,B,C,D\}} U_p(a,\mathcal{G}) = \frac{4}{1+2.5\sqrt{2}/2} > 1$. But this situation could not be reached from groups consisting of a single agent by individual rational decisions, since the best interest one single agent could get from grouping with another agent would be $\frac{2}{2.25} < 1$, so no agent would want to group with any other. With our algorithm, if we start with a group $\mathcal{G} = \{A\}$, due to making computations based on the potential maximum size of a group rather than its actual size, the individual interest of B and C for joining this group would be estimated at $\frac{4}{2.25} > 1$ and the group $\mathcal{G}$ will be updated to $\mathcal{G} = \{A,B,C\}$. On the second run of the inner loop, with the barycenter of $\mathcal{G}$ being at $[2.5/3, 2.5/3]$, the estimated interest for D to join $\mathcal{G}$ would be $\frac{4}{1+2.5\sqrt{2}/2} > 1$. We end up with $\mathcal{G} = \{A,B,C,D\}$.

4   Experimentation

In this section, we evaluate Algorithm 1, in its guaranteed convergence variant (noted $AUCCCR$) and see how much, in practice, it deviates from the Nash equilibrium property of the base variant.[1]

4.1   Experimental Set-Up

Competitors. We compare our algorithm with two reference clustering algorithms: OPTICS [2] and k-means [11] (with a k-means++ [3] initialization). All three algorithms are guaranteed to terminate.

Hyperparameters and Loss Functions. For all algorithms, we use the usual Euclidean distance for d in $\mathbb{R}^k$. For AUCCCR, we use $n(x) = \frac{1}{1+x}$ as the normalization function, and $v(x) = \sqrt{x}$ as the group value function (see Sect. 2.1). We

---
[1] Our code is available at https://gitlab.inria.fr/abouchra/distributed_neural_netw orks.

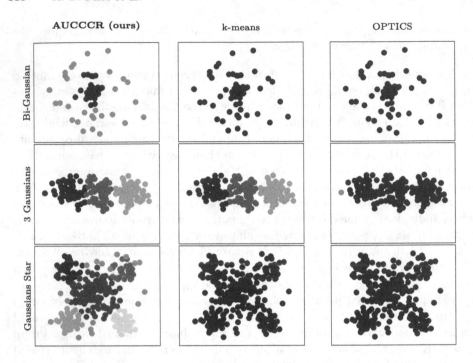

Fig. 1. Clusters found in various synthetic cases

further set the number of trials to $prc = 20$, and the momentum to $mmt = 5$. k-means' parameter k is searched for the same way as for AUCCCR. For OPTICS, which is not randomized and, as such, does not suffer from the same instability in its output, the procedure used to select the proper minimum reachability distance is equivalent but with $prc = mmt = 1$.

Datasets. We perform this evaluation on two kinds of datasets: synthetic datasets generated using three different mixtures of Gaussian distributions (a *bi-Gaussian*, a *3-Gaussian*, and a *Gaussian Star*), and two real datasets related to leisure travel, and group buying (*BuddyMove* [14], and *Wholesale* [9]). We describe each dataset in more detail in the relevant subsections below.

Metrics. We use the following three metrics to measure the performance of each algorithm:

– The losses of agents (summed among all agents).
– The share of agents having losses.
– The global utility of agents (sum of all utilities).

A good group formation algorithm (according to our definition of this problem) should deliver a close-to-maximum global utility, along with a low sum of agents' losses, and a low rate of losing agents (in the ideal case these last two metrics should have value 0).

In addition to the above metrics, we also display the results of example runs (in the case of random synthetic data, using the same data for all algorithms). In these, a color identifies a cluster, gray points represent isolated agents (or agents in a cluster of size 1). Results are averaged over 10 runs; error bars indicate the standard deviation.

4.2 Synthetic Data

We first evaluate our algorithm on $\mathbb{R}^2$ data generated using a combination of Gaussian distributions (Gaussian mixtures). We choose Gaussians because those distributions are usually considered good models of characteristic distributions in real populations. We generate a fixed number of points, using a mixture of Gaussian distributions that are combined according to some predetermined ratio (probability for points to be generated according to each distribution).

Gaussian Mixtures. We use three different Gaussian-mixtures generators:

- **Bi-Gaussian:** two $(0,0)$-centered Gaussian distributions with 1 and 8 as standard deviations and a 0.5–0.5 ratio (same probability for each distribution), 100 points.
- **3 Gaussians:** three Gaussian distributions, $(-6,0)$, $(0,0)$ and $(0,6)$-centered with 1.5 as standard deviations and a 0.33–0.33–0.33 ratio (same probability for each distribution) 200 points.
- **Gaussian Star:** five Gaussian distributions, $(0,0)$, $(-10,-10)$, $(-10,10)$, $(10,-10)$ and $(10,10)$-centered with 5 for the $(0,0)$ and 2 for others as standard deviations and a 0.5–0.125–0.125–0.125–0.125 ratio, 300 points.

Additionally, we use a single Gaussian with 2 as standard deviation for the size constraint test.

Results. Example runs for the three distributions are shown in Fig. 1. The performance of each algorithm in terms of global utility, sum of agents' losses, and rate of losing agents is shown in Figs. 2, 3 and 4.

In more detail, looking at the Bi-Gaussian mixture (Fig. 2), we see that AUC-CCR yields a lower global utility, but also that its output is very close to a Nash equilibrium (close to 0 losses), while k-means and OPTICS have more than 25% of agents experiencing losses. Looking at example runs (Fig. 1, first line), we observe that k-means and OPTICS tend to produce a single global cluster, maximizing its overall utility but causing losses for individual agents that lie far away from the barycenter. By contrast, our algorithm is able to take better care of individual interests, because it allows agents to stay on their own (by choosing $\perp$ as their group in the algorithm).

Turning to the 3-Gaussian mixture (Fig. 3), we note that AUCCCR and k-means both perform well in terms of global utility, better than OPTICS, likely due to OPTICS' density-centered design being ineffective for this kind of distribution. In particular, OPTICS experiences difficulties distinguishing close but

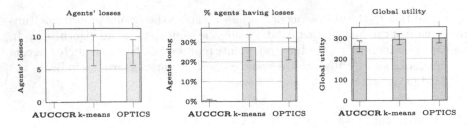

Fig. 2. Metrics in Bi-Gaussian for all algorithms

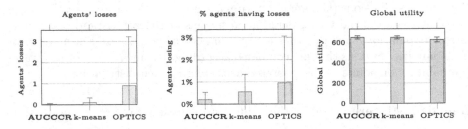

Fig. 3. Metrics in 3 Gaussians for all algorithms

clearly distinct distributions. Losses are overall limited, but AUCCCR clearly outperforms k-means and, even more markedly, OPTICS.

Finally, for the Gaussians Star (Fig. 4), all algorithms deliver similar results in terms of global utility (albeit slightly lower ones for AUCCCR on average). The gap is much larger on losses, with AUCCCR clearly leading. Looking at sample runs, AUCCCR seems to be able to identify relatively precisely the five Gaussians mixed in the input data, in contrast to k-means and OPTICS.

From these metrics we can conclude that AUCCCR causes very low individual losses for agents compared to both, k-means and OPTICS (the tendency is less marked in 3 Gaussians). We also see that the global utility achieved by AUCCCR is very close to what is obtained with k-means and OPTICS (slightly lower in Bi-Gaussian).

Size Constraint. To evaluate the ability of our algorithm to manage cases where very large clusters are not desirable (agents want to be in a sufficiently large group but not too large) we take a simple case, with a single Gaussian, in which our usual $v(x) = \sqrt{x}$ function would make a single cluster (with all agents in it) optimal. We then change $v(x)$ so that too large clusters are not optimal.

We test two different functions, both equal to $\sqrt{x}$ for $x \leqslant 20$, but for $x > 20$ the first function is constant $\sqrt{20}$ while the second is decreasing $\sqrt{\frac{20}{1+\frac{|20-x|}{20}}}$ (both functions are continuous). The number 20 is an arbitrary choice and can be considered as a target size for clusters.

Results are shown in Fig. 5. We can see that, with the proper $v(x)$ function, AUCCCR can integrate size constraints.

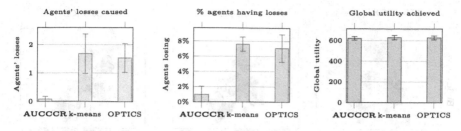

Fig. 4. Metrics in Gaussians Star for all algorithms

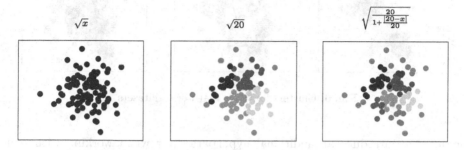

Fig. 5. Clusters with AUCCCR and different $v(x)$

4.3   Real Data

We now apply our algorithm and its two competitors on two real datasets: *BuddyMove* and *Wholesale*.

Datasets. The BuddyMove [14] dataset consists of statistics from users of a travel review website. For each user, the dataset provides the number of reviews written for each of 6 classes of destination (e.g. religious sites, parks, etc.). From these data, we derive for each user the share of reviews written for each destination class. This can be interpreted as the relative interest of a user for each kind of destination. This could help provide for instance recommendations to users on whom they should go with for group travels, based on the similarity of their preferences.

The Wholesale [9] dataset consists of statistics from customers of a wholesale vendor. For each customer, the dataset indicates the annual spending for each of 6 classes of products (e.g. fresh, frozen, etc.). From these data, we derive for each customer the share of reviews written for each product class. This result could for instance be used to provide recommendations to customers on whom they should collaborate with to make grouped orders more directly, removing the wholesale distributor from the circuit (short circuit distribution), based on which kinds of product they usually order.

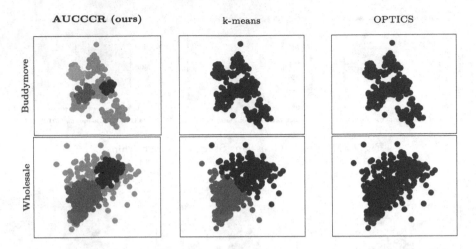

Fig. 6. Clusters found in real-world datasets

Scale. We introduce an additional hyperparameter when working with real datasets: *scale*. The scale describes how "far" a given distance is considered by the algorithm and can be seen as parameter of the projector function p we presented earlier. The distances naturally present in the dataset must be given an absolute cardinal signification for the algorithm, the right choice is up to the user, we variate it to see how this choice affects the results. For example, if the scale is 10, a distance of 1 will be computed as 10 for computation. As the scale grows, the distance between points will be considered longer by the algorithm.

Results. Sample runs for BuddyMove and Wholesale are presented in Fig. 6. Both datasets include 6 features, the views we provide are reduced to two dimensions. For this, we use the *scikit-learn* [13] Python library, which implements various algorithms that can perform non-linear dimension reduction. The specific algorithm we use is MDS (Multi-Dimensional Scaling) [5].

On BuddyMove, with AUCCCR, we obtain three clusters containing about 30 agents each, with 150 agents remaining alone. k-means and OPTICS put all agents in a single large cluster.

On Wholesale, with AUCCCR, we got one cluster of 100, one of 200 and 100 agents remaining alone. k-means gave two clusters of 200, OPTICS put everyone in a 400 cluster.

Metrics for the two dataset are presented in Figs. 7 (BuddyMove) and 8 (Wholesale). Note that in Fig. 7 the lines for k-means and OPTICS are superposed.

In these graphs, we see that AUCCCR maintains very low losses for the whole scale range while k-means and OPTICS both have higher losses that quickly increase as scale increases. In terms of global utility, AUCCCR is comparable

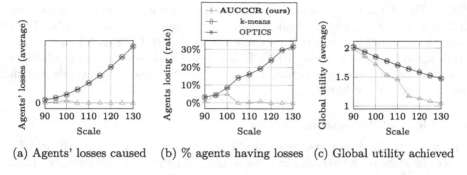

(a) Agents' losses caused   (b) % agents having losses   (c) Global utility achieved

Fig. 7. Metrics in Buddymove for different scales by different algorithms

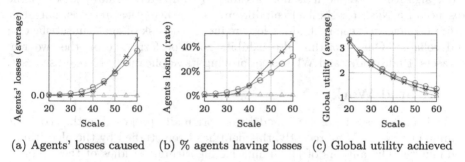

(a) Agents' losses caused   (b) % agents having losses   (c) Global utility achieved

Fig. 8. Metrics in Wholesale for different scales by different algorithms

to k-means and OPTICS for Wholesale and for low scale values on BuddyMove, but it achieves lower global utility for high scale values in this latter dataset.

In the details, for BuddyMove, k-means and OPTICS behave identically. For a low scale, users are considered close and all algorithms will detect a single cluster but, as the scale increases, the results start diverging. AUCCCR, contrary to its competitor, seeks to avoid individual losses; as we can see, losses are very low for AUCCCR, especially at large scales, while, for other algorithms, losses increase dramatically. For a scale of 130, AUCCCR has nearly 0 agents losses while k-means and OPTICS have 30%. This comes at the cost of a reduced global utility, which slowly drops until it nearly hits 1 (equivalent to everyone alone) at 130. For midrange values, we see that AUCCCR has nearly no losses with a still high average global utility (1.5 for AUCCCR, compared to 1.7 for its competitors; this lower value being due to the small size of clusters possible with minimal individual losses) while k-means and OPTICS cause between 15% to 20% of agents to experience losses.

For Wholesale, at low scale, users are considered close and all algorithms will detect a single cluster but, as the scale increases, the results become very different. AUCCCR, contrary to the reference algorithms, will prevent individual losses; as we can see, losses are close to 0 for AUCCCR, especially for large scales, while, for other algorithms, losses increase dramatically. For a scale of 60, AUCCCR has nearly 0 agents losses while k-means has >30% and OPTICS has

>40%. The global utility remains similar to that of k-means and greater than for OPTICS for the lower half of scale values, only higher values have a utility lower than that of OPTICS but still very close to both, OPTICS and k-means. For midrange values, we see that AUCCCR has nearly no losses with a global utility close to that of k-means and higher than OPTICS, while reference algorithms have around 10% of losses.

4.4   General Analysis of Results

On our synthetic data tests, we showed that our algorithm is more efficient for cluster identification than classical clustering algorithms. It also appeared that, as intended, our algorithm causes much lower individual losses to agent than other algorithms. While it is not possible to completely remove losses in the absence of a Nash (or similar) equilibrium for mathematical reasons, our algorithm manages to keep such losses low, reducing the risk of agent unsatisfaction and collapse. Our algorithm also exhibits good performance on the two real datasets, Buddymove and Wholesale, proving its applicability to real use cases.

5   Related Work

Hedonic games as a theoretical model were originally proposed in the economic community by [7]. More recently, this field has been studied by the algorithmic community. The authors of [4] summarize algorithmic studies of these games. Existing algorithmic solutions all rely on some kind of equilibrium (similar to Nash equilibrium) which may not exist in real applications.

In addition to those general researches on hedonic games, some researchers have considered using hedonic games to solve practical problems. For example, [15] worked on collaboration between roadside units in Intelligent Transportation Systems. [1] considered hedonic games in the context of Fog Computing. Application of these games to edge computing is also proposed in [18]. These games can also be applied to energy networks, as suggested by [12]. Each of these articles presents its own model and algorithm, to solve the specific problem they are considering.

The most well-known clustering algorithm is k-means [11]. A lot of variants have been proposed, notably k-means++ [3], which improves its initialization. There are also variants of k-means with size constraints [10,16] or density [6] but such variants can only find clusters with similar characteristics (size/density) leading to results unwanted for our application (the graphs in [10] show this clearly). Another well-known clustering algorithm is OPTICS [2], which can be seen as an improved DBSCAN [8]; both OPTICS and DBSCAN are density-based (k-means is distance-based).

6   Conclusion

In this work, we defined a new class of hedonic games for providing cooperation recommendation, and proposed an algorithm to solve such games. We experimentally compared our algorithm to classical clustering algorithms that could also have been used for the same recommendation problem, and showed that our algorithm yields results that exhibit close to no losses and are thus similar to a Nash equilibrium (which does not necessarily exist in the general case) than other algorithms.

References

1. Anglano, C., Canonico, M., Castagno, P., Guazzone, M., Sereno, M.: A game-theoretic approach to coalition formation in fog provider federations. In: FMEC 2018 (2018)
2. Ankerst, M., Breunig, M., Kriegel, H.P., Sander, J.: Optics: ordering points to identify the clustering structure. In: SIGMOD 1999 (1999)
3. Arthur, D., Vassilvitskii, S.: K-means++: the advantages of careful seeding. In: SODA 2007 (2007)
4. Aziz, H., Savani, R.: Hedonic games. In: Handbook of Computational Social Choice (2016)
5. Borg, I., Groenen, P.J.F.: Modern Multidimensional Scaling - Theory and Applications. Springer, New York (1997). https://doi.org/10.1007/0-387-28981-X
6. Davidson, I., Vassilvitskii, S.: Agglomerative hierarchical clustering with constraints: theoretical and empirical results. In: PKDD 2005 (2005)
7. Drèze, J.H., Greenberg, J.: Hedonic coalitions: optimality and stability. Econometrica **48**, 987–1003 (1980)
8. Ester, M., Kriegel, H.P., Sander, J., Xu, X.: A density-based algorithm for discovering clusters a density-based algorithm for discovering clusters in large spatial databases with noise. In: KDD 1996 (1996)
9. Fernandes Marques de Abreu, N.G.C.: Análise do perfil do cliente Recheio e desenvolvimento de um sistema promocional. Master's thesis, Instituto Universitário de Lisboa (2011)
10. Ganganath, N., Cheng, C.T., Tse, C.K.: Data clustering with cluster size constraints using a modified k-means algorithm. In: CyberC 2014 (2014)
11. Lloyd, S.P.: Least squares quantization in PCM. IEEE Trans. Inf. Theory **28**(2), 129–137 (1982)
12. Mei, J., Chen, C., Wang, J., Kirtley, J.: Coalitional game theory based local power exchange algorithm for networked microgrids. Appl. Energy **239**, 133–141 (2019)
13. Pedregosa, F., et al.: Scikit-learn: machine learning in Python. JMLR **12**, 2825–2830 (2011)
14. Renjith, S., Anjali, C.: A personalized mobile travel recommender system using hybrid algorithm. In: ICCSC 2014 (2014)
15. Saad, W., Han, Z., Hjorungnes, A., Niyato, D., Hossain, E.: Coalition formation games for distributed cooperation among roadside units in vehicular networks. IEEE J. Sel. Areas Commun. **29**(1), 48–60 (2010)
16. Tang, W., Yang, Y., Zeng, L., Zhan, Y.: Size constrained clustering with MILP formulation. IEEE Access **8**, 1587–1599 (2019)
17. von Neumann, J., Morgenstern, O.: Theory of Games and Economic Behavior. Princeton University Press, Princeton (1944)
18. Zhang, T., Chen, W., Yang, F.: Data offloading in mobile edge computing: a coalitional game based pricing approach. In: IEEE 2017 (2017)

Blockchain

Blockchain Using Proof-of-Interaction

Jean-Philippe Abegg[1,2(✉)], Quentin Bramas[1], and Thomas Noël[1]

[1] ICUBE, University of Strasbourg, Strasbourg, France
jp.abegg@unistra.fr
[2] Transchain, Strasbourg, France

Abstract. Proof-of-Work is originally a client-side puzzle proposed to prevent spam or denial of service attacks. In 2008, Satoshi Nakamoto used it as an election mechanism (or equivalently, to replace a centralized time server) in the first Blockchain: Bitcoin. In the same year, another spam prevention algorithm was proposed, based on a guided-tour puzzle, but received only little attention.

The main motivation of our work is to see if a Blockchain protocol can use the guided-tour puzzle like Bitcoin uses Proof-of-work.

In this paper we extend the guided tour puzzle to a new Puzzle called Proof-of-Interaction and we show how it can replace, in the Bitcoin protocol, the Proof-of-Work algorithm. We show that it uses a negligible amount of computational power compared to Bitcoin, and scales very well in term of number of messages. We analyze the security of our protocol and show that it is not subject to selfish mining. However, our protocol currently works only when the nodes in the network are known, but we discuss how this assumption could be weakened in future work.

Keywords: Blockchain · Proof of interaction · Distributed systems · Consensus algorithm

1 Introduction

A Blockchain is a Distributed Ledger Technology (DLT) *i.e.*, a protocol executed by a set of nodes to maintain a data-structure where data can only be appended in blocks. It is maintained in a distributed manner by many participants, who may not trust each-other, and some of which can be faulty or malicious. In order for this data-structure to be consistent among the participants, a protocol is used to ensure that every one agrees on the next block that is appended into the Blockchain.

The most famous example of such protocol, Bitcoin [18], uses the Proof-of-Work to elect a single participant that is responsible for appending the next block. In more details, Proof-of-work is a client-puzzle that is executed by all the nodes in the network. Finding a solution of the puzzle requires a large amount of computational power but is easily verifiable. The first node that finds a solution is the one that is allowed to append a block to the chain of block. The difficulty of the puzzle increases with the total computational power of the network in order

© Springer Nature Switzerland AG 2021
K. Echihabi and R. Meyer (Eds.): NETYS 2021, LNCS 12754, pp. 129–143, 2021.
https://doi.org/10.1007/978-3-030-91014-3_9

to limit the chance of having multiple concurrent elected nodes. This implies that the total power consumption increases linearly with the total computational power of the participants. According to the latest estimates, the Bitcoin network consumes more than the Czech Republic [7,11] (and just account for less than 70% of all the Proof-of-work based Blockchains).

There have been many attempts to avoid using Proof-of-work based agreement, but usually adding other constraints [22] (*eg.*, small number of nodes, hardware prerequisite, new security threats).

In this paper, we propose to use a new client-puzzle called Proof-of-Interaction to define a new energy-efficient Blockchain protocol.

Related Work. Proof-of-work [4] (PoW) is a method initially intended for preventing spamming attacks. It was then used in the Bitcoin protocol [18] as a way to prove that a certain amount of time has passed between two consecutive blocks. Another way to see the aim of the Proof-of-work in the Bitcoin protocol is as a leader election mechanism, to select who is responsible for writing the next block in the blockchain. This leader election has several important properties, including protection against Sybil attacks [9] and against denial-of-service [16]. Also, it has a small communication complexity. However, the computational race consumes a lot of energy. The majority of current Blockchain protocols uses Proof-of-Work, with different hashing functions [17].

In 2012, S. King and S. Nadal [14] proposed the Proof-of-Stake (PoS), an alternative for PoW. This leader election mechanism requires less computational power but has security issues [5,12] (*eg.*, Long range attack and DoS). Intel proposed another alternative to PoW, the Proof-of-Elapsed-Time (PoET) [1]. This solution requires Intel SGX as a trusted execution environment. Thus, Intel becomes a required trusted party to make the consensus work, which might imply security concerns [8] and is against blockchain idea to remove third parties. Other mechanisms where proposed such as Proof-of-Activity and Proof-of-Importance [3], which are hybrid protocols or protocols using properties from the network itself.

Vote-based protocols refer to the family of Byzantine Agreement protocols, such as PBFT [6]. Such protocols do not use client-puzzle, hence are energy-efficient. They can handle a large amount of transactions but must be executed in a known network, and do not scale well with the number of nodes due to their communication complexity.

The previous paper most related to our work was presented just prior the publication of Bitcoin in 2008, by M. Abliz and T. Znati [2]. They proposed *A Guided Tour Puzzle for Denial of Service Prevention*, which is another spam protection algorithm. This mechanism has not yet been used in the Blockchain context, and is at the core of our new Proof-of-Interaction. The idea was that, when a user wants to access a resource in a server that is heavily requested, the server can ask the user to perform a tour of a given length in the network. This tour consists of accessing randomly a list of nodes, own by the same provider as the server. After the tour, a user can prove to the server that it has completed the task and can then retrieve the resource. The way we generate our tour in our Proof-of-Interaction is based on the same idea. We generalized the approach of M. Abliz and T. Znati to work with multiple participants, and we made the tour length variable.

Contributions. The contribution of this paper is twofold. First, we propose a better alternative to Proof-of-Work, called Proof-of-Interaction, which requires negligible computational power. Second, we show how it can be used to create an efficient Blockchain protocol that is resilient against selfish mining, but assumes for now that the network is known.

Paper Structure. Section 2 presents the model and illustrates the problem with naive approaches. This also helps to understand how our protocol is built. Then we present the Proof-of-Interaction protocol, that could be used outside of the Blockchain context. Then, in Sect. 4 we explain how we use this proof mechanism to create a Blockchain protocol. In Sect. 5, we analyze the security properties of our protocol. Finally, we conclude and discuss possible extensions in Sect. 6.

2 Preliminaries

2.1 Model

The network, is a set $\mathcal{N}$ of n nodes that are completely connected. Each node has a pair of private and public cryptographic keys. Nodes are uniquely identified by their public keys (*i.e.*, the association between the public keys and the nodes is common knowledge). Each message is signed by its sender, and a node cannot fake a message signed by another (non-faulty) node.

We denote by $\text{sign}_u(m)$ the signature by node u of the message m, and $\text{verif}_u(s, m)$ the predicate that is true if and only if $s = \text{sign}_u(m)$. For now, we assume the signature algorithm is a deterministic one-way function that depends only on the message m and on the private key of u. This assumption might be very strong as, with common signature schemes, different signature could be generated for the same message, but there are ways to remove this assumption by using complex secret generation and disclosure schemes, not discussed in this paper, so that each signature is in fact a deterministic one-way function. The function H is a cryptographic, one-way and collision resistant, hash function [19].

As for the Bitcoin protocol, we assume the communication is partially synchronous *i.e.*, there is a fixed, but unknown, upper bound Δ on the time for messages to be delivered.

The size of a set S is denoted with $|S|$.

2.2 Guided Tour

The guided tour defined by M. Abliz and T. Znati [2] can be summarized as follow. When a resource server is under DOS attack, it responds to a given request by a random seed hash h_0, a set S of n servers and a length L. The client has to solve a puzzle in order to complete its request to the resource server. To solve the puzzle, the client makes L requests to the servers in S in a specific order. The index, in S, of the first server to request is deduced from h_0. Let $i_0 \in [0, n-1]$ such that $i_0 \equiv h_0 \mod n$. Then, the client sends message h_0 to the i_0-th server in S. The server responds with hash h_1. Then then client computes $i_1 \in [0, n-1]$ such that $i_1 \equiv h_1 \mod n$, and sends message h_1 to the i_1-th server in S, and so on. This continues until hash h_L is obtained. h_L is a proof that the tour as been completed, and is sent to the resource server to obtain the requested resource. Thanks to a secret shared among all the servers, the resource server is able to check that hash h_L is indeed the expected proof for the initial seed h_0. This idea is interesting because the whole tour depends only on the initial value, and cannot be performed in parallel because each hash h_i cannot be found until h_{i-1} is known. We then present a naive approach on how it can be used as a distributed client-puzzle.

2.3   Naive Approach

We give here a naive approach on how asking participants to perform a tour in the network can be used as a leader election mechanism to elect the node responsible for appending the next block in a Blockchain.

When a node u_0 wants to append a block to the blockchain, it performs a random tour of length L in the network retrieving signatures of each participants it visits. The first node u_1 to visit is the hash of the last block $h_0 = last\_block\_hash$ of the blockchain modulo n (if we order nodes by their public keys, the node to visit is the i-th with $i = h_0 \mod n$). u_1 responds with the signature $s_1 = \mathrm{sign}_{u_1}(h_0)$. The hash $h_1 = H(s_1)$, modulo n, gives the second node u_2 to visit, and so on. This idea is similar to the guided tour of M. Abliz and T. Znati [2], and here the whole tour depends only the hash of the last block. Given h_0, anyone can verify that the sequence of signatures $(s_1, s_2, \ldots, s_L)$ is a proof that the tour has been properly performed. If each node in the network performs a tour, the first node to complete its tour is elected broadcast its block, containing the proof, to the other nodes to announce it.

However, here, each node has to perform the same tour, which could be problematic. An easy fix is to select the first node to visit, not directly using the hash of the last block, but also based on the signature of the node initiating the tour, $h_0 = H(\mathrm{sign}_{u_0}(last\_block\_hash))$. Now, given h_0, the sequence of signatures $(\mathrm{sign}_{u_0}(h_0), s_1, s_2, \ldots, s_L)$ proves that the tour has been properly performed by node u_0. Each tour, performed by a given node, is unique, and a node cannot compute the sequence of signature other than by actually asking each node in the tour to sign a message. Indeed, the next hop of the tour depends on the current one.

Here, one can see that it could be a good idea to also make the tour dependent on the content of the block node u_0 is trying to append. Indeed, using only the last block to generate a new proof does not protect the content of the current block, *i.e.*, the same proof can be used to create two different blocks. To prevent this behavior, we can assume that $h_0 = H(\mathrm{sign}_{u_0}(last\_block\_hash) \cdot M)$ ($\cdot$ being the concatenation operator) where M is a hash of the content of the block node

u_0 is trying to append. In practice, it is the root of the Merkel tree containing all the transactions of the block. Here, the proof is dependent on the content of the block, which means that if the content of the block changes the whole proof needs to be computed again.

From there, we face another issue. Each node performs a tour of length L, so each participant will be elected roughly at the same time, creating a lots of forks. To avoid this, we can make the tour length variable. We found two ways to do so. The first one is not to decide on a length in advance, and perform the tour until the hash of k-th signature is smaller than a given target value, representing the difficulty of the proof. In this way, every interaction with another node during a tour can be seen as a tentative to find a good hash (like hashing a block with a given nonce in the PoW protocol). The target value can be selected so that the average length of the tour is predetermined. However doing so, since the proof does depend on the content of the block, u_0 can change the content of the block, by adding dummy transactions for instance, so that the tour stops after one hop[1]. The other way to make the tour length variable is to use a cryptographic random number generator, seeded with $\mathtt{sign}_{u_0}(last\_block\_hash)$, to generate the length L. Doing so, the length depends only on u_0 and on the previous block. Then a tour of length L is performed as usual.

To complete the scheme, we add other information to the message sent to the visited node so that they can detect if we try to prove different blocks in parallel. We also make u_0 sign each response before computing the next hop, so that the tour must pass through u_0 after each visit. Finally, we will see why it is important to perform the tour, not using the entire network, but only a subset of it.

3   The Proof-of-Interaction

In this section we define the most important piece of our protocol, which is, how a given node of the network generates a proof of interaction. Then, we will see in the next section how this proof can be used as an election mechanism in our Blockchain protocol.

3.1   Algorithm Overview

We present here two important algorithms. One that generates a Proof-of-Interaction (PoI), and one that checks the validity of a given PoI.

Generating a Proof-of-Interaction. Consider we are a node $u_0 \in \mathcal{N}$ that wants to generate a PoI. Given a fixed *dependency* value denoted d, the user u_0 wants to prove a *message* denoted m. The user has no control over d but can chose any message to prove.

---

[1] There are some ways to limit this attack, but we believe it will remain an important attack vector.

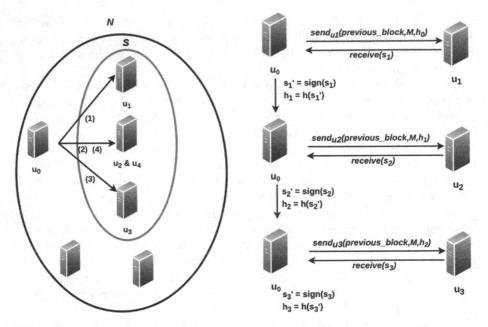

Fig. 1. u_0 interacts randomly with a subset S of the nodes

Fig. 2. u_0 interacts with a sequence of nodes to construct a PoI. In this example, the dependency is the hash of the previous block.

The signature by u_0 of the dependency d, denoted $s_0 = \texttt{sign}_{u_0}(d)$, is used to generate the subset S of $n_S = \min(20, n/2)$ nodes to interact with

$$S = \{S_0, S_2, \ldots, S_{n_S-1}\} = \texttt{createServices}\,(\mathcal{N}, s_0)\,.$$

S is generated using the pseudo-random Algorithm $\texttt{createServices}$, and depends only on d and on u_0. From s_0, we also derive the length of the tour $L = \texttt{tourLength}(D, s_0)$, where D is a probabilistic distribution that corresponds to the difficulty parameter. $\texttt{tourLength}$ is a random number generator, seeded with s_0 that generates a number according to D. Using D one can easily change the average length of the tour for instance.

Now u_0 has to visit randomly L nodes in S to complete the proof, as illustrated in Fig. 1. To know what is the first node u_1 we have to visit, we first hash the concatenation of s_0 with m to obtain $h_0 = H(s_0 \cdot m)$. This hash (modulo $|S|$) gives the index i in S of the node we have to visit, $i \equiv h_0 \bmod |S|$. So we send the tuple (h_0, d, m) to node $u_1 = S_i$, which responds by signing the concatenation, $s_1 = \texttt{sign}_{u_1}(h_0 \cdot d \cdot m)$.

To know what is the second node u_2 we have to visit, we sign and hash the response from u_1 to obtain $h_1 = H(\texttt{sign}_{u_0}(s_1))$, so that $u_2 = S_j \in S$ with $j = h_1 \bmod |S|$. Again, we send the tuple (h_1, d, m) to u_2, which responds by signing the concatenation, $s_2 = \texttt{sign}_{u_2}(h_1 \cdot d \cdot m)$. We sign and hash the response from u_1 to obtain $h_2 = H(\text{sign}_{u_0}(s_2))$ and find the next node we have to visit, and so on (see Fig. 2). This continues until we compute $\text{sign}_{u_0}(s_L)$, after the response of the L-th visited node.

The *Proof of Interaction* (PoI) with dependency d of message m by node u_0 and difficulty D is the sequence

$$(s_0, s_1, \text{sign}_{u_0}(s_1), s_2, \text{sign}_{u_0}(s_2), \ldots, s_L, \text{sign}_{u_0}(s_L)).$$

Checking a Proof-of-Interaction. To check if a PoI $(s_0, s_1, s'_1 \ldots, s_k, s'_k)$ from user u, is valid for message m, dependency d and difficulty D, one can first check if s_0 is a valid signature of d by u_0. If so, we can obtain the set $S = \text{createServices}(\mathcal{N}, s_0)$ of interacting nodes, the length $L = \text{tourLength}(D, s_0)$, and the hash $h_0 = H(s_0 \cdot m)$. From h_0 and S, we can compute what is the first node u_1 and check if s_1 is a valid signature from u_1 of $(h_0 \cdot d \cdot m)$, and if s'_1 is a valid signature of s_1 from u_0. Similarly, one can check all the signatures until s'_k. Finally, if all signatures are valid, and $k = L$, the PoI is valid.

3.2   Algorithm Details

The pseudo code of our algorithms are given below.

The algorithm `createServices` is straightforward. We assume that we have a random number generator (RNG)—defined by the protocol hence the same for all the nodes in the network—that we initialize with the given seed. The algorithm then shuffles the input array using the given random number generator. Finally, it simply returns the first n_S elements of the shuffled array.

Algorithm createServices: create a pseudo-random subset of nodes

Input: N, the set of nodes
h, a seed
Output: S, a subset of nodes
1 $RNG.seed(h)$
2 $S \leftarrow \text{shuffled}(N, RNG)$
3 $S \leftarrow S.slice(0, n_S)$
4 **return** S

The main part of the algorithm `generatePoI` consists in a loop, that performs the L interactions. The algorithm requires that each node in the network is executing the same algorithm (it can tolerates some faulty nodes, as explained later). The end of the algorithm shows what is executed when a node receives a message from another node. The procedure `checkMessage` may depends on what the PoI is used for. In our context, the procedure checks that the nodes that interacts with us does not try to create multiple PoI with different messages, and use the same dependency as everyone else. We will see in details in the next section why it is important.

Algorithm generatePoI: Program executed by u_0 to generate the PoI

Input: d, the dependency (hash of last block of the blockchain)
   m, the message (root of the merkle tree of the new block)
   D, difficulty of the PoI
   N, the set of nodes in the network
Output: P, a list of signatures $\{s_0, s_1, s_1', s_2, s_2', \ldots, s_k, s_k'\}$

1 $P \leftarrow []$
2 $s_0 \leftarrow \text{sign}_{u_0}(d)$
3 $S \leftarrow \text{createServices}(N, s_0)$
4 $L \leftarrow \text{tourLength}(D, s_0)$
5 $P.append(s_0)$
6 $current\_hash \leftarrow H(s_0 \cdot m)$
7 **for** L *iterations* **do**
8     $\quad next\_hop \leftarrow current\_hash \% |S|$
9     $\quad s \leftarrow sends_{S_{next\_hop}}(current\_hash, d, m)$
10    $\quad P.append(s)$
11    $\quad s \leftarrow \text{sign}_{u_0}(s)$
12    $\quad P.append(s)$
13    $\quad current\_hash \leftarrow H(s)$
14 **return** P

15 **When Receive** (h, d, m) from u **do**
16 **if** $checkMessage(u, h, d, m)$ **then**
17    $\quad$ **Reply** $\text{sign}_{u_0}(h \cdot d \cdot m)$

Algorithm checkMessage: Check the message received from node u

---

Input: u, the sender of the request
   h, difficulty of the PoI
   d, the dependency (hash of last block of the blockchain)
   m, the message (root of the merkle tree of the new block)
Output: whether to accept or not the request

1 **if** *d is the hash of the latest block of one of the longest branches* **then**
2     $\quad$ **if** *Received[(u, d)] exists and is not equal to m* **then**
3        $\quad\quad$ penalties (u)
4        $\quad\quad$ **return** false
5     $\quad$ Received$[(u, d)] = m$
6     $\quad$ **return** true
7 **else**
8     $\quad$ **if** *unknown d* **then**
9        $\quad\quad$ **Ask** block d
10       $\quad\quad$ **return** false

The algorithm checkPoI that checks the validity of a PoI is checking that each signature from the proof is valid and respects the proof generation algorithm.

Algorithm checkPoI: Program executed by anyone to check the validity of a PoI

Input: P, a proof-of-interaction
u, creator of the proof
d, the dependency (hash of last block of the blockchain)
m, the message (root of the merkle tree of the new block)
D, difficulty of the PoI
N, the set of nodes in the network
Output: whether P is a valid PoI or not

1   if *not* $verif_u(P[0],d)$ then
2 | return *false*

3   $S \leftarrow$ createServices$(N, P[0])$;
4   $L \leftarrow$ tourLength$(D, P[0])$;
5   if $L * 2 + 1 \neq |P|$ then
6 | return *false*

7   $current\_hash \leftarrow H(P[0] \cdot m)$;
8   for $i = 0$; $i < L$; $i + +$ do
9 | $next\_hop \leftarrow current\_hash \% |S|$;
10 | if *not* $verif_{S_{next\_hop}}(P[2 * i + 1], current\_hash \cdot d \cdot m)$ then
11 | | return *false*
12 | if *not* $verif_u(P[2 * i + 2], P[2 * i + 1])$ then
13 | | return *false*
14 | $current\_hash \leftarrow H(P[2 * i + 2])$;

15 return *true*

Proof-of-Interactions Properties. Now we show that the Proof-of-Interaction has several properties that are awaited by client-puzzle protocols [21].

Computation guarantee: The proof can only be generated by making each visited node sign a particular message in the correct order. The sequence of visited node depends only on the initiator node, on the dependency d, and on the message m, and cannot be known before completing the tour. Furthermore, a node knows the size of his tour before completing it, which means that the node knows before doing his tour how much messages it needs to exchange and how much signatures it will do to have a correct proof.

Non-parallelizability: A node cannot compute a valid PoI for a given dependency d and message m in parallel. Indeed, in order to know what is the node of the i-th interaction, we need to know h_{i-1}, hence we need to know s_{i-1}. s_{i-1} is a signature from u_{i-1}. So we can interact with u_i only after we receive the answer from u_{i-1} *i.e.*, interactions are sequential.

Granularity: The difficulty of our protocol is easily adjustable using the parameter D. The expected time to complete the proof is $2 \times mean(D) \times Com$ where Com is the average duration of a message transmission in the network, and $mean(D)$ is the mean of the distribution D.

Efficiency: Our solution is efficient in terms of computation for all the participants. The generation of one PoI by one participant requires $mean(D)$ hashes and $mean(D)$ signatures in average for the initiator of the proof, and $mean(D)/n$ signatures in average for another node in the network. The verification requires $2D + 1$ signature verification and $mean(D)$ hashes in average. The size of the proof is also linear in the difficulty, as it contains $2mean(D) + 1$ signatures.

4   Blockchain Consensus Using PoI

In this section we detail how we can use the PoI mechanism to build a Blockchain protocol. The main idea is to replace, in the Bitcoin protocol, the Proof-of-work by the Proof-of-interactions, with some adjustments. We prove in the next section that it provides similar guarantees to the Bitcoin protocol.

Block Format. First, like in the Bitcoin protocol, transactions are stored in blocks that are chained together by including in each block, a field containing the hash of the previous block. In Bitcoin, a block includes a nonce field so that the hash of the block is smaller than a target value (hence proving that computational power has been used) whereas in our protocol, the block includes a proof of interaction where the dependency d is the hash of the previous block, and the message m is the root of the Merkel tree storing the transactions of the current block. Like for the transactions, the block header could contains only the hash of the PoI, and the full proof can be stored in the block data, along with the sequence of transactions.

Block Generation. Now we explain how the next block is appended in the blockchain. Like in Bitcoin, each participant gathers a set of transactions (not necessarily the same) and when the last block is received, wants to append a new block to the blockchain. To do so, each node tries to generate a PoI with the hash of the last block as dependency d, the root of the Merkel tree of the transactions of their own block as message m, and using the last block difficulty D. We assume the difficulty D is characterized by its mean value $mean(D)$, which is the number that is stored into the block. Like in Bitcoin, the difficulty can be adjusted every given period, depending on the time it takes to generate the last blocks.

Participants have no choice over d so the length of their tour, and the subset S of potential visited nodes is fixed for each participants (one can assume that it is a random subset). Each participant is trying to complete its PoI the fastest as possible, and the first one that completes it, has a valid block. The valid block is broadcasted into the network to announce to everyone that one have completed a PoI for its new block. When a node receives a block from another node, it checks if all the transactions are valid and then checks if the PoI is valid. If so, it appends the new block to its local blockchain and starts generating a PoI based on this new block.

First, one can see that this could lead to forks, exactly like in the Bitcoin protocol, where different part of the networks try to generate PoI with different dependencies. Thus, the protocol dictates that only one of the longest chain should be used as a dependency to generate a PoI. This is defined in the procedure `checkMessage`. When a node receives a message from another node, it first checks if the dependency matches the latest block of one of the longest chain. If not, the request is ignored.

Incentives. Like in Bitcoin, we give incentives to nodes that participate to the protocol. The block reward (that could be fixed, decreasing over time, or just contains the transactions fees) is evenly distributed among all the participants of the PoI of the block. This implies that, to maximize their gain, nodes should answer as fast as possible to all the requests from the other nodes currently generating their PoI, to increase their chance of being part of the winning block.

Also, it means that we do not want to answer a request for a node that is not up to date *i.e.*, that is generating a PoI for a block for which there is already a valid block on top, or for a block in a branch that is smaller than longest one.

Preventing Double-Touring Attacks. What prevents a node to try to generate several PoI using different variation of its block? If a node wants to maximize its gain (without even being malicious, but just rational) it can add dummy transactions to its current block to create several versions of it. Each version can be used to initiate the generation of a PoI using different tours. However, he has to send the message m every times he interacts with another node. If the length of the tour is long enough, the probability that two different tours intersect is very high. In other words, a node that receives two messages from the same node, with the same dependency d, but different values of m will raise the alarm. To prevent *double-touring*, it is easy to add an incentive to discourage nodes from generating several blocks linked to the same dependency. To do so, we assume each participant has locked a certain amount of money in the Blockchain, and if a node u has a proof that another node has created two different blocks with the same dependency (*i.e.*, previous block), then the node u can claim as reward the locked funds of the cheating node. In addition, it can have other implications such as the exclusion of the network. We assume that the potential loss of being captured is greater than the gain (here the only gain would be to have a greater probability to append its own block).

Difficulty Adjustment. The difficulty could be adjusted exactly like in Bitcoin. The goal is to chose the difficulty so that the average time B to generate a block is fixed. Here, the difficulty parameter D gives a very precise way to obtain a delay B between blocks and to limit the probability of fork at the same time. If *Com* denotes the average duration of a transmission in the network, then we want the expected shortest tour length among the participants to be $\lceil B/Com \rceil$.

For instance, it is known that the average minimum of n independent random variables uniformly distributed on the interval (a, b) is

$$\frac{b + na}{n + 1}.$$

Thus, if D is the uniform distribution between 1 and $\lceil B/Com \rceil (n + 1) - 1$, then the length of the shortest tour among all the participants will be $\lceil B/Com \rceil$ in average.

Every given period (*eg.*, 2016 blocks as in Bitcoin), the difficulty could be adjusted using the duration of the last period (using the timestamps included in each block) to take into account the possible variation of Com, so that the average time to generate a block remains B.

Communication Complexity. A quick analysis shows that each node sends messages sequentially, one after receiving the answer of the other. At the same time, it answers to signature requests from the other nodes. In average, a node is part of n_S tours. Hence the average number of messages per unit of time is constant *i.e.*, $n_S + 1$ every Com. Then, the total amount of messages, per unit of time, in the whole network is linear in n.

5 Security

This section discusses about common security threats and how our PoI-based Blockchain handles them. We assume that honest nodes will always follow our algorithms but an attacker can have arbitrary behavior, while avoiding receiving any penalty (which could remove him from the network). We assume that an attacker can eavesdrop every messages exchange between two nodes but he can not change them. Also, assume that an attacker A cannot forge messages from another honest node B.

Crash Faults. A node crashes when it completely stops its execution. The main impact is that it does not respond to the sign requests of other nodes. This can be an issue because at each step of the PoI generation, the initiator node could wait forever the response of a crashed node. Crashed nodes are handled by the fact that a node only has to interact with a subset S of the whole network $\mathcal{N}$, computed using the service creation function, `createServices`. Hence, if a node crashes, only a fraction of the PoI that are being generated will be stuck waiting for it. All the nodes whose Service sets S do not contains crash faults are able to generate their PoI entirely. Since each set S is of size $n_S = \min(n/2, 20)$, we have that, if half of the nodes crash, the probability a given set S contains a crashed node is $1 - \left(\frac{1}{2}\right)^{n_S}$. So that the probability p that at least one set S contains only correct nodes is

$$p = 1 - \left(1 - \left(\frac{1}{2}\right)^{n_S}\right)^n$$

One can see that the probability p tends quickly (exponentially fast) to 1 as n tends to infinity. For small values of n, the probability is greater than a fixed non-null value. In the rare event that all the sets S contain at least a crashed node, then the protocol is stuck until some crashed nodes reboot and are accessible again.

Finally, we recall that honest nodes are incentivized to answer, because they get a reward when they are included in the next block's PoI. Hence, honest nodes will try be back again as fast as possible.

Selfish mining. Selfish mining [10] is an attack where a set of malicious nodes collude to waste honest nodes resources and get more reward. It works as follow. Once a malicious node finds a new block, it only shares it with the other malicious nodes. All malicious nodes will be working on a private chain without revealing their new block, so that honest nodes are working on a smaller public branch *i.e.*, honest nodes are wasting resources to find blocks on a useless branch. When honest nodes find a block, the malicious nodes might reveal some of their private blocks to discard honest blocks and get the rewards.

In Bitcoin, selfish mining is a real concern as attackers having any fraction of the whole computational power could successfully use this strategy [20].

Interestingly, our PoI-based Blockchain is less sensible to such attack. Our algorithm gives a protection by design. Indeed, when generating a PoI, a node has to ask to a lots of other nodes to sign messages containing the hash of the previous block, forcing it to reveal any private blocks. Other nodes in the network will request the missing block before accepting to sign the message. In other words, it is not possible to generate a PoI alone. Moreover, if a node is working on a branch that is smaller than the legitimate chain and ask for the signature of an honest node, the latter will tell the former to update its local Blockchain, thus preventing him from wasting resources.

Shared Mining. During the PoI, a node will most of the time be waiting for the signature of another node. So the network delay has the highest impact on the block creation time. To remove this delay, a set of malicious nodes can share theirs private keys between each other and try to create a set S where every nodes are malicious. If one malicious node of the pool succeeds, it can compute the proof locally without sending any messages. It will generate the PoI faster than honest nodes and have a high chance to win.

We defined earlier that each node of the network is known. Which mean that each node is a distinct entity. For this attack to succeed, entities need to share their private keys. This is a very risky move because once you give your private key to someone, he can create transactions in your name without your authorization. This risk alone should discourage honest nodes to do it, even if they want to maximize their gain.

We can still assume that a small number of malicious nodes do know each other and collude to perform this attack. We show now that this attack is hard to perform. S only depends on the previous block and on the identity of the initiator of the proof, so the nodes have no control over it. S consists of n_S nodes randomly selected among the network. So if there are F malicious friends on the network, there is on average the same fraction (n/F) of malicious friends in S as in $\mathcal{N}$. However, the probability for the tour to contains only malicious friend is very low. Indeed, with F malicious friends on the network, the probability that the entire tour consists of malicious friends is $(n/F)^{mean(D)}$ in average.

When a malicious node initiates a PoI for a given message, it can see whether the tour contains an honest node or not, so it might be tempted to change the content (by reordering the transaction or inserting dummy transactions) of its block until the tour contains only his malicious friends. However, even if there is a fraction $(n/F) = 0.1$ of malicious friends in the network (hence in S), and if $mean(D) = 100$, for instance, then the probability that a given tour contains only malicious friends is 10^{-100}. To find a tour with only malicious friends, an initiator would have to try in average 10^{100} different block content, which is not feasible in practice.

6   Conclusion and Possible Extensions

We have presented an new puzzle mechanism that requires negligible work from all the participants. It asks participants to gather sequentially a list of signatures from a subset of the network, forcing them to wait for the response of each visited node. This mechanism can be easily integrated into a Blockchain protocol, replacing the energy inefficient Proof-of-work. The resulting Blockchain protocol is efficient and more secure than the Bitcoin protocol as it is not subject to selfish-mining. Also, it does not have the security issues found in usual PoW replacements such as Proof-of-stack or Proof-of-elapsed time. However, it currently works only in networks where participants are known in advance. The design of our Blockchain protocol makes it easy to propose a possible extension to remove this assumption.

The easiest way to allow anyone to be able to create blocks, is to select as participants the n nodes that locked the highest amount of money. This technique is similar to several existing blockchain based on protocols that work only with known participants (such as Tendermint [15] using an extension of PBFT [6]) or where the nodes producing blocks are reduced for performance reasons (such as EOS [13] where 21 producer nodes are elected by votes from stakeholders). We believe a vote mechanism from stakeholders can elect the set of participants executing our protocol. The main advantage with our solution is that the number of participants can be very high, especially compared to previously mentioned protocols.

References

1. Poet 1.0 specification. https://sawtooth.hyperledger.org/docs/core/releases/1.2.4/ architecture/poet.html
2. Abliz, M., Znati, T.: A guided tour puzzle for denial of service prevention. In: 2009 Annual Computer Security Applications Conference, pp. 279–288. IEEE (2009)
3. Alsunaidi, S.J., Alhaidari, F.A.: A survey of consensus algorithms for blockchain technology. In: 2019 International Conference on Computer and Information Sciences (ICCIS), pp. 1–6. IEEE (2019)

4. Back, A., et al.: Hashcash-a denial of service counter-measure (2002)
5. Bonnet, F., Bramas, Q., Défago, X.: Stateless distributed ledgers. arXiv preprint arXiv:2006.10985 (2020)
6. Castro, M., Liskov, B.: Practical byzantine fault tolerance and proactive recovery. ACM Trans. Comput. Syst. (TOCS) **20**(4), 398–461 (2002)
7. CBECI: Cambridge bitcoin electricity consumption index (2020). https://www.cbeci.org
8. Chen, L., Xu, L., Shah, N., Gao, Z., Lu, Y., Shi, W.: On security analysis of Proof-of-Elapsed-Time (PoET). In: Spirakis, P., Tsigas, P. (eds.) Stabilization, Safety, and Security of Distributed Systems, SSS 2017. LNCS, vol. 10616, pp. 282–297. Springer, Cham (2017). https://doi.org/10.1007/978-3-319-69084-1_19
9. Douceur, J.R.: The Sybil attack. In: Druschel, P., Kaashoek, F., Rowstron, A. (eds.) Peer-to-Peer Systems, IPTPS 2002. LNCS, vol. 2429, pp. 251–260. Springer, Heidelberg (2002). https://doi.org/10.1007/3-540-45748-8_24
10. Eyal, I., Sirer, E.G.: Majority Is not enough: bitcoin mining is vulnerable. In: Christin, N., Safavi-Naini, R. (eds.) Financial Cryptography and Data Security, FC 2014. LNCS, vol. 8437, pp. 436–454. Springer, Heidelberg (2014). https://doi.org/10.1007/978-3-662-45472-5_28
11. Gallersdörfer, U., Klaaßen, L., Stoll, C.: Energy consumption of cryptocurrencies beyond bitcoin. Joule (2020)
12. Gaži, P., Kiayias, A., Russell, A.: Stake-bleeding attacks on proof-of-stake blockchains. In: 2018 Crypto Valley Conference on Blockchain Technology (CVCBT), pp. 85–92. IEEE (2018)
13. IO, E.: Eos. IO technical white paper. EOS. IO (2017). https://github.com/EOSIO/Documentation. Accessed 18 December 2017
14. King, S., Nadal, S.: Ppcoin: Peer-to-peer crypto-currency with proof-of-stake. self-published paper, August 19 (2012)
15. Kwon, J.: Tendermint: Consensus without mining. Draft v. 0.6, fall 1(11) (2014)
16. Mirkovic, J., Dietrich, S., Dittrich, D., Reiher, P.: Internet denial of service: attack and defense mechanisms (Radia Perlman Computer Networking and Security). Prentice Hall PTR (2004)
17. Mukhopadhyay, U., Skjellum, A., Hambolu, O., Oakley, J., Yu, L., Brooks, R.: A brief survey of cryptocurrency systems. In: 2016 14th annual conference on privacy, security and trust (PST), pp. 745–752. IEEE (2016)
18. Nakamoto, S.: Bitcoin: a peer-to-peer electronic cash system (2008)
19. Preneel, B.: Analysis and design of cryptographic hash functions. Ph.D. Thesis, Katholieke Universiteit te Leuven (1993)
20. Sapirshtein, A., Sompolinsky, Y., Zohar, A.: Optimal selfish mining strategies in bitcoin. In: Grossklags, J., Preneel, B. (eds.) FC 2016. LNCS, vol. 9603, pp. 515–532. Springer, Heidelberg (2017). https://doi.org/10.1007/978-3-662-54970-4_30
21. Tritilanunt, S., Boyd, C., Foo, E., González Nieto, J.M.: Toward non-parallelizable client puzzles. In: Bao, F., Ling, S., Okamoto, T., Wang, H., Xing, C. (eds.) Cryptology and Network Security, pp. 247–264. Springer, Berlin Heidelberg, Berlin, Heidelberg (2007)
22. Wang, W., Hoang, D.T., Hu, P., Xiong, Z., Niyato, D., Wang, P., Wen, Y., Kim, D.I.: A survey on consensus mechanisms and mining strategy management in blockchain networks. IEEE Access **7**, 22328–22370 (2019)

LighTx: A Lightweight Proof-of-Bandwidth Transactions Transfer System

Imane El Abid[1]([✉]), Yahya Benkaouz[2], and Ahmed Khoumsi[3]

[1] Mohammed VI Polytechnic University, Benguerir, Morocco
imane.elabid@um6p.ma
[2] LCS, Faculty of Sciences, Mohammed V University, Rabat, Morocco
yahya.benkaouz@um5.ac.ma
[3] University of Sherbrooke, Quebec, Canada
Ahmed.Khoumsi@usherbrooke.ca

Abstract. The race to solve the so-called *Blockchain trilemma* (i.e., decentralization, scalability, and security) has resulted in a multitude of solutions, each providing at most two of the three features. Moreover, existing blockchain systems still represent several technical hurdles in terms of computation effectiveness and energy consumption, especially in large-scale networks. In this paper, we design LighTx a cost-effective and scalable transaction transfer system that aims to reach agreement in public peer-to-peer networks at a low cost.

LighTx leverages a *Byzantine Reliable Broadcast* (BRB) primitive to transmit and validate transactions in a logarithmic communication cost with respect to the number of nodes. We additionally deploy a *Proof-of-Bandwidth-based reputation system* to mitigate the threats enforced by Sybil attack in public networks. We assess the performance of our system and demonstrate a considerable transaction rate of hundreds of transactions per second and low latency in the order of few seconds while providing defense against Byzantine adversaries.

Keywords: Byzantine reliable broadcast · Proof-of-bandwidth · Reputation score · Sybil attack

1 Introduction

In 2008, Satoshi Nakamoto presented Bitcoin [31], a digital cryptocurrency implemented as a decentralized public peer-to-peer network. At its core, Blockchain was designed to transfer transactions while preventing double-spending and was soon after used to manage other kinds of data for different applications. Blockchain achieves consensus among peers via cryptographic puzzles solving known as Proof-of-Work (PoW) that challenges a computationally bounded adversary from gaining control over the network.

© Springer Nature Switzerland AG 2021
K. Echihabi and R. Meyer (Eds.): NETYS 2021, LNCS 12754, pp. 144–160, 2021.
https://doi.org/10.1007/978-3-030-91014-3_10

Despite being a major asset, PoW by its very nature incorporates a serious handicap; the excessive energy consumption induced by mining activity and the time required for transaction validation results in high costs and low scalability. On the other hand, decisions are only made by miners dedicating a massive computing power for the mining activity. To fasten transactions validation, miners gather in cooperating mining pools, leading the system to be partially centralized. Moreover, popular blockchain systems lack fairness among users as they allow unfair participation opportunities for miners with limited resources. Decisions are made by users with high computational power using PoW [31] or important stake deposit in the Proof-of-Stake (PoS) [35].

Two known approaches emerged to resolve these concerns and improve the performance of the inherent agreement protocol. One approach focuses on variants of Nakamoto consensus [1,4,35], while the other adopts voting-based consensus protocols such as [2,9–11,26]. Both approaches succeeded to some extent to eliminate the issue of high energy consumption posed by PoW-based protocols and the constraints of stakes holding in [35], due to miners selection via competitive voting mechanisms. However, these approaches face other challenges such as significant communication overhead and Sybil attack vulnerability.

System openness and decentralization are tied one to another. Decentralization improves security by avoiding central entities prone to become a single point of failure while storing critical data. By contrast, openness comes with new challenges, mainly the Sybil attack [13]. Sybil attack refers to an adversary that forges a large number of fake identities to undermine network services. Sybil defense involves rigorous verification of the genuine identities of nodes and determines if we can trust their participation to propagate and validate transactions, in order to maintain the system robustness.

This work aims to create a lightweight version of blockchain that meets application-level demands without a central party or synchrony assumptions. We seek to demonstrate a scalable and reliable system in public networks by deploying low-cost techniques and with no need for computation-intensive consensus protocols. We suggest LIGHTX, a cost-effective and scalable transaction logging system that makes use of BRB primitive [19] to concurrently and securely commit transactions. Inheriting the properties of the underlying broadcast abstraction and supplied with identification and routing mechanisms, LIGHTX disposes of consensus complexities and converges to a trade-off between reliability and scalability guarantees to fit into application demands. We further deploy for public environments, a lightweight Proof-of-bandwidth (PoB) jointly with a reputation system to provide Sybil defense. This is achieved by leveraging the concept of trust transitivity in peer-to-peer systems inspired by [21].

The rest of the paper is organized as follows: In Sect. 2, we present an overview of the existing consensus families. In Sect. 3, we describe the system and threat models. Then, in Sect. 4, we describe our permissionless LIGHTX that builds upon a BRB abstraction jointly with a *PoB-based reputation system* to detect Sybil nodes and mitigate their impact. In Sect. 5, we report the experimental results. We discuss related work in Sect. 6. Finally, we conclude in Sect. 7.

2   Background

Consensus protocol, as one of the core components of blockchain, directly affects the scalability of the system. In this section, we review the guarantees offered by both deterministic and probabilistic consensus and then we give an overview of Byzantine Reliable Broadcast and mention its limitations.

2.1   Deterministic vs. Probabilistic Consensus

Deterministic algorithms are restricted by impossibility results [16] and lower bound limits [27] with the presence of faulty processes and asynchrony assumptions. An alternative that makes it possible to achieve consensus is to enable randomization among nodes. Randomized consensus ensures rather probabilistic *liveness* properties (every operation eventually completes), but guarantees the same *safety* properties (nothing wrong happens) [3]. Namely, randomized algorithms can be much simpler and provably faster and provide a better defense to thwart an adversarial behavior [24].

Reliable broadcast abstraction [8] presents in the same context a convenient tool to reliably propagate a message through a network of nodes, such that all nodes deliver the message even with the existence of faulty nodes. In real terms, existing broadcast protocols make use of quorum-based techniques. Conceptually, quorum schemes require the confirmation of the majority of participants to deliver a message (i.e., the values proposed by participating nodes have to intersect in at least one correct node to enable the system to make a decision) [5,28,29]. Their communication complexity is quadratic [5,20], while optimized versions were barely able to reach a linear complexity [6].

2.2   Byzantine Reliable Broadcast Abstraction

BRB Specifications. Byzantine Reliable Broadcast (BRB) [19] is a broadcast primitive that falls in the category of randomized algorithms. BRB aims to transmit a message over a network of nodes, such that the same version of the message is delivered by all correct nodes. By making use of BRB abstraction, we achieve consensus at logarithmic communication cost. Being a probabilistic approach helps bypassing the impossibility result of agreement in the presence of one faulty process [16] by guaranteeing probabilistic properties rather than deterministic ones. Moreover, the peer sampling method used by BRB to sample communicating peers enables each node to pick a sample of peers (following Poisson distribution) to collect feedback for message delivery, instead of expecting confirmation from a majority quorum.

This sampling approach is considered a key feature for a scalable design as it allows reduced messages exchange to validate a transaction. The samples are significantly smaller in size compared to quorums. The importance of this feature becomes evident as the network grows bigger and the difference between quorums and samples sizes gets flagrant. That results also in a low latency due to the minimal messages exchange. Indeed, a node does not have to wait for the responses of the majority to deliver or discard a message. Rather, it communicates only with some selected peers that will reflect the status of the network. Also, unlike in PoW and PoS systems where validators are privileged based on their dedicated resources (E.g., computation power, stake), adopting BRB enables all nodes to be potential validators. Nodes collect confirmation from their communication samples to validate a transaction, giving equal chances to all users to participate in the agreement process.

BRB represents an appealing workaround to replace consensus and solve the double-spending problem by ensuring three major properties: (1) *Validity*: If a correct sender sends a message, all correct processes eventually deliver the broadcast message. (2) *Consistency*: Every correct process delivers the same message if it delivers one at all and (3) *Totality*: If any correct process delivers some message, every correct process eventually delivers that message.

BRB Sub-protocols. BRB abstraction consists of three nested sub-protocols. First *Murmur* that propagates a transaction m across the network via a generated Erdös-Rényi random graph [15]. Upon delivery of m by *Murmur*, comes *Sieve* that ensures the consistency of the broadcast transaction. *Sieve* delivers m only when enough nodes have witnessed the same m. Then *Contagion* provides secure delivery by all correct nodes through a feedback mechanism that first verifies if enough nodes are ready for delivering m, before effectively delivering it [19].

Murmur represents the starting of the broadcast operation by creating an instance of a gossip-based algorithm that distributes a message m (broadcast by the sender) over the network. Upon initialization, each node subscribes to a randomly picked *gossip sample*. The sender then signs m and sends it to members of its *gossip sample*. Upon receipt of a gossip message m, the recipient verifies the identity of the sender by verifying the signature and delivers the message, then in turn, dispatches it to members of its *gossip sample* (Fig. 1). During this gossip process, nodes may receive several copies of the broadcast message m from different nodes as they are included in multiple *gossip samples* of many other nodes. Murmur delivers only the first received gossip message and discard the others.

Sieve represents an instance of consistent broadcast. This means that Sieve guarantees that each correct node delivers the same message, if it delivers one at all, even with the presence of a faulty sender. Upon initialization, each node subscribes to a randomly picked *echo sample*. Sieve proceeds when a message m is delivered by the lower layer (i.e.,Murmur); a node p_i echoes the gossiped message m to all nodes in its subscription set. When p_i collects enough *echoes* associated to m from its *echo sample*, it delivers m (Fig. 2).

Contagion. Besides consistency property inherited from Sieve, Contagion complements the previous blocks by ensuring *totality*. These combined properties imply the equivalent of agreement.

Upon initialization, each node subscribes to its *ready* and *delivery samples*. When a message m is delivered by the lower layer (i.e., Sieve), Contagion sends

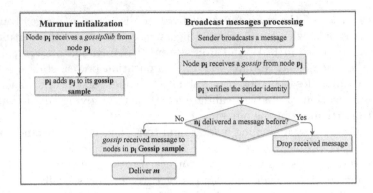

Fig. 1. Murmur

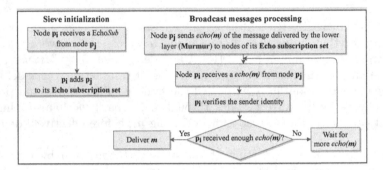

Fig. 2. Sieve

a *ready* message to nodes of its subscription set and waits to receive feedback messages from its *ready* and *delivery samples*. When node p_i receives enough confirmations for a message m from its ready sample, it becomes ready for m too. Contagion delivers the first message m for which a node p collected enough *ready* messages from its *delivery sample* (Fig. 3).

BRB Limitations. As we described the broadcast primitive, we identified two main constraints we had to address for building our system. First, the broadcast algorithm generates one broadcast instance to deliver a single transaction. In BRB, every submitted transaction is carried by a single broadcast instance. This helps you to differentiate between multiple send operations and inconsistent message dissemination (in the case of a Byzantine sender). However, this feature imposes a constraint in concurrent environments where all nodes attempt to submit multiple transactions. When concurrent transactions are issued, nodes receive concurrently different BRB messages of all broadcast transactions without being able to identify their broadcast instances. Therefore, it becomes challenging to handle exchanged messages in each broadcast transaction and associate them with their corresponding transactions.

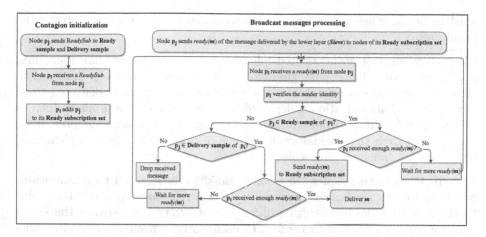

Fig. 3. Contagion

On the other hand, BRB requires additional mechanisms to prove robustness in high churn networks. BRB tends to be more suitable for private networks [19] and claims that in an open environment with slow churn, the issue could be approached in the same fashion as in private networks to deal with the effect of openness. Nevertheless, in open environments, robustness challenges get even harder, through the exposure to dynamic malicious users. A Sybil attacker can spread and vote for inconsistent transactions. This requires proper inspection of the real identity of nodes in order to preserve system robustness in networks with churn.

3 Problem Definition

3.1 System Model

We consider a public asynchronous fully connected message-passing system that provides direct communication between each pair of nodes. The term node and process are used interchangeably. The system builds upon the following assumptions:

Nodes. Nodes are denoted $p_1, .. , p_n$. We assume that each node is associated to an account and each account has a single owner. We identify a transaction issued by p_i to p_j as a transfer of a digital asset from the account of p_i to the account of p_j. We consider a straightforward probability sampling strategy, by deploying a sampling oracle Ω that, at each execution, generates randomly selected samples (i.e. subsets with parametrized sizes) of nodes among the total population of nodes. All nodes have direct access to the sampling service to pick their samples. We leverage existing encryption and authentication primitives: digital signature, SSL certificates and hash functions, which are known to be hard to break. Then we assume that adversaries cannot break cryptographic primitives.

Links. We consider authenticated reliable point-to-point communication links [8]. In principle, reliable links guarantee delivery and no message duplication. Given any message m sent from a node p_i to a node p_j, m is delivered to p_j exactly once. Authentication is deployed to ensure no creation property (i.e., to prevent malicious nodes from creating and propagating a message and pretending it originates from the original sender). Such a link abstraction is implemented by lower-level transport protocols.

Churn. Participating nodes are recommended to dedicate all their bandwidth resources for the application, especially during assessment phase. This is due to the fact that the bandwidth will be shared arbitrarily between the set of running applications on their devices. Consequently, it results in undesired fluctuated measurements caused by this bandwidth usage. Thus, we can track the bandwidth changes and assert that they reflect the behavior of the nodes in the application. We also assume that nodes have bounded bandwidth resources, but they are not required to be homogeneous in terms of bandwidth capacity, since the system only relies on detecting bandwidth fluctuations and not the bandwidth capacity.

Nodes newly joining the network are not assigned any reputation score; reputation scores are only assigned after interacting with new nodes and recording their bandwidth measurements. New-comers (includes recently active nodes) get their scores either by referring to previously calculated global scores (measured by the network) or by proceeding to a new PoB phase.

We also assume a set of high-ranked nodes, we call *Pre-trusted*, that a node can trust their opinion about the network status. Members of the *Pre-trusted* set can be chosen on different bases. For instance, choosing the first nodes joining the network [21], or dynamically choosing highly reputable nodes based on their history [25,37]. We opt for a hybrid approach to assess bandwidth metric via both proactive and reactive assessment. First, proactively, by launching the *PoB* mechanism periodically (The period τ is a system parameter that is determined at application level). And reactively, by responding to the network changes, in terms of network size and transaction submission rate. During a measurement phase, we consider the system to be of static size.

3.2   Threat Model

We consider that nodes may experience Byzantine failures (i.e. nodes fail in an arbitrary manner that may deviate in any conceivable way from the algorithm [8]). Byzantine nodes are represented by f computationally unbounded malicious nodes that may collude to interfere with the algorithm. We also assume that Byzantine nodes cannot compromise authentication primitives.

Scenario (1): We define a Byzantine node as a node that acts maliciously by presenting conflicting or inconsistent messages with the objective to carry out a double-spending. Double-spending occurs when a transaction uses the same input as another transaction that has already been spent. Concretely, the transaction is spent twice.

Scenario (2): We consider Sybil adversaries who declare multiple fake identities in the network and thus gain a majority opinion to send out inconsistent trans- actions. A Sybil attacker may also collude with other Byzantine nodes on the network, who may falsely report each other's scores. When dealing with power- ful adversaries (e.g., nationwide or global adversaries), we need to deploy more advanced probing techniques that enable nodes to quantify the dedicated band- width resources of their peers and thus detect fluctuations caused by adversarial behavior. At the current stage, the analysis of appropriate probing techniques has not yet been studied.

4 LighTx

4.1 Beyond BRB

The design of LIGHTX draws primarily on BRB abstraction [19]. We customize BRB via setting up a routing mechanism for messages identification management to support concurrency in transactions submission. When concurrent transac- tions are broadcast, nodes need to link the issued BRB messages to their corre- sponding transactions. For this purpose we identify broadcast channels for each created transaction and create a routing entity at each node to map each broad- cast instance to its corresponding message. Algorithm 1 describes the routing mechanism in LIGHTX.

The sender s initiates a broadcast operation by creating a broadcast instance brb identified with a unique broadcast channel identifier brb_{id}, then adds the new pair $\{brb_{id} : \boldsymbol{brb}\}$ to the routing table (line 3). Next, the sender broadcasts the transaction labelled with its broadcast channel ID using the BRB primitive as described in [19] (line 4). The broadcast channel ID is generated by combining the broadcast instance hash code with the current timestamp. When a broad- cast operation is triggered, the broadcast instance is created, by the sender s, associated with its broadcast channel ID. The broadcast channel ID is unique across all nodes.

Upon receipt of any BRB message, the receiver consults its routing table. If the routing table contains the broadcast channel ID of the received BRB mes- sage, the corresponding broadcast instance handles the BRB message (line 3). Otherwise, a new broadcast instance will be added to the routing table to han- dle messages originating from this broadcast channel (line 5). Once the broad- cast channels are identified, the message processing continues following the *BRB* primitive (line 3).

4.2 Proof-of-Bandwidth Scheme

To resolve the issues posed by Sybil attack while maintaining the major fea- ture BRB offers which is scalability, we propose a lightweight PoB-based rep- utation system to prevent malicious peers from performing a Sybil attack and

Algorithm 1. LIGHTX ROUTING

1: Initiate **BRB** broadcast:
2:        create broadcast instance **brb** with broadcast channel ID $< brb_{id} >$
3:        $routingTable \leftarrow \{brb_{id} : brb\}$
4:        **brb**.broadcast $< tx|brb_{id} >$ ▷ The sender

1: Upon receipt of **BRB** message with brb_{id} ▷ All nodes
2:        $brb \leftarrow Routing(brb_{id})$
3:        proceed **BRB** broadcast process with broadcast instance **brb**

1: **function** $Routing(brb_{id})$
2:        **if** $routingTable$ contains brb_{id} **then**
3:                **return** getValue(brb_{id})
4:        **else**
5:                $routingTable$.add(brb_{id} : new **brb**)
6:                **return** getValue(brb_{id})
7:        **end if**
8: **end function**

---

thus, manipulating transactions validation. Unlike the famed Proof-of-Work [31], nodes are not exhausted by investing a huge amount of energy and CPU power to prove their identities. We utilize bandwidth measurement, then inject the collected measurements into a reputation system. The reputation scores of the network are aggregated until they converge to a global score in few rounds. The concept drives from the intuition that transmitted transactions should be proportional to the used bandwidth resources. A genuine identity should reflect stable bandwidth measurement records, whilst a node running multiple fake identities will have fluctuating bandwidth measurements caused by flow interference, high load and medium occupancy.

Bandwidth Measurements. As a first step, we set up a bandwidth evaluation method, to measure the available bandwidth of peers. Over multiple rounds, we perform bandwidth measurements and track bandwidth changes of the evaluated nodes. Multiple bandwidth estimation techniques and tools exist in the literature depending on application usage. Most of the bandwidth estimation techniques are based on probe messages. We adopt the one-way-delay method [23] as it represents a low overhead compared to other existing methods [22,33].

Receivers may misbehave by faking timestamps of the probe messages. However, given that the adversary does not know about the measured round trip time (RTT) on the sender side. Also, given that probe messages are sent on various throughput speed, then inducing random timestamps to the probe reply comes against the peer itself. Therefore, measuring a stable bandwidth capacity would be preferred.

Local Scores Assignment. Bandwidth measurements are performed over r rounds. Each node keeps track of the r bandwidth values collected over r rounds in a vector BW for each of its assessed peers. When the measurement phase is finished, peers initially assign local scores to their peers. Each assessed node will be assigned an initial local score $w_{ij}^{(0)}$. We set $w_{ij}^{(0)}$ to be equal to the mean μ_{BW_0} over all the assessed peers in the first measurement round. Then, scores are updated based on the variance of the measurements taken over r rounds and the maximum assessed bandwidth of each peer. If the variance of the measurements is negligible, the initial scores are maintained. When the fluctuations are considerable, the new score takes into account the variance of bandwidth measurement and the maximum measured bandwidth. That implies that nodes with high bandwidth are heavily penalized when they show some anomalies, since they are more likely to have a big impact on the system if they are malicious. We can translate the score update by Expression 1.

$$w_{ij} = \begin{cases} w_{ij}^{(0)} - \sigma_{BW_j}.BW_{j_{max}} : & if \quad \sigma_{BW_j} > \alpha \\ \quad w_{ij}^{(0)} & : \quad otherwise \end{cases} \tag{1}$$

Where w_{ij} is the updated local score, $w_{ij}^{(0)}$ is the initial local score, σ_{BW} is the variance of bandwidth measurements measured over r rounds, BW_{max} is the maximum bandwidth capacity measured of node p_j and α is fluctuations tolerance factor which indicates the degree of bandwidth fluctuations that can be allowed without penalty. Algorithm 2 describes the *Proof-of-Bandwidth* scheme and the local scores update during assessment phase.

4.3   Reputation System

Score Aggregation. To support the *Proof-of-Bandwidth* mechanism we setup a reputation system to evaluate every single node on the network. Our approach is inspired by EigenTrust [21] that builds upon the notion of transitive trust. Nodes consider the opinion of remote nodes they trust, by collecting their local reputation scores. Based on the remote evaluation, each node aggregates the reputation score from its own local scores and the scores received from remote nodes, then shares it back with nodes that trust him. By iteratively repeating this process for a certain number of rounds, it converges to the *eigenvector* of the score matrix. Each node in the system converges to a global reputation score, based on which they can make a decision on the malicious nodes.

As all nodes do not have the same measurement records, their local scores can widely vary. The local scores should be normalized to limit the impact of an adversary arbitrarily giving high local reputation scores to other malicious peers, or a low reputation score to honest ones and poison the view of the network. Therefore, we first normalize the local reputation scores in a range between zero and one. We define m_{ij} as the normalized score assigned by node p_i to node p_j in Eq. 2.

$$m_{ij} = \frac{max(w_{ij}, 0)}{\sum max(w_{ij}, 0)} \tag{2}$$

To aggregate scores, each node requests the peers it evaluated for their local scores, weights their opinions and take them into consideration to compute the global score value by iterating over the matrix of normalized scores M.

Algorithm 2. PROOF-OF-BANDWIDTH

1: **Parameters.** α : Fluctuation tolerance factor
2:                    R : Number of rounds

1: **procedure** $Probe()$
2:     **for** each node p_i **do**
3:         **for** r rounds **do**
4:             Send $ProbeMessage$ to nodes of **BRB** samples
5:         **end for**
6:     **end for**
7: **end procedure**

1: **function** $ScoreUpdate(\sigma_{BW}, BW_{max})$
2:     **if** $\sigma_{BW} > \alpha$ **then**
3:         **return** $\omega_{ij}^{(0)} - \sigma_{BW}.BW_{max}$
4:     **else**
5:         **return** $\omega_{ij}^{(0)}$
6:     **end if**
7: **end function**

1: **Upon receipt of a** $ProbeMessage$
2: **if** $BW < R$ **then**
3:     $BW.add(BW_i)$
4: **else**
5:     $BW_{max} = Max(BW)$
6:     Calculate variance σ_{BW} of BW
7:     $\omega_{ij}^{(0)} = \mu_{BW_0}$
8:     $\omega_{ij} = ScoreUpdate(\sigma_{BW}, BW_{max}, \alpha)$
9: **end if**

In the first iteration, we compute the score vector m' by multiplying the local score matrix M with the remote score vector m (Eq. 3). Then, for each iteration, p_i updates its global score vector $m'' = M.m' = M^{(2)}.m$ to $m^{(n)} = M^{(n)}.m$ until no changes are observed when updating the global score vector. At this stage, the network converges to one global vector which corresponds to the principal eigenvector of matrix M.

$$m' = M.m \tag{3}$$

Byzantine Resilient Score Aggregation. To avoid the impact of *sinks* (i.e. nodes who share very low scores about other nodes, influencing their global scores), we use vector $\vec{\rho}$, that is a distribution over the set of pre-trusted peers P (a priori trusted set of nodes). We define ρ_i the elements of vector $\vec{\rho}$ a follows:

$$\rho_i = \begin{cases} \frac{1}{|P|} : & if \quad i \in P \\ 0 : & otherwise \end{cases} \tag{4}$$

The use of vector $\vec{\rho}$ is claimed by [7,21] to lead to faster convergence, so we use the vector $\vec{\rho}$ as the start vector for the score aggregation. We also include $\vec{\rho}$ in the later computation to weight pre-trusted peers by a damping factor

d [7] (as described by Eq. 5). Introducing this damping factor d aims to break malicious collectives, by injecting some randomness into the data received from other peers when measuring global scores. Since all the scores should be summed together, the global scoring measured from other peers is "reduced" by the factor d to prevent the other nodes from inflating the scores. The $1 - d$ factor is to compensate the probability of the first part, by adding some confidence to the set of pre-trusted nodes.

$$M^{(k+1)} = d.Mm^{(k)} + (1 - d)\overrightarrow{\rho} \tag{5}$$

We describe the distributed version of the score aggregation in Algorithm 3.

Algorithm 3. REPUTATION SYSTEM

1: **Parameters.** A_i : Nodes evaluated by node p_i
2:                              B_i : Nodes who evaluated node p_i
3:                              ϕ : Convergence precision
4:                              d : Damping factor

5: **for** Each node p_i **do**
6:         Query all nodes $p_j \in A_i$ for m_{ji} and $m_j^{(0)} = \rho_j$
7:         **repeat**
8:             $m_i^{(k+1)} = d(m_{1i}m_1^{(k)} + m_{2i}m_2^{(k)} + .. + m_{ni}m_n^{(k)}) + (1 - d)\rho_i$
9:             send $m_{ij}m_i^{(k+1)}$ to all nodes $p_j \in B_i$
10:            wait for all nodes $p_j \in A_i$ to return their $m_{ji}m_j^{(k+1)}$
11:        **until** $|m_i^{(k+1)} - m_i^{(k)}| < \phi$
12: **end for**

---

5 Evaluation

In this section, we assess the performance of our system. Our prototype was implemented in Java language. The code is available in https://github.com/ImaneElabid/LighTx

Experimental Setup. All experiments are conducted on a computer running Windows 10 OS, with Intel(R) Xeon(R) W-2123 CPU 3.60 GHz and 32.0G of RAM. For the sake of simplicity, we withdraw link authentication property to dispose of authentication certificate setups. We use therefore, TCP as transport protocol. In what follows, we detail the circumstances of the experiment and analyze the results. In the correct execution model, a sender s broadcasts a transaction and distributes it among the network. We assume that all nodes of the network behave correctly.

Communication Complexity. All samples sizes are set to be equal and logarithmic in size with respect to the number of nodes. With this setting, we vary the number of nodes from 10 to 250 nodes and we measure the number of received messages per process for each broadcast instance. Figure 4a depicts that contrarily to quorum-based BRB, the number of messages of LIGHTX using

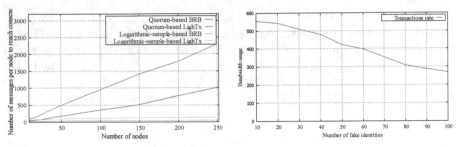

(a) Number of messages per node to reach consensus in sample-based vs. quorum-based LIGHTX.

(b) Transaction rate as a function of the network size.

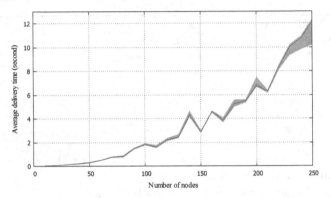

(c) Average latency as a function of the network size.

Fig. 4. LIGHTX performance efficiency.

sample-based BRB does not increase with the total number of nodes. This is because each node communicates only with members of its samples (of logarithmic size). Even with such a minimal message exchange, LIGHTX achieves agreement.

Latency. For the system latency, we measure the time elapsed between the broadcast and the delivery operation. We plot the delivery time averaged over multiple runs. Figure 4c shows that latency increases with the number of nodes. For example, we see that it takes on average 12 seconds to deliver a consistent transaction over a network of 250 nodes. Note that part of the latency augmentation is generally due to external causes, such as concurrency and multi-threading issues related to hardware limitation. In our simulation environment, as the system grows bigger, concurrency issues become glaring. We presume that these external delays can be considerably reduced in a distributed architecture.

Data Rate. The obtained transaction rate is shown in Fig. 4b where we note that the application reaches a throughput of more than 500 tps when the number of nodes is relatively small (e.g., < 30). Transaction rate decreases with the number of nodes, but remains clearly higher compared to Bitcoin which can only handle around 7 tps [12] or Ethereum that reaches 20 tps [36].

Adversarial Execution. In a centralized model of our reputation system, the network converges to global trusts within few rounds (about 5 rounds, see Fig. 5a) the decentralized version of the algorithm. In this simulation, in a network of 100 nodes having equal bandwidth resources, we assess bandwidth usage to validate one transaction. We record the bandwidth measurements of all nodes on the network, while varying k_{id} the number of identities created by a Sybil attacker. Figure 5b shows the difference between honest nodes bandwidth records and a Sybil node records as we increase the fake identities k_{id} of a Sybil attacker. We note that as k_{id} grows, the ability to detect Sybil nodes increases, as they become overloaded with the different flows they receive and that lowers their transmission rate. Concretely, a Sybil node p_k should run a large number of fake identities k_{id} to take over the network. p_k in this case, communicates simultaneously with $S_{avg}.k_{id}$ (where S_{avg} is the average size of communication samples).

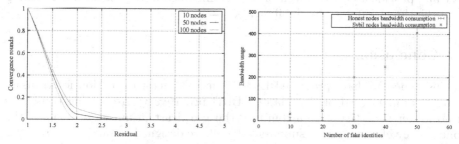

(a) Convergence rounds in the centralized version of the reputation system.

(b) Correct vs. Sybil nodes bandwidth consumption for transaction transfer.

Fig. 5. LIGHTx PoB-based reputation system.

6   Related Work

With the wide adoption of Blockchain technology, the necessity to conceive alternatives to optimize Blockchain systems in various aspects, such as performance, energy-efficiency and security has become of paramount importance. Lately, new families of consensusless algorithms are emerging. AT2 [18] proposes multiple protocols for assets transfer based on BRB abstraction to build an asset transfer system abandoning the expenses of consensus. AT2 also refers to the idea of a *Proof-of-Bandwidth* protocol but does not describe any comprehensive practical solution. [34] combines traditional consensus and Nakamoto's consensus [31] to come up with a cost-effective protocol that makes use of gossip protocols with recurrent sampling. [34] however, has a communication complexity of $O(kn)$ where $(k << n)$ is a security parameter.

Other alternatives opting for randomization to achieve agreement, such as HoneyBadgerBFT [30] and BEAT [14], stress on the security aspect by setting extensive use of cryptographic operations (i.e., erasure coding and threshold encryption). They manage indeed to reduce communication complexity (linear at the best case) and enhance system security. However, they result in a non-negligible computation overhead. Algorand [17] is also a recent cryptocurrency that confirms transactions with latency on the order of a minute while scaling to thousands of users. Algorand uses a new Byzantine Agreement (BA) protocol to reach consensus. The consensus protocol used by Algorand consists of a synchronous protocol that incorporates a PoS system with a Byzantine fault tolerance agreement.

Solutions such as off-chain blockchain [32] have emerged to solve the scalability problem. The Lightning Network is an off-chain [32] that establishes payment channels for micro-payments that do not require to be included in the blockchain one by one. Consequently, payment networks can achieve higher transaction throughput and lower latency than blockchain. However, they also suffer from security challenges and impractical trust assumptions.

None of the aforementioned achieve consensus in logarithmic communication cost and provide a low computation and communication overhead when dealing with the threat enforced by Sybil attack.

7 Conclusion

In this paper, we have presented LIGHTx a cost-effective solution for transactions recording based on an adapted BRB primitive where solving consensus is no longer necessary. By customizing the broadcast primitive to support the concurrency of transactions submission, we have designed a transaction transfer system, inheriting BRB scalability and robustness. We also detailed our proposal to enable LIGHTx in open environments while alleviating the impact of Sybil nodes. Our approach leveraging a Proof-of-Bandwidth based reputation system demonstrates appealing results for adversaries detection. In future work, we intend to formally analyze the practical use of LighTx as a cryptocurrency system.

Acknowledgement. El Abid Imane acknowledges Karim Boubouh and Meryem Janati Idrissi (Mohammed VI Polytechnic University) for the fruitful scientific discussions and all their suggestions and reviews that have contributed to the improvement of this work.

References

1. Abraham, I., Malkhi, D., Nayak, K., Ren, L., Spiegelman, A.: Solida: a blockchain protocol based on reconfigurable Byzantine consensus. In: 21st International Conference on Principles of Distributed Systems (OPODIS 2017), Leibniz International Proceedings in Informatics (LIPIcs), vol. 95, pp. 25:1–25:19 (2018)
2. Androulaki, E., et al.: Hyperledger fabric: a distributed operating system for permissioned blockchains. In: Proceedings of the Thirteenth EuroSys Conference, pp. 1–15 (2018)
3. Aspnes, J.: Notes on theory of distributed systems. arXiv preprint arXiv:2001.04235 (2020)

4. Bentov, I., Gabizon, A., Mizrahi, A.: Cryptocurrencies without proof of work. In: Clark, J., Meiklejohn, S., Ryan, P.Y.A., Wallach, D., Brenner, M., Rohloff, K. (eds.) FC 2016. LNCS, vol. 9604, pp. 142–157. Springer, Heidelberg (2016). https://doi.org/10.1007/978-3-662-53357-4_10

5. Bracha, G.: Asynchronous Byzantine agreement protocols. Inf. Comput. **75**(2), 130–143 (1987)

6. Bracha, G., Toueg, S.: Asynchronous consensus and broadcast protocols. J. ACM **32**(4), 824–840 (1985)

7. Brin, S., Page, L.: The anatomy of a large-scale hypertextual web search engine. In: Proceedings of the Seventh International Conference on World Wide Web 7, WWW7, pp. 107–117. Elsevier Science Publishers B. V., NLD (1998)

8. Cachin, C., Guerraoui, R., Rodrigues, L.: Introduction to Reliable and Secure Distributed Programming, 2nd edn. Springer Publishing Company, Incorporated, Heidelberg (2011). https://doi.org/10.1007/978-3-642-15260-3

9. Castro, M., Liskov, B.: Practical Byzantine fault tolerance. In: Proceedings of the Third Symposium on Operating Systems Design and Implementation, OSDI '99, pp. 173–186. USENIX Association, USA (1999)

10. Crain, T., Gramoli, V., Larrea, M., Raynal, M.: DBFT: efficient leaderless Byzantine consensus and its application to blockchains. In: 2018 IEEE 17th International Symposium on Network Computing and Applications (NCA), pp. 1–8 (2018). https://doi.org/10.1109/NCA.2018.8548057

11. Crain, T., Gramoli, V., Larrea, M., Raynal, M.: DBFT: Efficient leaderless byzantine consensus and its application to blockchains. In: 2018 IEEE 17th International Symposium on Network Computing and Applications (NCA), pp. 1–8. IEEE (2018)

12. Croman, K., et al.: On scaling decentralized blockchains. In: Clark, J., Meiklejohn, S., Ryan, P.Y.A., Wallach, D., Brenner, M., Rohloff, K. (eds.) FC 2016. LNCS, vol. 9604, pp. 106–125. Springer, Heidelberg (2016). https://doi.org/10.1007/978-3-662-53357-4_8

13. Douceur, J.R.: The Sybil attack. In: Druschel, P., Kaashock, F., Rowstron, A. (eds.) IPTPS 2002. LNCS, vol. 2429, pp. 251–260. Springer, Heidelberg (2002). https://doi.org/10.1007/3-540-45748-8_24

14. Duan, S., Reiter, M.K., Zhang, H.: Beat: Asynchronous BFT made practical. In: Proceedings of the 2018 ACM SIGSAC Conference on Computer and Communications Security, CCS '18, pp. 2028–2041. Association for Computing Machinery, New York (2018). https://doi.org/10.1145/3243734.3243812

15. Erdős, P., Rényi, A.: On the evolution of random graphs. Publ. Math. Inst. Hung. Acad. Sci **5**(1), 17–60 (1960)

16. Fischer, M.J., Lynch, N.A., Paterson, M.S.: Impossibility of distributed consensus with one faulty process. J. ACM **32**(2), 374–382 (1985). https://doi.org/10.1145/3149.214121

17. Gilad, Y., Hemo, R., Micali, S., Vlachos, G., Zeldovich, N.: Algorand: scaling byzantine agreements for cryptocurrencies. In: Proceedings of the 26th Symposium on Operating Systems Principles, SOSP '17, pp. 51–68. Association for Computing Machinery, New York (2017). https://doi.org/10.1145/3132747.3132757

18. Guerraoui, R., Kuznetsov, P., Monti, M., Pavlovic, M., Seredinschi, D.A.: AT2: asynchronous Trustworthy Transfers (2018)

19. Guerraoui, R., Kuznetsov, P., Monti, M., Pavlovic, M., Seredinschi, D.A.: Scalable Byzantine Reliable Broadcast. In: 33rd International Symposium on Distributed Computing (DISC 2019). Leibniz International Proceedings in Informatics (LIPIcs), vol. 146, pp. 22:1–22:16. Schloss Dagstuhl-Leibniz-Zentrum fuer Informatik (2019)

20. Imbs, D., Raynal, M.: Trading off t-resilience for efficiency in asynchronous Byzantine reliable broadcast. Parallel Process. Lett. **26**(4), 1650017 (2016)
21. Kamvar, S.D., Schlosser, M.T., Garcia-Molina, H.: The eigentrust algorithm for reputation management in P2P networks. In: Proceedings of the 12th International Conference on World Wide Web, WWW '03, pp. 640–651. Association for Computing Machinery, New York (2003)
22. Karame, G., Gubler, D., Čapkun, S.: On the security of bottleneck bandwidth estimation techniques. In: Chen, Y., Dimitriou, T.D., Zhou, J. (eds.) SecureComm 2009. LNICST, vol. 19, pp. 121–141. Springer, Heidelberg (2009). https://doi.org/10.1007/978-3-642-05284-2_8
23. Koitani, K., Hasegawa, G., Murata, M.: End-to-end measurement of hop-by-hop available bandwidth. In: Proceedings of the 2014 IEEE 28th International Conference on Advanced Information Networking and Applications, AINA '14, pp. 17–24. IEEE Computer Society (2014)
24. Kortsarts, Y., Rufinus, J.: Randomized algorithms. J. Comput. Sci. Coll. **20**(5), 66–67 (2005)
25. Kurdi, H.A.: Honestpeer. J. King Saud Univ. Comput. Inf. Sci. **27**(3), 315–322 (2015)
26. Kwon, J.: Tendermint: Consensus without mining. Draft v. 0.6, fall **1**(11) (2014)
27. Lamport, L., Shostak, R., Pease, M.: The byzantine generals problem. ACM Trans. Program. Lang. Syst. **4**(3), 382–401 (1982)
28. Malkhi, D., Mansour, Y., Reiter, M.K.: On diffusing updates in a byzantine environment. In: Proceedings of the 18th IEEE Symposium on Reliable Distributed Systems, SRDS '99, p. 134. IEEE Computer Society, USA (1999)
29. Malkhi, D., Merritt, M., Rodeh, O.: Secure reliable multicast protocols in a wan. In: Proceedings of the 17th International Conference on Distributed Computing Systems (ICDCS '97), ICDCS '97, p. 87. IEEE Computer Society, USA (1997)
30. Miller, A., Xia, Y., Croman, K., Shi, E., Song, D.: The honey badger of BFT protocols. In: Proceedings of the 2016 ACM SIGSAC Conference on Computer and Communications Security, CCS '16, pp. 31–42. Association for Computing Machinery, New York (2016)
31. Nakamoto, S.: Bitcoin: a peer-to-peer electronic cash system. Technical report. www.bitcoin.org
32. Poon, J., Dryja, T.: The bitcoin lightning network: scalable off-chain instant payments (2016)
33. Prasad, R., Dovrolis, C., Murray, M., Claffy, K.: Bandwidth estimation: metrics, measurement techniques, and tools. IEEE Netw. **17**(6), 27–35 (2003)
34. Rocket, T.: Snowflake to avalanche: a novel metastable consensus protocol family for cryptocurrencies. Technical report (2018)
35. Wood, G., et al.: Ethereum: a secure decentralised generalised transaction ledger. Ethereum Project Yellow Paper **151**(2014), 1–32 (2014)
36. Xu, X., et al.: The blockchain as a software connector. In: 2016 13th Working IEEE/IFIP Conference on Software Architecture (WICSA), pp. 182–191 (2016). https://doi.org/10.1109/WICSA.2016.21
37. Zhou, R., Hwang, K.: Powertrust: a robust and scalable reputation system for trusted peer-to-peer computing. IEEE Trans. Parallel Distrib. Syst. **18**(4), 460–473 (2007)

Efficient and Secure TSA for the Tangle

Quentin Bramas[(⊠)]

ICUBE, Strasbourg University, CNRS, Strasbourg, France
bramas@unistra.fr

Abstract. The Tangle is the data structure used to store transactions in the IOTA cryptocurrency. In the Tangle, each block has two parents. As a result, the blocks do not form a chain, but a directed acyclic graph. In traditional Blockchain, a new block is appended to the heaviest chain in case of fork. In the Tangle, the parent selection is done by the Tip Selection Algorithm (TSA). In this paper, we make some important observations about the security of existing TSAs. We then propose a new TSA that has low complexity and is more secure than previous TSAs.

1 Introduction and Background

A Distributed Ledger Technology (DLT) is a distributed protocol executed by a set of nodes to maintain an append-only data structure. In Bitcoin, the data-structure is a chain of blocks, containing transactions. Blocks are appended one after the other to form a chain. Each block requires some amount of computational power, *called weight*, to be created. In Bitcoin (and other Proof-of-Work Blockchains), a new block is added to the heaviest branch *i.e.*, the branch that maximizes the sum of the weights of the blocks it contains. This behavior is at the core of the security of Bitcoin.

In this brief announcement, we are interested in the data structure called the *Tangle*, used to store transactions in the IOTA cryptocurrency, and especially in the algorithm used to append new data. We make some important observations about the security of such algorithms and how previous algorithms do not satisfy them. We then propose a new algorithm that is more secure and more efficient than previous solutions.

The Tangle. *The Tangle* is a data-structure where each block of transactions, called *site*, is linked to two previous sites (using hash pointers), called *parents*. The *genesis* site is the only site without parents. Thus, sites form a Directed Acyclic Graph (DAG) of sites. A site is said to *confirm* all its ancestors in the Tangle. A *tip* of the Tangle is a site which has no child *i.e.*, which is not confirmed by any site.

We consider a network composed of connected nodes that generate and broadcast new sites. Each node has a local copy of the tangle that is updated when a new site is appended.

In order to append a transaction in the Tangle, a node must perform a Proof-of-Work *i.e.*, solving a cryptographic puzzle requiring a certain amount of

K. Echihabi and R. Meyer (Eds.): NETYS 2021, LNCS 12754, pp. 161–166, 2021.
https://doi.org/10.1007/978-3-030-91014-3_11

computational power. The *weight* of a site represents this work and we assume each site has a weight of 1. Then, the *cumulative weight* of a site is defined [6] as the sum of its own weight with the weight of its descendants (the sites that confirm it).

Tips Selection Algorithm (TSA). When a site is added to the Tangle, its parents are selected by a *Tip Selection Algorithm* (TSA). The TSA must select two tips (unconfirmed sites) that are not conflicting (informally, two transactions are conflicting if accepting both would produce a double spend). The TSA is a fundamental component of the protocol because it implicitly indicates how the nodes agree on the current state of the Tangle. Indeed, if two tips are conflicting, the TSA indicates which one is considered correct (and should be extended by appending a new site to it) or orphaned (by ignoring it).

Since each node in the network maintains its own version of the Tangle, a site can end up having multiple children. Indeed, due to the latency in the network, the TSA could chose a site which is a tip locally, but that is already confirmed in another version. The Tangle whitepaper [6] presents two TSAs[1]:

- Uniform TSA: Each parent is chosen uniformly at random among all the tips.
- Markov Chain Monte Carlo (MCMC): the selection of each parent is done by using a random walk. A walker starts from a given site (e.g., the genesis), moves from site to child site, and stops when it reaches a tip. The probability of moving to a child site, depends on its cumulative weight (see [6] for more details).

The Double Spending Attack. In the Tangle, an attacker that wants to double spend must generate two conflicting transactions and first broadcast only one of them. When the first transaction is considered well-confirmed (*i.e.*, the honest nodes think the probability to reverse it is small enough), then the attacker can broadcast the second transaction and append a lots of sites (forming *a parasite chain* [2,6]) so that the first transaction is discarded.

2   Related Work and Motivations

Related Work. The Uniform TSA was initially proposed for its simplicity. One of its advantages is that tips are quickly confirmed [5,6]. However, it is easy to see that it offers no protection against double spending attacks. Indeed, an attacker just has to generate more tips than the current number of honest tips to have a higher probability to be selected by honest nodes. Hence, even very old transactions could be canceled easily.

The MCMC algorithm was the first to offer protection against double spending attacks. Indeed, the older a transaction is, the harder it is to cancel it [1,6]. However, the MCMC requires computing the cumulative weight of every sites in

---

[1] A third one is briefly presented but is actually just a variation of the MCMC that we present here.

the tangle (which has worst-case quadratic complexity in the number of sites), and its security depends on a parameter α which also influences the number of tips that are left behind [5] (*i.e.*, tips that are never confirmed). In other words, better security implies less stability, and usability.

The efficiency and the security of the MCMC has been improved with $MCMC_{rw}$ [1], by using a simpler version of the cumulative weight (which has linear complexity in the number of tips). $MCMC_{rw}$ obtains a better trade-off security/stability than standard MCMC. G-IOTA [3] and E-IOTA [4] are two extensions of IOTA that proposed mechanisms to limit the number of left-behind tips, while still using MCMC for its security.

All the previously proposed TSAs mixes in the same algorithm the security and the stability aspects. Our goal is to give an algorithm that separates these two aspects.

The last version of the IOTA whitepaper [7] has a similar approach. It proposes to use a completely distinct algorithm to resolve conflicts so that the TSA is not concerned by the security aspect. However, the security of the proposed consensus algorithm has not yet been formally studied. Our goal is to improve previously defined TSAs, using the same model as the original Tangle whitepaper, which has been formally defined [2,6].

Motivations. Our motivation comes from three important observations.

Observation 1. *If, between two conflicting transactions, one is considered malicious[2] with higher probability than the other, does it make sense to choose the malicious transaction as parent with non-zero probability ?*

Regardless of the algorithm used to compare conflicting transactions, we believe a transaction that is considered malicious should never be selected as parent, even with small probability. Otherwise, a fraction of the honest nodes will support the malicious transactions and help the adversary. So, we think a secure TSA should resolves conflicts in a **deterministic** manner, using another algorithm that we call the *Conflict Resolving Algorithm* (CRA).

Observation 2. *The uniform random tip selection is the algorithm that offers the best confirmation time and produces the smallest number of tips on average. However it offers poor security guarantees.*

The main reason Uniform random TSA is not used in practice is because it offers poor security guarantees. Indeed, it is very easy for a malicious node to generate a small number of transactions to give a high probability for an old transaction to be selected as parent. However, when there is no conflicts, transactions are confirmed very quickly and no transaction is left over. Thus, there is no issue in using the uniform random TSA, after that the set of non-conflicting tips has been deterministically selected.

---

[2] Here malicious just means that it conflicts with a transaction that is considered correct.

Observation 3. *MCMC offers good security guarantees at the price of slower confirmation time and higher number of tips on average.*

Again, if an algorithm provides a good way to discriminate conflicting transactions, then there in no reason not to use it for this purpose. Then, another algorithm can be used to randomly select parents among the non-conflicting remaining tips efficiently. The security of MCMC is due to the fact that a random walker has a greater probability to move towards sites with higher cumulative weight. However, we think there is no need to do it for all the sites, but instead, it should be done only when comparing conflicting sites.

3   A New Secure TSA: The Two-Step TSA

Model. Given a Tangle, S denotes the set of sites. For any subset C of sites, we say that C is *conflict-free* if all the sites in C are pairwise non-conflicting. We now give a more precise definition of tips that takes into account conflicts. We say a conflict-free set C is a *set of tips*, if there is no sites $s \in S$ and $t \in C$ such that s confirms t and $C \cup \{s\}$ is conflict-free. This means that, if a tip in C is confirmed by some site in $s \in S$, then s does conflict with another site in C. For a site s, $w(s)$ denotes its cumulative weight.

The 2-Steps TSA. Our 2-Step TSA first resolves conflicts between sites and then dispatch parents among conflict-free sites.

Our Conflict Resolver Algorithm (CRA) takes a Tangle and returns a maximal conflict-free set of tips C such that, for any pair of conflicting sites s_1 and s_2, if s_1 is confirmed by some site in C, then $w(s_1) \geq w(s_2)$ *i.e.*, the conflict-free set of tips that confirms only the heaviest site in case of conflicts, and is maximal in the sense that no more site can be added to the set without creating conflicts.

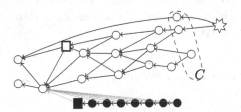 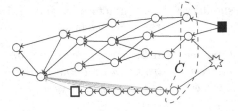

Fig. 1. In this example, the white square is considered correct and the black one is discarded.

Fig. 2. In this example, the new site can merge both branches.

Our Tip Dispatcher Algorithm (TDA) takes a set of conflict-free tips C and returns two tips p_1 and p_2 selected uniformly at random among C, with p_1 and p_2 distinct if $|C| \geq 2$.

In Fig. 1 we see a tangle and two conflicting transactions (the two squares). Our CRA first discriminates between the two and considers the white square to be correct and discards the black sites (all the sites confirming the black square). The output of the CRA is the set C containing three conflict-free tips. Then our TDA dispatches the two parents without discriminating between the old and the recent sites. The goal of the TDA is to confirm as many sites as possible, reducing the number of left-over sites.

Security. Using our algorithm, if the honest nodes agree on a conflict-free set of tips, then they all extend the tangle in the same way, increasing the weight of the same set of sites. In other words, for any discarded site, there is a site, considered correct, whose weight increases for each new honest site.

This implies that, if an adversary wants to discard a site that is considered correct, it has to generate sites at a higher rate than the honest nodes (which is a necessary assumption anyway [2]). It means that, like in Bitcoin, the probability of creating a successful double spending attack on a site decreases exponentially fast with its weight.

This property is not obtained by previous TSAs. For instance, if a parasite chain has probability 1/3 to be selected by the MCMC, then an honest node will append one third of its transactions in the parasite chain (assuming they are independent), which is not the intended behavior. In addition, 1/3 of the honest transactions globally will end up selecting the parasite chain as the correct one. A third of the honest nodes plus the malicious node then represents half of the computational power, so that it becomes even easier for the malicious node to increase the probability of selecting its parasite chain.

Using our TSA, the parasite chain is never selected if its cumulative weight is smaller than another branch of the tangle. This implies that a malicious node that wants to double spend has to create a parasite chain on its own and is not helped by honest nodes.

Another interesting property of our TSA is that it does not automatically consider correct a site that is located on the main branch. Instead, it compares conflicting sites independently on where they are located on the tangle. Doing so, we can confirm a separate branch that could look like a parasite chain, but is in fact older and might contain honest transactions as well, for instance if it was generated offline. Indeed, we do not want to discard an entire chain just because a conflicting site appears on top of the main chain. Figure 2 illustrates the situation. We see that the white square has a greater cumulative weight compared to the black square, so only the site confirming the black square (there is no such site in this example) are discarded, creating two tips (the parents of the black-square site). We then have a chance to merge the two branches with a new site (the star-shaped one) using our TDA.

In this situation the MCMC would choose the main branch with greater probability and would almost never merge both branches since the MCMC would never stops its random walk to a parent of black square because it is not a tip.

Performances. Despite using the cumulative weight, which is computed in $\Theta(n)$ time for a given site, our algorithm can have constant complexity in most situations.

After receiving the Tangle from its peers, a node can compute the conflict-free set of tips C with the CRA, while storing the cumulative weight of each site for later use. After that, every time the node has to generate a site s, the TSA will return two parents p_1 and p_2 among C and there is no need to run the CRA again for the next site as the new conflict-free set of tips is simply $C \cup \{s\} \setminus \{p_1, p_2\}$. Similarly, for each incoming site s, if s confirms a site in C, then we know s is considered correct and we can update C by adding s and removing the confirmed tips. So if all the nodes are honest, after the first run of the CRA, every execution of the TSA has constant-time complexity.

However, if an incoming site confirms a site s_m, considered malicious, and conflicting a site s_c considered correct, then we can increment the weight of s_m by one and compare it to the previously computed weight of s_c. If the weight of s_m is still smaller than the weight of s_c, we can safely ignore the new site as running the CRA again will not change our current conflict-free set of sites C. If the weight of s_m becomes greater than the previously computed weight of s_c, then we have to update the weight of s_c and do the comparison again. We believe other optimizations could be performed in this case as well.

Concluding Remarks. We propose a new paradigm for constructing secure and efficient TSAs. We observed that existing TSAs can be improved by splitting the parent selection into a conflict resolving phase and tip dispatcher phase. We believe this work will open new research on the security and the performances of TSAs.

References

1. Attias, V., Bramas, Q.: How to choose its parents in the tangle. In: Atig, M.F., Schwarzmann, A.A. (eds.) NETYS 2019. LNCS, vol. 11704, pp. 275–280. Springer, Cham (2019). https://doi.org/10.1007/978-3-030-31277-0_18
2. Bramas, Q.: The Stability and the Security of the Tangle (April 2018). https://hal.archives-ouvertes.fr/hal-01716111
3. Bu, G., Gürcan, Ö., Potop-Butucaru, M.: G-iota: fair and confidence aware tangle. In: IEEE INFOCOM 2019-IEEE Conference on Computer Communications Workshops (INFOCOM WKSHPS), pp. 644–649. IEEE (2019)
4. Bu, G., Hana, W., Potop-Butucaru, M.: Metamorphic iota. arXiv preprint arXiv:1907.03628 (2019)
5. Kusmierz, B., Sanders, W., Penzkofer, A., Capossele, A., Gal, A.: Properties of the tangle for uniform random and random walk tip selection. In: 2019 IEEE International Conference on Blockchain (Blockchain), pp. 228–236. IEEE (2019)
6. Popov, S.: The tangle. white paper (2016). https://iota.org/IOTA_Whitepaper.pdf
7. Popov, S., et al.: The coordicide (2020)

Verification

Separating Map Variables in a Logic-Based Intermediate Verification Language

Daniel Dietsch, Matthias Heizmann, Jochen Hoenicke, Alexander Nutz$^{(\boxtimes)}$, and Andreas Podelski

University of Freiburg, Freiburg, Germany
`alex@certora.com`

Abstract. In SMT solver based verification, the program to be verified is often given in an intermediate verification language such as Boogie. We present a program transformation that aims at splitting mathematical arrays (i.e., maps, which are typically used to model arrays and specifically the heap) into different partitions, so that the resulting verification conditions are easier to solve (due to the need of fewer case splits when analysing the effect of reads and writes over the same array). Our method takes the similar role of classical preprocessing steps based on alias analysis; the difference is that it works on any (mathematical) map, as opposed to a data structure that is known to present a chunk of memory managed by some compiler. Having to forfeit the benefits of general assumptions about memory (e.g., allocate-before-use), we need to deal with additional difficulties but obtain a more general technique. In particular, our technique can be applied to arbitrary programs in the intermediate verification language, including programs that are not directly derived from a program in a production-type programming language, like C or Java. We have implemented a prototypical version of the program transformation in order to demonstrate that it can lead to up to exponential reductions in execution time for the Ultimate software verification tool, despite the cost of performing the initial static analysis.

1   Introduction

We present a program transformation to split the map variables in a program into as many different maps as possible (given the results of an initial static analysis which conservatively approximates the program's semantics). The benefit of the program transformation is that the verifier can be freed of some of the burden of proving independence of map operations since map operations on different maps are (trivially) independent.

Our method is designed as a source-to-source transformation of programs in a logics-based intermediate verification language. By "logics-based" we mean in particular that the native data types of the language represent mathematical objects that are to be reasoned about using formal logics, as opposed to the

© Springer Nature Switzerland AG 2021
K. Echihabi and R. Meyer (Eds.): NETYS 2021, LNCS 12754, pp. 169–186, 2021.
https://doi.org/10.1007/978-3-030-91014-3_12

data types of production-type languages, like C or Java, which represent chunks of memory of the computer that the program is executed on.

Translation into such a language is a common intermediate step taken by symbolic verifiers, e.g., Dafny, Ultimate Automizer, Ultimate Taipan, SMACK, SeaHorn, JayHorn, and many more [2,3,8,10,11,13]. This step allows the verifier to separate the modelling of the input language (often a production-type language like C or Java) from the application of verification techniques that are independent of the particularities of the input language. To guarantee this strict separation of concerns, it is essential to make the intermediate language fully self-contained, in particular no hidden assumptions may be carried over from the source language, since they would break the interchangeabilty of techniques that can be applied to programs in the intermediate language. A prominent example for programs in an intermediate verification language that have no direct correspondence to the verification input anymore are *path programs*, as presented and used in [1] and [7]. That our technique consumes programs in an intermediate verification language is a crucial difference to techniques that achieve a similar splitting of maps but depend on assumptions from the semantics of their input language, like the partitioned memory model presented by Wang et al. whose static analysis is specific to C programs [19], as well as other approaches that separate the heap of a C program in order to speed up verification [9,18,19].

Map data types in particular are a common tool for modelling dynamic memory in languages like C and Java. For instance, the verifiers in the Ultimate framework model the heap of a C program as one large integer-indexed map. Since this means that all heap accesses in the C program are translated to accesses of the same map variable in the intermediate program, very long chains of store operations occur naturally and induce a significant burden on the reasoning engine inside the verifier. This becomes problematic when there are many reads over such long store chains, since for inferring the outcome of each read operation, the verifier needs to determine which of the stores can be responsible for the value of the map at the read position. This leads to a lot of case splits which can slow down verification massively, as shown for example by Wang et al. [19] along with the effectiveness of partitioning memory to alleviate the problem.

Structure of the Paper. Since the correctness of our program transformation is justified by the independence of groups of map accesses, we will first give a formal definition of when two accesses are semantically independent. Based on this, we will define an interface to a static analysis such that the static analysis can be used to conservatively infer independence of map accesses for a given program. Based on such an analysis result, we will define the program transformation itself and show its correctness through a proof of bisimulation between the input program and the output program. We will use a scalable benchmark program to give an example of the reduction in verification time that that the transformation can obtain, despite the cost of performing the initial static analysis.

Our technical contributions are as follows.

- We formally introduce the independence property which enables the desired program transformation (in the context for of the intermediate programming language).
- We define an instrumentation of a program with auxiliary variables such that an existing static analysis can infer the independence property.
- We define a program transformation that takes as input a program and the inferred independence property and returns a new program. In the new program statements use different map variables according to the inferred independence property. We prove of the transformation by showing bisimulation-equivalence with the original program.
- We present an experiment for a scalable benchmark program.

2   Example

The left hand side of Fig. 1 shows an example program given in the Boogie [12] verification language. While the program models a program in the C programming language we want to stress that our technique cannot rely on any metainformation specific to C, like the meaning of the `malloc` procedure, or the absence of access to uninitialized memory cells. Map semantics in Boogie follow McCarthy's theory of arrays [14], which is also used in SMT solvers.

The example program is artificial. Its purpose is to necessitate a large number of non-interference checks in a program of minimal size. So the main obstacle to verification is the necessity of proving non-interference between the map updates.

In the example, dynamically allocated memory is modeled by the two map variables `mem` and `valid`. The map `mem` stores the contents of the memory. The map `valid` stores which memory cells are allocated. C's malloc function is modeled by the procedure `malloc`, which returns a memory location that is not currently in use. (For simplicity we assume that all memory blocks are of size 1).

The procedure `main` starts by allocating two pointers and storing them to variables p and q. The contents of both memory locations p and q are initialized to 0. Then, the value at location p is incremented nondeterministically often, and the value at location q is decremented nondeterministically often. The `assert` statements express that, at the end of the program the values in memory at p and q contain a non-negative or a non-positive value respectively.

As an intermediate goal to correctness, a solver must prove that the operations on memory cells p and q do not interfere. A typical CEGAR-based, or bounded model checking-based, solver will need to do this for every spurious counterexample.

Our technique provides a preprocessing such that the solver can instead prove correctness of the transformed program on the right hand side of Fig. 1. In the transformed example, the map `mem` has been replaced by two maps `mem_1` and `mem_2`. Memory accesses at p are modeled by accessing `mem_1`, memory accesses at q are modeled by accessing `mem_2`. That way the solver does not need to

```
var mem : [int] : int;              var mem_1, mem_2 : [int] : int;
var valid : [int] : bool;           var valid : [int] : bool;

procedure main() {                  procedure main() {
  var p, q : int;                     var p, q : int;

  call p := malloc();                 call p := malloc();
  call q := malloc();                 call q := malloc();

  mem[p] := 0;                        mem_1[p] := 0;
  mem[q] := 0;                        mem_2[q] := 0;

  while (*) {                         while (*) {
    if (*) {                            if (*) {
      mem[p] := mem[p] + 1;               mem_1[p] := mem_1[p] + 1;
    } else {                            } else {
      mem[q] := mem[q] - 1;               mem_2[q] := mem_2[q] - 1;
    }                                   }
  }                                   }

  assert mem[p] >= 0;                 assert mem_1[p] >= 0;
  assert mem[q] <= 0;                 assert mem_2[q] <= 0;
}                                   }

procedure malloc() returns (ptr : int);   procedure malloc() returns (p : int);
ensures !old(valid)[ptr];           ensures !old(valid)[ptr];
ensures valid == old(valid)[ptr:=true];   ensures valid == old(valid)[p := true];
```

Fig. 1. Example of a program and its transformation. The program serves also as the basis of our scalable benchmark suite. The value of the variable mem is a mathematical map. It is used to model the memory. The program transformation makes the independence of the two statements in the loop apparent. Intuitively, the two statements use the map mem differently. The transformation introduces *diffent maps for different uses*.

prove non-interference between the increment and decrement operations for each spurious counterexample, which typically results in a dramatic speedup.

3   Preliminaries

We fix our notation for syntax and semantics of map manipulating programs.

Program Syntax. We distinguish two types of variables, *map variables* and *base variables*. Map variables are named $a, b, \ldots$. We use $i, j, \ldots$ for base variables that are used as map indices in the current context and $x, y, \ldots$ for all-purpose base variables. We use constant (or literal) expressions named $lit, lit_1, lit_2, \ldots$. We use a special variable pc $\in$ Variables called the *program counter*. We use typewriter font for program variables (e.g., i, x) and italics for mathematical variables (e.g., i, x).

Expressions in our programs can have one of three types.

Expressions of base type: $e_{base} ::= \text{lit} \mid \text{x} \mid \text{a[i]}$

Expressions of map type: $e_{map} ::= \text{a} \mid \text{a[i:=x]} \mid (\text{const lit})$

Boolean expressions: $e_{bool} ::= \text{x==y} \mid \,! e_{bool} \mid e_{bool} \text{ \&\& } e_{bool} \mid e_{bool} \text{ || } e_{bool}$

The set of all *commands* is generated by the following grammar. We refer to this set by Command. We refer to individual commands by the letters c, c'.

$$c ::= \texttt{x:=}e_{\text{base}} \mid \texttt{a:=}e_{\text{map}} \mid \texttt{havoc x} \mid \texttt{havoc a} \mid \texttt{assume } e_{\text{bool}}$$

The commands assignments, as well as the havoc and assume commands which are common in verification languages. The set of *program locations*, Loc, is a set of distinct identifiers $\{\ell, \ell', \ell_0, \ell_1, \ldots\}$. A *statement* is a triple of a source program location, a command, and a target program location, i.e., Statement $=$ Loc $\times$ Command $\times$ Loc. We use the letter σ for statements. Let $\sigma = (\ell, c, \ell')$ be a statement, then we refer to the *source location* of σ by $src(\sigma)$. In contexts where the locations are not important we omit them from the statement and write only the command. We call statements whose command is of the form a:=a[i:=x] *map write statements*, and we call statements whose command is of the form x:=a[i] *map read statements*. To highlight that a statement's command is a map write (read), we name the statement σ_{wr} (σ_{rd}).

For our program transformations, we will also use *sequential composition statements* of the form $(\ell, \sigma_1; \sigma_2; \ldots; \sigma_n, \ell')$, where $\sigma_1, \sigma_2, \ldots$ are commands. The meaning of this notation is that the transformed program contains the locations $\ell_1, \ell_2, \ldots$, which appear nowhere else in the program, and the statements (ℓ, σ_1, ℓ_1), (ℓ, σ_2, ℓ_1), $\ldots$, $(\ell_{n-1}, \sigma_n, \ell)$.

A *program* P is given as a control flow graph whose edges are statements. Formally: $P = (\text{Loc}, \Sigma, \ell_0)$, where Loc is a set of locations, $\Sigma \subseteq$ Statement is a set of statements, and $\ell_0 \in$ Loc is the initial location. For technical reasons we do not allow incoming control flow edges at the initial location. A program P induces a set of *program variables*, Var, which are all the variables that occur in any of the statements of P. We sometimes refer to only the basic variables $\text{Var}_{base} \subseteq$ Var or only the map variables $\text{Var}_{map} \subseteq$ Var. We call the subset of Σ that contains all the map write (read) statements Σ_{wr} (Σ_{rd}). From now on we assume the program P is given as described here.

We do not allow equating maps in assume statements (assume a==b). In our experience this restriction does not matter in practice. Furthermore, we only allow equalities between (base) variables, not between expressions. This is not a proper restriction.

We will abbreviate a:=a[i:=x] as a[i]:=x. We may omit the case when the store is over a different map, like a:=b[i:=x], from case distinctions, since it can be simulated by a map update followed by a map assignment; in this case a:=b followed by a[i]:=x. Also, we omit chains of stores applied to one map variable; again this omission does not change the expressiveness of the programming language.

Program Semantics. For simplicity of presentation we consider only two sorts, namely some fixed non-empty set that is the base sort, and the set of all functions from that set into itself, this is the sort of our map variables. We denote the base sort Sort and the map sort Sort $\rightarrow$ Sort.

A *state* in our program is a mapping from program variables to values from our set of sorts. The base variables, like x and i are assigned values from Sort.

The map variables, like a, are assigned values from sort $\mathsf{Sort} \to \mathsf{Sort}$. The Boolean sort $\{\mathsf{true}, \mathsf{false}\}$ occurs only during evaluation of Boolean expressions. The program counter variable pc is a special case, its value denotes the location $\ell \in \mathsf{Loc}$ that the execution is currently in.

We use the (semantic) map update operator $\cdot[\cdot \mapsto \cdot]\colon (\mathsf{Sort} \to \mathsf{Sort}) \times \mathsf{Sort} \times \mathsf{Sort} \to (\mathsf{Sort} \to \mathsf{Sort})$: Let a be a map, then $a[i \mapsto x]$ is the map that returns the value $a(j)$ for all arguments $j \neq i$ and the value x for the argument i.

For expressions e we give an evaluation function $\cdot[\![\cdot]\!]\colon \mathsf{States} \times \mathsf{Expressions} \to (\mathsf{Sort} \cup (\mathsf{Sort} \to \mathsf{Sort}))$, which, given a valuation of the variables, assigns a value to e: Every literal has one value in Sort it is associated with; the literal evaluates to that value regardless of state. A variable is evaluated by looking up its value in the state. A map variable's value is a map, a map access at some index evaluates to the application of the evaluated map value to the evaluated index value. The semantics of the store operator is given as the above-mentioned map update operator. A constant map expression with some argument lit evaluates to a map whose value is lit at every position. The Boolean operators are evaluated as usual. Formally:

$$s[\![\mathtt{lit}]\!] \overset{\mathrm{def}}{=} \mathtt{lit} \qquad\qquad s[\![\mathtt{v}]\!] \overset{\mathrm{def}}{=} s(\mathtt{v})$$

$$s[\![\mathtt{a[i]}]\!] \overset{\mathrm{def}}{=} s[\![\mathtt{a}]\!](s[\![\mathtt{i}]\!]) \qquad s[\![\mathtt{a[i:=x]}]\!] \overset{\mathrm{def}}{=} s[\![\mathtt{a}]\!][s[\![\mathtt{i}]\!] \mapsto s[\![\mathtt{x}]\!]]$$

$$s[\![\mathtt{(const\ lit)}]\!] \overset{\mathrm{def}}{=} \lambda x.\,\mathtt{lit} \qquad s[\![\mathtt{e==e'}]\!] \overset{\mathrm{def}}{=} \begin{cases} \mathtt{true} & \text{if } s[\![\mathtt{e}]\!] = s[\![\mathtt{e'}]\!] \\ \mathtt{false} & \text{otherwise} \end{cases}$$

Note in particular that equality is defined both for expressions of both the base sort and the map sort, where the semantic equality of maps is defined as usual, i.e., two maps are equal when they map every input element to the same output element.

The *concrete post* operator $\mathsf{post}\colon 2^{\mathsf{States}} \times \mathsf{Statement} \to 2^{\mathsf{States}}$ is given as follows.

$$\mathsf{post}(S, (\ell, \mathtt{x:=}e_{\mathsf{base}}, \ell')) \overset{\mathrm{def}}{=} \{s[\mathtt{pc} \mapsto \ell'][\mathtt{x} \mapsto s[\![e_{\mathsf{base}}]\!]] \mid s \in S, s(\mathtt{pc}) = \ell\}$$

$$\mathsf{post}(S, (\ell, \mathtt{a:=}e_{\mathsf{map}}, \ell')) \overset{\mathrm{def}}{=} \{s[\mathtt{pc} \mapsto \ell'][\mathtt{a} \mapsto s[\![e_{\mathsf{map}}]\!]] \mid s \in S, s(\mathtt{pc}) = \ell\}$$

$$\mathsf{post}(S, (\ell, \mathtt{havoc\ x}, \ell')) \overset{\mathrm{def}}{=} \{s[\mathtt{pc} \mapsto \ell'][\mathtt{x} \mapsto v] \mid s \in S, s(\mathtt{pc}) = \ell, v \in \mathsf{Sort}\}$$

$$\mathsf{post}(S, (\ell, \mathtt{havoc\ a}, \ell')) \overset{\mathrm{def}}{=} \{s[\mathtt{pc} \mapsto \ell'][\mathtt{a} \mapsto v] \mid s \in S, s(\mathtt{pc}) = \ell,$$
$$v \in \mathsf{Sort} \to \mathsf{Sort}\}$$

$$\mathsf{post}(S, (\ell, \mathtt{assume\ e}, \ell')) \overset{\mathrm{def}}{=} \{s[\mathtt{pc} \mapsto \ell'] \mid s \in S, s(\mathtt{pc}) = \ell, s[\![\mathtt{e}]\!] = \mathtt{true}\}$$

$$\mathsf{post}(S, (\ell, \mathtt{c\ ;\ c'}, \ell')) \overset{\mathrm{def}}{=} \mathsf{post}(\mathsf{post}(S, (\ell, \mathtt{c}, \ell)), (\ell, \mathtt{c'}, \ell'))$$

An *execution* e is a sequence of statements and states in alternation, i.e.,

$$e = s_0.\,\sigma_0.\,\dots\,.\,\sigma_{n-1}.\,s_n.$$

Every execution starts in an initial state, i.e., a state s_0 where the program counter pc is assigned the initial location ℓ_0. Furthermore, the sequence must be consecutive, i.e., for all i from 0 to $n-1$, the state s_{i+1} must be contained in the set of post states of the state s_i under the statement σ_i, i.e.,

$$s_{i+1} \in \mathsf{post}(\{s_i\}, \sigma_i).$$

Note that the common condition that for two consecutive statements σ_i, σ_{i+1} in an execution the source location of σ_{i+1} must match the destination location of σ_i is, maybe not obviously, implied by the above condition due to the program counter value being part of our program states. The empty executions are special cases; an empty execution s_0 consists of an initial state only. We can write every non-empty execution as $e.\sigma.s$ where e is an execution. We denote the set of all executions Executions.

The *reachable states* are all states s such that there is an execution that ends in s.

$$Reach \overset{\text{def}}{=} \{s \mid \exists e \in \mathsf{Executions}.\, e = e's\}$$

4   Dependency Analysis

Our program transformation is based on an analysis of the dependencies between the statements in the program P. In this section, we describe a property that makes explicit which map update statements may be responsible for the value of a map at some index at some program location. For this, we introduce the relation LstWr (read: "last writes") that contains for a potential read in the program all the map updates that are relevant for that read in some execution of the program. Note that the relation LstWr is an adaptation of the reaching definitions relation, which is textbook knowledge [15] – but only for programs without maps.

Last Write Relation. LstWr The relation LstWr $\subseteq \Sigma_{\mathsf{wr}} \times \Sigma_{\mathsf{rd}}$ relates all map write statements σ_{wr} to all the map read statements σ_{rd} such that σ_{wr} is responsible for the value that is read in σ_{rd} in some execution.

Definition 1 (Last Writes Relation LstWr). *The Last Write relation* LstWr $\subseteq \Sigma_{\mathsf{wr}} \times \Sigma_{\mathsf{rd}}$ *contains a pair* $(\sigma_{\mathsf{wr}}, \sigma_{\mathsf{rd}})$, *where the command in* σ_{wr} *is of the form* a[i]:=x, *and the command in* σ_{rd} *is of the form* y:=b[j], *whenever there is an execution e and a value v such that v is written by* σ_{wr} *and is read by* σ_{rd}, *i.e., if e fulfills the following linear time property.*

$$\Diamond\,(\boldsymbol{pc} = src(\sigma_{\mathsf{wr}}) \wedge \boldsymbol{x} = v \wedge \Diamond\,(\boldsymbol{pc} = src(\sigma_{\mathsf{rd}}) \wedge b\,[j] = v))$$

In this definition we assume that every value that is written to a map during an execution is unique; this can be accommodated by providing each value with a timestamp. Furthermore, in this definition a and b may or may not refer to the same program variables, the same holds for, i and j and x and y.

Alternative Characterisation of the Last Writes Relation. LstWr We provide an alternative characterisation of the Last Writes relation LstWr. This characterisation will lead to an instrumentation of the program that will allow us to compute an relation $\mathsf{LstWr}^{\#}$ that overapproximates the Last Writes relation.

We next define the function lw which, given a position i, given a map a, and given an execution e, returns the write statement σ_{wr} that is responsible for the value that the map a has at position i in the last state of the execution e. For technical reasons we will use the symbol $\bot$ (to cater for the case where the map a has not been written at position i in execution e).

Formally, we define the function $\mathsf{lw} : \mathsf{Var}_{\mathsf{map}} \times \mathsf{Sort} \times \mathsf{Executions} \rightarrow \Sigma_{\mathsf{wr}} \cup \{\bot\}$ by induction over the length of the execution e. (As explained above, an execution of length 0 is of the form s_0 where s_0 is an initial state, and an execution of length $n+1$ is of the form $e.\sigma.s$ where σ is a statement and s is a state.)

$$\mathsf{lw}(\mathsf{a}, j, s_0) \overset{\mathsf{def}}{=} \bot$$

$$\mathsf{lw}(\mathsf{a}, j, e.\,\mathsf{havoc}\ \mathsf{a}.\,s) \overset{\mathsf{def}}{=} \bot$$

$$\mathsf{lw}(\mathsf{a}, j, e.\,\mathsf{a}{:}{=}(\mathsf{const}\ \mathsf{lit}).\,s) \overset{\mathsf{def}}{=} \bot$$

$$\mathsf{lw}\big(\mathsf{a}, j, e.\,\mathsf{a[i]}{:}{=}\mathsf{x}.\,s\big) \overset{\mathsf{def}}{=} \begin{cases} \mathsf{a[i]}{:}{=}\mathsf{x} & \text{if } s[\![\mathsf{i}]\!] = j \\ \mathsf{lw}(\mathsf{a}, j, e) & \text{if } s[\![\mathsf{i}]\!] \neq j \end{cases}$$

$$\mathsf{lw}(\mathsf{a}, j, e.\,\mathsf{a}{:}{=}\mathsf{b}.\,s) \overset{\mathsf{def}}{=} \mathsf{lw}(\mathsf{b}, j, e)$$

$$\mathsf{lw}(\mathsf{a}, j, e.\,\sigma.\,s) \overset{\mathsf{def}}{=} \mathsf{lw}(\mathsf{a}, j, e) \text{ if } e.\,\sigma.\,s \text{ matches none of the above}$$

Intuitively, the definition of $\mathsf{lw}(\mathsf{a}, j, e)$ traces the value of the map a at index j back within the execution e until it hits the map write statement that is responsible for the fact that a has that value at position j at the end of e. This write statement is returned by lw. If the execution consists only of an initial state s_0, or the last statement was a havoc statement with argument a, or when a has been set to a constant map by the last statement, then no value in a depends on a map write statement, so lw returns the symbol $\bot$. If the last statement in the execution has been a write to map a, then LstWr checks whether the write was at position j. If that is the case, the last write is returned, otherwise lw recurses on the prefix of the execution where the write statement and its successor state have been dropped. If the last statement in the execution assigned another map b to a, the lw recurses on the execution prefix, and it looks for writes on b instead of writes on a. Otherwise, the last statement in the execution had no influence on values in a, so it is evaluated recursively on the prefix without the last statement and state.

As above, the Last Writes relation LstWr relates all the write statements σ_{wr} to all the read statements σ_{rd}, such that there is an execution where σ_{wr} is responsible for the value that σ_{rd} reads. From the function lw we build the explicit characterization of the relation $\mathsf{LstWr} \subseteq \Sigma_{\mathsf{wr}} \times \Sigma_{\mathsf{rd}}$ as follows.

$$\mathsf{LstWr} \overset{\text{def}}{=} \{(\sigma_{\mathsf{wr}}, \sigma_{\mathsf{rd}}) \mid \sigma_{\mathsf{rd}} = (\ell, \mathtt{x} \colon = \mathtt{a[i]}, \ell')$$
$$\wedge \exists e.\, s \in \mathsf{Executions}.\, s[\![\mathtt{pc}]\!] = \ell \wedge s[\![\mathtt{i}]\!] = i \wedge \mathsf{lw}(\mathtt{a}, i, e.\, s) = \sigma_{\mathsf{wr}}$$
$$\wedge\, \sigma_{\mathsf{wr}} \neq \bot\}$$

5   Computing Dependencies

In this section, we present an instrumentation of the program P such that the Last Writes relation LstWr can be expressed in terms of the set of reachable states of the instrumented program P_{LstWr}.

5.1   Instrumentation

We introduce an auxiliary map variable $\mathtt{a\text{-}lw}$ for every map-variable $\mathtt{a}$ that occurs in the program P. The values of the maps that are assigned to $\mathtt{a\text{-}lw}$ are not values from our base sort Sort, but instead are symbols that refer to write statements that occur in P.

Intuitively, the transformation is designed in such a way that the fresh $\mathtt{lw}$-maps capture the results of the lw-function for each program location. We construct the transformation in three steps. We begin by defining a transformer $\tau^c_{\mathsf{LstWr}} \colon \mathsf{Command} \to \mathsf{Command}$ for some commands whose transformation result does not depend on their location in the program.

If the command c is a havoc to map variable $\mathtt{a}$, or if c assigns a constant map to $\mathtt{a}$, then $\mathtt{a\text{-}lw}$ is assigned a constant map that contains the symbol $\bot$ at all positions. This represents that no write statement has an influence on any value in the map $\mathtt{a}$ after the command c has been executed. If c assigns the value of a map variable to another map variable, then the analogous assignment is done on the respective $\mathtt{lw}$-maps. This expresses that all map write statements that have an influence on $\mathtt{a}$ also have an influence on $\mathtt{b}$ after the command c has been executed. In all other cases, the transformation τ^c_{LstWr} leaves the command c unchanged.

$$\tau^c_{\mathsf{LstWr}}(\mathtt{havoc\ a}) \overset{\text{def}}{=} \mathtt{havoc\ a;\ a\text{-}lw} \colon = (\mathtt{const}\ \bot)$$

$$\tau^c_{\mathsf{LstWr}}(\mathtt{a} \colon = (\mathtt{const\ lit})) \overset{\text{def}}{=} \mathtt{a} \colon = (\mathtt{const\ lit});\ \mathtt{a\text{-}lw} \colon = (\mathtt{const}\ \bot)$$

$$\tau^c_{\mathsf{LstWr}}(\mathtt{b} \colon = \mathtt{a}) \overset{\text{def}}{=} \mathtt{b} \colon = \mathtt{a};\ \mathtt{b\text{-}lw} \colon = \mathtt{a\text{-}lw}$$

$$\tau^c_{\mathsf{LstWr}}(c) \overset{\text{def}}{=} c \text{ where none of the other cases apply}$$

From τ^c_{LstWr} we construct the transformer $\tau^\sigma_{\mathsf{LstWr}} \colon \mathsf{Statement} \to \mathsf{Statement}$, which transforms the map write statements. Whenever a map variable $\mathtt{a}$ is written to at index $\mathtt{i}$, then $\mathtt{a\text{-}lw}$ is written at the same index, but with a special value that identifies the updating statement. Statements that are not map write statements are left unchanged by $\tau^\sigma_{\mathsf{LstWr}}$.

$$\tau^\sigma_{\mathsf{LstWr}}(\sigma_{\mathsf{wr}}) \stackrel{\text{def}}{=} (\ell,\ \mathtt{a[i]:=x;\ a-lw[i]:=}\sigma_{\mathsf{wr}}, \ell')$$
$$\text{where } \sigma_{\mathsf{wr}} = (\ell,\ \mathtt{a[i]:=x}, \ell')$$
$$\tau^\sigma_{\mathsf{LstWr}}((\ell, c, \ell')) \stackrel{\text{def}}{=} (\ell, \tau^c_{\mathsf{LstWr}}(c), \ell') \text{ where } (\ell, c, \ell') \notin \Sigma_{\mathsf{wr}}$$

From the above-described transformers, we can now construct the statement tranformer used for the actual instrumentation, $\tau_{\mathsf{LstWr}} \colon$ Statement $\to$ Statement. The statement originating at the initial location ℓ_0 constitutes a special case. Since at the initial location no map writes have been executed, we set every $\mathtt{lw}$-variable to a constant map containing the symbol $\bot$. (Note that we assume that the initial location has no incoming statements.) The transformation for all other statements just uses the above-defined $\tau^\sigma_{\mathsf{LstWr}}$.

$$\tau_{\mathsf{LstWr}}((\ell_0, c, l)) \stackrel{\text{def}}{=} \mathtt{a-lw:=(const\ \bot)};$$
$$\cdots$$
$$\mathtt{z-lw:=(const\ \bot)};$$
$$c'$$
$$\text{where } \tau^\sigma_{\mathsf{LstWr}}((\ell_0, c, l)) = (\ell_0, c', \ell) \text{ and } \mathsf{Var}_{\mathsf{map}} = \{\mathtt{a}, \ldots, \mathtt{z}\}$$
$$\tau_{\mathsf{LstWr}}((\ell, c, \ell')) \stackrel{\text{def}}{=} \tau^\sigma_{\mathsf{LstWr}}((\ell, c, \ell')) \text{ where } \ell \neq \ell_0$$

We are now ready to define the instrumented program P_{LstWr}. We define the instrumented program P_{LstWr} through applying the transformation function τ_{LstWr} to each statement in Σ. Formally:

$$P_{\mathsf{LstWr}} \stackrel{\text{def}}{=} (\mathsf{Loc}, \{\tau_{\mathsf{LstWr}}(\sigma) \mid \sigma \in \Sigma\}, \ell_0)$$

We can now express the Last Write relation LstWr through the set of reachable states of the instrumented program P_{LstWr}.

Proposition 1. *The Last Writes relation LstWr, as defined in Sect. 4, is identical to the relation that relates a map write statement σ_{wr} to a map read statement σ_{rd} of the form $(\ell, \mathtt{x:=a[i]}, \ell')$ if there is a state s in the set of reachable states of the instrumented program P_{LstWr} such that the program counter $\mathbf{pc}$ points to the source location of σ_{rd}, ℓ, and the value that s assigns to the map read expression $\mathtt{a-lw[i]}$ is the write statement σ_{wr}. Formally:*

$$\mathsf{LstWr} = \{(\sigma_{\mathsf{wr}}, \sigma_{\mathsf{rd}}) \mid \sigma_{\mathsf{rd}} = (\ell, \mathtt{x} := \mathtt{a[i]}, \ell')$$
$$\wedge\, \exists s \in Reach(P_{\mathsf{LstWr}}).\, s[\![\mathbf{pc}]\!] = \ell \wedge s[\![\mathtt{a}]\!]\text{-}\mathtt{lw}[i] = \sigma_{\mathsf{wr}}$$
$$\wedge\, \sigma_{\mathsf{wr}} \neq \bot\}$$

The proof can be given as an induction over the length of executions and is this is relatively straightforward.

We state the following lemma for later reference (proof of Theorem 1 in Sect. 6).

Lemma 1. *P and P_{LstWr} are bisimulation-equivalent.*

The proof of this lemma is obvious from the fact that the additional commands introduced by the transformation are ghost code.

5.2 Computing an Overapproximation of the Last Writes Relation LstWr

We have seen that the relation LstWr can be expressed through the set of reachable states of the instrumented program P_{LstWr}. The set of reachable states is not computable in general. Thus, we apply a static analysis that computes an overapproximation of the set of reachable states.

The static analysis must be able to handle programs that manipulate maps. An example is a static analysis based on the Map Equality Domain [4]. This domain is useful to infer equalities and disequalities between expressions which can involve maps.

We have implemented an extension of the Map Equality Domain. The extensions supports constraints of the form $x \in \{\text{lit}_1, \text{lit}_2\}$ which allows us to succinctly express constraints like $\texttt{a-lw[i]} \in \{\sigma_1, \sigma_2\}$. Here, σ_1 and σ_2 are literals (referring to the corresponding statements). All literals are pairwise different. Thus, these constraints allow us to infer constraints like $\texttt{a-lw[i]} \neq \sigma_3$. Such constraints are crucial to infer independence of statements.

From now on, we use $\text{LstWr}^{\#}$ to refer to the overapproximation of the relation LstWr computed by applying the above-described static analysis to the instrumented program P_{LstWr}. The static analysis always computes an overapproximation of the set of reachable states of P_{LstWr}. Thus, the relation $\text{LstWr}^{\#}$ is an overapproximation of the Last Writes relation LstWr. We state the following remark for later reference (in Lemma 2).

Remark 1. The relation $\text{LstWr}^{\#}$ is an overapproximation of the Last Write relation LstWr, i.e.,

$$\text{LstWr}^{\#} \supseteq \text{LstWr}.$$

6 Program Transformation

In this section we introduce the program transformation that transforms the program P, given the relation $\text{LstWr}^{\#}$, which approximates the Last Write relation LstWr of program P.

6.1   Computing a Partition of the Map Write Statements

First, we define the relation $R \subseteq \Sigma_{\mathsf{wr}} \times \Sigma_{\mathsf{wr}}$ that relates all write statement that map influence the same read statement. Two write statements σ_{wr} and σ'_{wr} are related by R if there exists a read statement σ_{rd} such that the relation $\mathsf{LstWr}^{\#}$ relates both σ_{wr} to σ_{rd} and σ'_{wr} to σ_{rd}. Formally:

$$R \stackrel{\mathrm{def}}{=} \{(\sigma_{\mathsf{wr}}, \sigma'_{\mathsf{wr}}) \mid \exists \sigma_{\mathsf{rd}} \in \Sigma_{\mathsf{rd}}. \, \mathsf{LstWr}^{\#}(\sigma_{\mathsf{wr}}, \sigma_{\mathsf{rd}}) \wedge \mathsf{LstWr}^{\#}(\sigma_{\mathsf{wr}}, \sigma_{\mathsf{rd}})\}$$

Based on the relation R, we define the relation $r \subseteq \Sigma_{\mathsf{wr}} \times \Sigma_{\mathsf{wr}}$ as the smallest equivalence relation that contains the relation R. This equivalence relation r induces a partition over the set Σ_{wr}, i.e., a set $\mathcal{W} \subseteq 2^{\Sigma_{\mathsf{wr}}}$ of subsets of the set Σ_{wr} such that the disjoint union of the subsets is identical to the original set Σ_{wr}. Thus, the set $\mathcal{W}$ consists of disjoint subsets $\{W_1, \ldots, W_n\}$ of the set of all write statements Σ_{wr}. The partition $\mathcal{W}$ has the property that for every two blocks W_1 and W_2 in $\mathcal{W}$, we know that if we take one write statement σ_{wr} from W_1 and another write statement σ'_{wr} from W_2, then σ_{wr} and σ'_{wr} are independent in the sense that they never have an influence on the same read statement.

For technical reasons, we add a the singleton consisting only of the symbol $\perp$ to $\mathcal{W}$. Its use will become clear in the next subsection.

6.2   Program Transformation

We introduce a map variable a_W for each $W \in \mathcal{W}$. If for example the write statements $a[i]:=x$ and $a[j]:=y$ appear in different blocks W_1 and W_2, then we will replace the map variable a with two different variables a_W_1 and a_W_2 in these statements accordingly. (There is a subtle point regarding the fact that W is a mathematical object while a variable name consists of characters, which we neglect here.)

We use the notation $\mathsf{LstWr}^{\#-1}[\sigma_{\mathsf{rd}}]$ to denote the preimage of $\mathsf{LstWr}^{\#}$ with respect to some read statement $\sigma_{\mathsf{rd}} \in \Sigma_{\mathsf{rd}}$, i.e.,

$$\mathsf{LstWr}^{\#-1}[\sigma_{\mathsf{rd}}] \stackrel{\mathrm{def}}{=} \{\sigma_{\mathsf{wr}} \mid (\sigma_{\mathsf{wr}}, \sigma_{\mathsf{rd}}) \in \mathsf{LstWr}^{\#}\}.$$

The transformation updates the statements of program P using the transformation $\tau: \mathsf{Statement} \to \mathsf{Statement}$ as described in the following. The transformation result $\tau(\sigma)$ depends on the statement type of σ. If σ writes to map variable a, it is transformed to a statement that does the same update to map variable a_W, i.e., to the map variable corresponding to the block in the partition $W \in \mathcal{W}$ that contains σ. If σ reads from a map variable a, there are two cases. Either $\mathsf{LstWr}^{\#}$ at the read location yields the empty set. This means that it is guaranteed that the read position has never been written to in any execution that reaches σ. In this case, σ is transformed to a read from the map variable $a\_\{\perp\}$ instead of a. Otherwise, by construction of the partition $\mathcal{W}$, LstWr must yield a set that falls completely into a block W in the partition $\mathcal{W}$. In that case, σ is transformed to a read from the map variable a_W instead of a. If σ assigns a map variable a to a map variable b, it is transformed to a series of assignments that assign for each block in the partition $W \in \mathcal{W}$ the variable a_W to the variable b_W. A havoc to a map variable a is translated to havoc on all variables a_W for every block W in the partition $\mathcal{W}$, followed by an assume statement that ensures that all maps a_W have been set to the same value. In all other cases, the transformation leaves σ unchanged. Formally:

$$\tau((\ell, \texttt{a[i]:=x}, \ell')) \overset{\text{def}}{=} (\ell, \texttt{a\_}W\texttt{[i]:=x}, \ell')$$
$$\text{where } (\ell, \texttt{a[i]:=x}, \ell') \in W$$

$$\tau((\ell, \texttt{x:=a[i]}, \ell')) \overset{\text{def}}{=} (\ell, \texttt{x:=a\_}\{\bot\}\texttt{[i]}, \ell')$$
$$\text{if } \mathsf{LstWr}^{\#-1}[(\ell, \texttt{x:=a[i]}, \ell')] = \emptyset$$

$$\tau((\ell, \texttt{x:=a[i]}, \ell')) \overset{\text{def}}{=} (\ell, \texttt{x:=a\_}W\texttt{[i]}, \ell')$$
$$\text{if } \mathsf{LstWr}^{\#-1}[(\ell, \texttt{x:=a[i]}, \ell')] \neq \emptyset$$
$$\text{and } \mathsf{LstWr}^{\#-1}[(\ell, \texttt{x:=a[i]}, \ell')] \subseteq W$$

$$\tau((\ell, \texttt{b:=a}, \ell')) \overset{\text{def}}{=} (\ell, \texttt{b\_}W_1\texttt{:=a\_}W_1; \ \ldots; \ \texttt{b\_}W_n\texttt{:=a\_}W_n, \ell')$$
$$\text{where } \mathcal{W} = \{W_1, \ldots, W_n\}$$

$$\tau((\ell, \texttt{havoc a}, \ell')) \overset{\text{def}}{=} (\ell, \texttt{havoc a\_}W_1; \ \ldots; \ \texttt{havoc a\_}W_n;$$
$$\texttt{assume a\_}W_1\texttt{==} \ \ldots \ \texttt{==a\_}W_n, \ell')$$
$$\text{where } \mathcal{W} = \{W_1, \ldots, W_n\}$$

$$\tau((\ell, \texttt{a:=(const lit)}, \ell')) \overset{\text{def}}{=} (\ell, \texttt{a\_}W_1\texttt{:=(const lit)}; \ \ldots;$$
$$\texttt{a\_}W_n\texttt{:=(const lit)}, \ell')$$
$$\text{where } \mathcal{W} = \{W_1, \ldots, W_n\}$$

$$\tau(\sigma) \overset{\text{def}}{=} \sigma \text{ if } \sigma \text{ matches none of the above cases}$$

We construct the transformed program P' by replacing all statements σ in P by their transformed version $\tau(\sigma)$. Formally:

$$P' \overset{\text{def}}{=} (\mathsf{Loc}, \{\tau(\sigma) \mid \sigma \in \Sigma\}, \ell_0)$$

6.3   Correctness of the Transformation

In this subsection, we show that the transformation is correct, i.e., that the program P and the transformed program P' are bisimulation-equivalent. Given Lemma 1 it is sufficient to prove the following Lemma.

As an aside: it does not seem obvious to us how to give a bisimulation between the programs P and P' directly.

Lemma 2. *The programs P_{LstWr} and P' are bisimulation-equivalent.*

Proof. We define a bisimulation relation $\sim$ between P_{LstWr} and P' as follows.

The states $s \in \mathsf{States}_P$ and $t \in \mathsf{States}_{P'}$ are bisimilar, i.e., $s \sim t$, iff

$$\forall x \in \mathsf{Var}_{base}. \qquad\qquad\qquad\qquad s[\![x]\!] = t[\![x]\!] \qquad\qquad (1)$$

and

$$\forall \mathtt{a} \in \mathsf{Var}_{map}. \forall \mathtt{i} \in \mathsf{Var}_{base}.$$
$$(s[\![\mathtt{a\text{-}lw[i]}]\!] = \bot \implies \qquad \forall W \in \mathcal{W}. s[\![\mathtt{a[i]}]\!] = t[\![\mathtt{a\_W[i]}]\!]) \qquad (2a)$$
$$\wedge\, (\exists W \in \mathcal{W}. s[\![\mathtt{a\text{-}lw[i]}]\!] \in W \implies \qquad s[\![\mathtt{a[i]}]\!] = t[\![\mathtt{a\_W[i]}]\!]) \qquad (2b)$$

That $\sim$ is a bisimulation can be proven via induction over the types of program statements. For a detailed proof, we refer to the longer preprint version of this paper [5]. □

Theorem 1 (Bisimulation). *P and P' are bisimulation-equivalent.*

Proof. This follows by transitivity of bisimulation-equivalence from Lemmas 1 and 2. □

7   Implementation in Ultimate

We implemented our program transformation in the Ultimate program analysis framework[1]. The intermediate representation we support is the most expressive one used by Ultimate, namely the so-called *interprocedural control flow graph* (short: ICFG). In this section we elaborate on how our approach can be lifted from the minimal program representation we chose for our formal presentation to Ultimate's program representation.

Multidimensional Maps. In order to support maps of higher dimensions, we adapt the relation LstWr and the corresponding analysis. This is done by having not one but several lw-maps for each map variable in the original program. For an n-dimensional map variable a we would introduce n lw-maps a-lw-1 to a-lw-n where a-lw-1 is one-dimensional a-lw-2 is two-dimensional and so forth.

Procedures. In order to support procedures, two features are relevant: Map-valued parameters must be passed between procedures, and it must be possible to compute procedure summaries that describe the effect of a procedure on global map variables (in fact having one of these features would be enough in terms of expressiveness, but Ultimate supports both). Both of these features are enabled by our support for (by-value) assignments between maps.

8   Experiments on a Scalable Benchmark Suite

In this section, we will investigate whether our approach is applicable in principle. We will use a small benchmark suite which is specifically tailored to expose the case split explosion problem.

---

[1] https://github.com/ultimate-pa/ultimate.

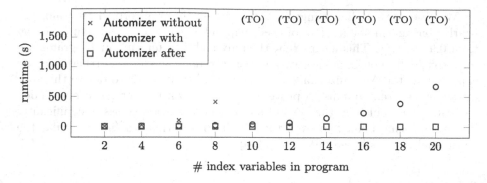

Fig. 2. The Ultimate Automizer toolchain *without* and *with* the program transformation as a preprocessing step, and the Ultimate Automizer toolchain in isolation applied *after* the program transformation, on a benchmark suite whose programs are scaled-up versions of the example program in Sect. 2. The timeout (TO) is set to 1800 s.

We obtain the bechmark suite by scaling up the example program from Sect. 2. The example program manipulates the map variable *mem* on the two *index* variables p and q. We obtain new program by adding another two variables and adding the corresponding statements which manipulate the map variable *mem* on two new variables in the same way as the existing statements do for p resp. q. We can iterate the process and thus obtain a scalable benchmark suite whose programs have 2, 4, 6, ... index variables.

Setup. We ran our experiments on a dedicated benchmarking system, each benchmark task was limited to 2 CPU cores at 2.4 GHz and 20 Gigabytes of RAM. We ran two toolchains and took three measurements. One toolchain, called "Automizer without", is the standard verification toolchain of the program verifier Ultimate Automizer. The toolchain computes an ICFG from the input program and then run's Automizer's verification algorithm on the ICFG. The second toolchain, called "Automizer with", applies our transformation after computing the interprocedural control flow graph and before running Automizer's verification algorithm. A third kind of measurements, denoted "Automizer after", are the timings of only the verification algorithm in the toolchain "Automizer with", i.e., how long the verification of the transformed program takes.

Results. In Fig. 2 we display the results of our experimental evaluation. The x-axis of the plot represents the different example programs, identified by the number of map index variables. The y-axis represents the time taken by each toolchain. We ran three toolchains: The Ultimate Automizer program verifier, Ultimate Automizer where before the verification run, the transformation is applied, and a toolchain where Automizer was run on the already transformed programs.

We observe that the timings of Automizer on the transformed programs are nearly constant in the number of used map index variables – the timings range from 0.9 s to 8.8 s. This means that the only real difficulty in our programs lies in deriving the non-interferences between the map accesses. Furthermore, we can see that the Automizer fails to scale well when it needs to derive the non-interferences itself: It fails to prove all examples with 10 or more map index variables. The toolchain that includes our transformation shows a significantly improved scaling behaviour even though the transformation (in particular the static analysis it is based on) is not cheap.

9   Related Work

There are several works resembling our approach of splitting map variables in that they propose computing non-interference properties between memory regions to simplify the verification conditions that are handed to an SMT solver. Rakamaric and Hu [18], as well as Wang et al. [19] propose a memory model that uses maps which are separated according to the results of an up-front alias analysis. Gurfinkel and Navas [9] propose a related but different memory model. In their setting, the heap state is passed between procedures through local map variables. They propose a memory model with a partitioning that is context-sensitive to improve precision. In contrast to our work, these papers all rely on C semantics for their input program, so they do not apply to arbitrary map manipulating programs.

Our relation LstWr and the corresponding property is reminiscent of a large field of work that is concerned with inferring guarantees about data dependencies between program parts in the presence of arrays, e.g., [6,16,17]. These papers propose various approaches of finding data dependencies in programs with arrays in different precisions, for different fragments and for different applications. None of them is aimed at symbolic program verification as our work is. To our knowledge, our property is the only one that accounts for maps, the crucial difference being the presence of by-value assignments of map variables.

10   Conclusion

We have investigated the theoretical foundations for a novel research question which may be relevant for the practical potential of intermediate verification languages. The question concerns a preprocessing step for intermediate verification languages which takes the similar role that alias analysis plays in the verification for programming languages. We have presented a preliminary solution in the form of a program transformation. We have integrated the program transformation into a toolchain. A preliminary experimentation on a small (and somewhat artificial) shows that the program transformation can be effective.

References

1. Beyer, D., Henzinger, T.A., Majumdar, R., Rybalchenko, A.: Path invariants. In: PLDI, pp. 300–309. ACM (2007)
2. Carter, M., He, S., Whitaker, J., Rakamaric, Z., Emmi, M.: SMACK software verification toolchain. In: ICSE (Companion Volume), pp. 589–592. ACM (2016)
3. Dietsch, D., et al.: Ultimate taipan with dynamic block encoding - (competition contribution). In: Beyer, D., Huisman, M. (eds.) TACAS 2018. LNCS, vol. 10806, pp. 452–456. Springer, Cham (2018). https://doi.org/10.1007/978-3-319-89963-3_31
4. Dietsch, D., Heizmann, M., Hoenicke, J., Nutz, A., Podelski, A.: The map equality domain. In: Piskac, R., Rümmer, P. (eds.) VSTTE 2018. LNCS, vol. 11294, pp. 291–308. Springer, Cham (2018). https://doi.org/10.1007/978-3-030-03592-1_17
5. Dietsch, D., Heizmann, M., Hoenicke, J., Nutz, A., Podelski, A.: Different maps for different uses. A program transformation for intermediate verification languages. CoRR, abs/1901.01915 (2019)
6. Feautrier, P.: Dataflow analysis of array and scalar references. Int. J. Parallel Prog. **20**(1), 23–53 (1991)
7. Greitschus, M., Dietsch, D., Podelski, A.: Loop invariants from counterexamples. In: Ranzato, F. (ed.) SAS 2017. LNCS, vol. 10422, pp. 128–147. Springer, Cham (2017). https://doi.org/10.1007/978-3-319-66706-5_7
8. Gurfinkel, A., Kahsai, T., Navas, J.A.: SeaHorn: a framework for verifying C programs (competition contribution). In: Baier, C., Tinelli, C. (eds.) TACAS 2015. LNCS, vol. 9035, pp. 447–450. Springer, Heidelberg (2015). https://doi.org/10.1007/978-3-662-46681-0_41
9. Gurfinkel, A., Navas, J.A.: A context-sensitive memory model for verification of C/C++ programs. In: Ranzato, F. (ed.) SAS 2017. LNCS, vol. 10422, pp. 148–168. Springer, Cham (2017). https://doi.org/10.1007/978-3-319-66706-5_8
10. Heizmann, M., et al.: Ultimate automizer and the search for perfect interpolants. In: Beyer, D., Huisman, M. (eds.) TACAS 2018. LNCS, vol. 10806, pp. 447–451. Springer, Cham (2018). https://doi.org/10.1007/978-3-319-89963-3_30
11. Kahsai, T., Rümmer, P., Sanchez, H., Schäf, M.: JayHorn: a framework for verifying Java programs. In: Chaudhuri, S., Farzan, A. (eds.) CAV 2016. LNCS, vol. 9779, pp. 352–358. Springer, Cham (2016). https://doi.org/10.1007/978-3-319-41528-4_19
12. Leino, K.R.M.: This is boogie 2. Technical report, Microsoft Research (2008). https://www.microsoft.com/en-us/research/publication/this-is-boogie-2-2/
13. Leino, K.R.M.: Dafny: an automatic program verifier for functional correctness. In: Clarke, E.M., Voronkov, A. (eds.) LPAR 2010. LNCS (LNAI), vol. 6355, pp. 348–370. Springer, Heidelberg (2010). https://doi.org/10.1007/978-3-642-17511-4_20
14. McCarthy, J.: Towards a mathematical science of computation. In: IFIP Congress, pp. 21–28 (1962)
15. Nielson, F., Nielson, H.R., Hankin, C.: Principles of Program Analysis. Springer, Heidelberg (1999)
16. Paek, Y., Hoeflinger, J., Padua, D.A.: Efficient and precise array access analysis. ACM Trans. Program. Lang. Syst. **24**(1), 65–109 (2002)
17. Pugh, W., Wonnacott, D.: An exact method for analysis of value-based array data dependences. In: Banerjee, U., Gelernter, D., Nicolau, A., Padua, D. (eds.) LCPC 1993. LNCS, vol. 768, pp. 546–566. Springer, Heidelberg (1994). https://doi.org/10.1007/3-540-57659-2_31

18. Rakamarić, Z., Hu, A.J.: A scalable memory model for low-level code. In: Jones, N.D., Müller-Olm, M. (eds.) VMCAI 2009. LNCS, vol. 5403, pp. 290–304. Springer, Heidelberg (2008). https://doi.org/10.1007/978-3-540-93900-9_24
19. Wang, W., Barrett, C., Wies, T.: Partitioned memory models for program analysis. In: Bouajjani, A., Monniaux, D. (eds.) VMCAI 2017. LNCS, vol. 10145, pp. 539–558. Springer, Cham (2017). https://doi.org/10.1007/978-3-319-52234-0_29

Petri Net Invariant Synthesis

Peter Chini[1]([✉]) and Florian Furbach[2]

[1] TU Braunschweig, Brunswick, Germany
p.chini@tu-braunschweig.de
[2] Uppsala University, Uppsala, Sweden
florian.furbach@it.uu.se

Abstract. We study the synthesis of inductive half spaces (IHS). These are linear inequalities that form inductive invariants for Petri nets, capable of disproving reachability or coverability. IHS generalize classic notions of invariants like traps or siphons. Their synthesis is desirable for disproving reachability or coverability where traditional invariants may fail.

We formulate a CEGAR-loop for the synthesis of IHS. The first step is to establish a structure theory of IHS. We analyze the space of IHS with methods from discrete mathematics and derive a linear constraint system closely over-approximating the space. To discard false positives, we provide an algorithm that decides whether a given half space is indeed inductive, a problem that we prove to be coNP-complete. We implemented the CEGAR-loop in the tool INEQUALIZER and our experiments show that it is competitive against state-of-the-art techniques.

1 Introduction

A major task of today's program verification is to formulate and prove safety properties. Such a property describes the desirable and undesirable behavior of a program, often expressed in terms of safe and unsafe states. A safety property is satisfied if all executions of a program explore only safe states. Phrased differently, it is violated if an unsafe state is reachable via an execution. Testing reachability is usually a rather complex problem and often undecidable [8,28,51,53].

To restore decidability, the behavior of a program is often over-approximated. Intuitively, an over-approximation describes a property that holds for all reachable states but fails for unsafe states. Hence, over-approximations act like a separator between reachable and unsafe states and provide a proof for the non-reachability of the latter. Computing over-approximations is often achieved by generating a type of invariant [4,18,29,50]. The challenge is to find a type that admits an efficient generation and is expressive enough to separate reachable from unsafe states. Inductive invariants are a prominent example [3,10,26]. If an inductive invariant holds for some state, then it also holds for any successor after a step of an execution. Hence, if it is satisfied initially, it holds for all reachable states.

© Springer Nature Switzerland AG 2021
K. Echihabi and R. Meyer (Eds.): NETYS 2021, LNCS 12754, pp. 187–205, 2021.
https://doi.org/10.1007/978-3-030-91014-3_13

We generate inductive invariants for Petri nets, a well-established model of concurrent programs [43,45]. Here, safety verification is usually expressed in terms of the Petri net reachability or coverability problem. The former is known to be ACKERMANN-complete [11,12,36,37], the latter is EXPSPACE-complete [6,38,47]. Despite the ongoing algorithmic development, in particular for coverability [24,30,31,48,54], computational requirements of solving both problems often exceeds practical limits. This has led to the development of classic Petri net invariants like *traps, siphons*, or *place invariants* [45] that may help to solve both problems more efficiently. Typically, these invariants are based on linear dependencies of places or transitions and can be synthesized easily by incorporating tools and solvers from linear programming.

The trade-off for the efficient synthesis is that the expressiveness of classic invariants is limited and often not sufficient to prove non-reachability of a marking. We study *inductive half spaces* (IHS) [49,52], a type of invariants with increased expressiveness. IHS generalize many of the classical Petri net invariants [52] and preserve their *linear nature*. For instance, any trap or siphon can be expressed as an IHS. However, the synthesis of IHS remained an open problem. An IHS consist of a tuple (k, c), where k is a vector over the places of the Petri net and c is an integer. The corresponding *half space* is a subset of the space of markings, containing all markings m satisfying the inequality $k \cdot m \geq c$. It is called *inductive* if the markings that are in the half space do not leave it after firing a transition.

Our main contribution is a method for the synthesis of inductive half spaces. More precise, we compute IHS that separate an initial marking m_0 from a final marking m_f, proving the latter non-reachable. The task is formalized in the *linear safety verification problem* LSV(R). Given m_0 and m_f, it asks for an IHS (k, c) such that $k \cdot m_0 \geq c$ and $k \cdot m_f < c$. The problem was first considered in [49] for continuous Petri nets. The synthesis of IHS is much easier in the continuous case. In fact, an entire subclass we call *non-trivial* inductive half spaces does not occur in this setting. So far, LSV(R) has not been considered in its full generality and decidability is still unknown.

We provide a semi-decision procedure for LSV(R) using *counter example guided abstraction refinement* (CEGAR) [9], a state-of-the-art technique in program verification. We illustrate the approach in Fig. 1. Suppose we are given a Petri net N, an initial marking m_0, and a marking m_f for which we want to disprove reachability from m_0. Our approach attempts to synthesize an IHS that separates m_f from the reachable markings of N. It begins by constructing a formula ϕ of linear constraints

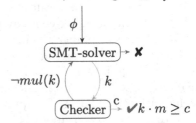

Fig. 1. The CEGAR loop.

from the given information and passes it to an SMT-solver. Roughly, ϕ describes necessary conditions for solutions of LSV(R). For each solution (k, c) of LSV(R), the vector k is a solution of ϕ. Hence, if the SMT-solver does not find a solution to ϕ, then no separating IHS exists. Otherwise we find a vector k of a half space candidate. The candidate is inductive for each transition t separately: there exists a $c_t \in \mathbb{Z}$ such that (k, c_t) is inductive for t. However, there might be no common $c \in Z$ such that (k, c) is inductive for all transitions. In order to synthesize such a c, we developed a *constant generation algorithm* (CGA). If CGA is successful, we have found a separating IHS (k, c). Otherwise, k does not admit a suitable constant c and we apply a refinement. We set $\phi = \phi \wedge \neg mul(k)$, where $\neg mul(k)$ is a linear constraint that excludes all multiples of k and repeat the above process. Note that the loop may not terminate. But if it does, we obtain an answer to LSV(R).

To realize the CEGAR-loop, we make the following main contributions.

- We develop a structure theory of inductive half spaces. It decomposes the space of IHS into *trivial* and *non-trivial* half spaces. While the synthesis of trivial IHS is simple, the synthesis of non-trivial ones is challenging. By employing techniques from discrete mathematics, we can determine necessary conditions for non-trivial IHS and construct the required formula ϕ.
- We present two algorithms: the *inductivity checking algorithm* (ICA) and the *constant generation algorithm* (CGA). The former determines whether a given half space is inductive, a problem that we prove coNP-complete. This answers an open question from [52]. ICA combines structural properties of IHS with dynamic programming. CGA synthesizes a constant c for a solution k of ϕ such that (k, c) is an IHS. CGA is an instrumentation of ICA. Its termination argument is an interesting connection between IHS and the *Frobenius number*.
- We implemented the CEGAR-loop in the tool INEQUALIZER. Employing it, we disproved reachability and coverability for a benchmark of widely used concurrent programs. The results are compared to algorithms implemented in MIST [44] and show INEQUALIZER to be competitive.

Related Work. The reachability problem of Petri nets is a central problem of theoretical computer science. Its complexity was finally resolved after 45 years: it is ACKERMANN-complete. The upper bound is due to Leroux and Schmitz [37]. The authors refined classical upper bounds like Kosaraju's [33], Mayr's [39,40], and Lambert's [34]. Hardness was first considered by Lipton [38]. He proved reachability EXPSPACE-hard. Czerwinski et al. [11] improved the lower bound to non-elementary. The gap was closed by Leroux, Czerwinski, and Orlikowski [12,36].

Many safety verification tasks can be phrased in terms of coverability. The EXPSPACE-completeness of coverability was determined by Rackoff [47] and Lipton [38]. Despite this, efficient algorithms keep getting developed [31]. Modern approaches are based on forward or backward state space exploration [21,23,30, 32,54]. A method that has drawn interest are unfoldings [15,16,35,41]. Notably, Abdulla et al. [1] solve coverability by constructing an unfolding that represents backwards reachable states. They analyze it using an SMT-formula.

Profiting from advances in SMT-solving, deriving program properties by constraint solving has become popular [1,14,26,27]. In [46], ranking functions are synthesized by solving linear inequalities. Synthesis methods for Petri nets involving SMT-solving often simplify by using continuous values. In [14], Esparza et al. generate inductive invariants disproving co-linear properties. IHS were first considered by Sankaranarayanan et al. [49]. The authors synthesize IHS over continuous Petri nets. Compared to these, we generate a larger class of invariants: *non-trivial* IHS do not occur in [49] but are necessary for discrete nets (see Fig. 2). The discrete structure of IHS was examined by Triebel and Sürmeli [52]. The authors show that IHS generalize notions like traps, siphons, and place invariants.

Outline. In Sect. 2, we introduce necessary notions around Petri nets. The structure of IHS is examined in Sect. 3. In Sect. 4, we formulate the SMT-formula in the CEGAR loop. The algorithms ICA and CGA are given in Sect. 5. Experimental results are presented in Sect. 6. For brevity, we omit a number of formal proofs. They can be found in the extended version of the paper [7].

2   Linear Safety Verification

We introduce the linear safety verification problems for Petri nets. They formalize the question of whether there exists an inductive half space which disproves reachability or coverability of a certain marking. To this end, we formally introduce half spaces and the necessary notions around Petri nets.

Petri Nets. A *Petri net* is a tuple $N = (P, T, F)$, where P is a finite set of *places*, T is a finite set of *transitions*, and $F : (P \times T) \cup (T \times P) \to \mathbb{N}$ is a *flow function*. We denote the number of places $|P|$ by n. The places are numbered. For convenience, we use a place p_i and their numeric value i interchangeably: Given a vector $x \in \mathbb{N}^n$, we denote its i-th component as both $x(i)$ and $x(p_i)$. For a transition $t \in T$, we define vectors $t^-, t^+ \in \mathbb{N}^n$. The i-th component of t^-, with $p_i \in P$, is defined to be $F(p_i, t)$, written $t^-(i) = t^-(p_i) := F(p_i, t)$. Similarly, $t^+(i) = t^+(p_i) := F(t, p_i)$. The vector t^Δ captures the difference $t^\Delta := t^+ - t^-$.

The semantics of a Petri net N is defined in terms of markings. A *marking* m is a vector in $\mathbb{N}^n$. Intuitively, it puts a number of *tokens* in each place. A marking is said to *enable* a transition t if $m(p) \geq t^-(p)$ for each place $p \in P$, written $m \geq t^-$. The set of all markings that enable t is called the *activation space* of t and is denoted by $\mathrm{Act}(t)$. Note that $\mathrm{Act}(t) = \{t^- + v \mid v \geq 0\}$. If $m \in \mathrm{Act}(t)$, then t can be *fired*, resulting in the new marking $m' = m + t^\Delta$. This constitutes the *firing relation*, written as $m[t\rangle m'$. We lift the relation to sequences of transitions $\sigma = t_1 \ldots t_k \in T^*$ where convenient, writing $m[\sigma\rangle m'$. A marking m_f is called *reachable* from a marking m_0 if there is a sequence of transitions σ such that $m_0[\sigma\rangle m_f$. We use $post^*(m_0)$ to denote the markings reachable from m_0 and $pre^*(m_f)$ are the markings from which m_f is reachable. The *upward closure* of m_f is $\uparrow m_f = \{m \in \mathbb{N}^n \mid m \geq m_f\}$. A marking m_f is *coverable* from m_0, if there is a sequence of transitions σ and an $m \in \uparrow m_f$ such that $m_0[\sigma\rangle m$.

(Inductive) Half Spaces. We describe sets of markings by means of half spaces. Let $N = (P, T, F)$ be a Petri net, $k \in \mathbb{Z}^n$ a vector, and $c \in \mathbb{Z}$ an integer. The *half space* defined by k and c is $\mathrm{Sol}(k, c) = \{m \in \mathbb{Z}^n \mid k \cdot m \geq c\}$. Here, $k \cdot m = \sum_{p \in P} k(p) \cdot m(p)$ is the usual scalar product. We also refer to the tuple (k, c) as half space. Note that we could also define half spaces via $k \cdot m \leq c$. This is of course equivalent since $k \cdot m \geq c$ if and only if $-k \cdot M \leq -c$. We are interested in half spaces that are inductive in the sense that they cannot be left by firing transitions. A half space (k, c) is *t-inductive* if for any $m \in \mathrm{Act}(t) \cap \mathrm{Sol}(k, c)$ we have $m + t^{\Delta} \in \mathrm{Sol}(k, c)$. A half space (k, c) is *inductive* if it is t-inductive for all $t \in T$. We use IHS as a shorthand for inductive half space.

A half space (k, c) is not t-inductive if and only if it contains a marking m with $k \cdot m \geq c$ that enables t, i.e. $m \geq t^-$, and from which we leave the half space by firing t: $k \cdot (m + t^{\Delta}) < c$. Since $m \geq t^-$ if and only if there is an $x \in \mathbb{N}^n$ with $m = t^- + x$, we can state inductivity in terms of an infeasibility requirement:

Theorem 1. *A half space (k, c) is t-inductive iff there is no vector $x \in \mathbb{N}^n$ with*

$$c \leq k \cdot x + k \cdot t^- < c - k \cdot t^{\Delta}.$$

Theorem 1 provides a way of disproving inductivity of a half space by finding a suitable vector x. It is a key ingredient of CGA and ICA.

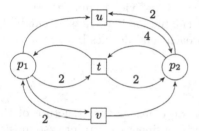

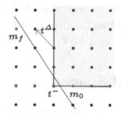

Fig. 2. Petri net with places p_1, p_2 and transitions u, t, v. Edges are entries of the flow function F. We omit the label 1.

Fig. 3. Geometric interpretation of the half space (k, c) in $\mathbb{Z}^2$. It is inductive and separates m_0 from m_f. (Color figure online)

Example. We provide some geometric intuition. Consider the Petri net in Fig. 2. Focus on transition t. The vectors describing t are $t^- = (2, 1)$ (incoming edges), $t^+ = (1, 2)$ (outgoing edges), and $t^{\Delta} = (-1, 1)$. The activation space of t is $\mathrm{Act}(t) = \{(2, 1) + (x, y) \mid x, y \in \mathbb{N}\}$. It is visualized by the yellow area in Fig. 3. Let $m_0 = (3, 1)$ and $m_f = (0, 4)$. Consider the half space defined by $k = (3, 2)$ and $c = 9$. In Fig. 3, it is indicated by the diagonal line $k \cdot x = c, x \in \mathbb{R}^2$. The set of integer vectors above it is $\mathrm{Sol}(k, c)$. Clearly, $m_0 \in \mathrm{Sol}(k, c)$ and $m_f \notin \mathrm{Sol}(k, c)$, the half space separates the markings. The markings in $\mathrm{Act}(t) \cap \mathrm{Sol}(k, c)$ are colored blue in Fig. 3. The half space is t-inductive: if $m \in \mathrm{Act}(t) \cap \mathrm{Sol}(k, c)$, firing t does not lead to a marking below the line. As we will see in Sect. 3, (k, c) is also u and v-inductive. Hence, it proves non-reachability of m_f from m_0.

Linear Safety Verification. Our goal is to find inductive half spaces that disprove reachability or coverability. Given a Petri net N and two markings m_0, m_f, we study two corresponding algorithmic problems: the *linear safety verification problem* LSV(R) for reachability and its coverability variant LSV(C).

LSV(R): Is there an IHS (k, c) with $m_0 \in \text{Sol}(k, c)$ and $m_f \notin \text{Sol}(k, c)$?
LSV(C): Is there an IHS (k, c) with $m_0 \in \text{Sol}(k, c)$ and $\uparrow m_f \cap \text{Sol}(k, c) = \emptyset$?

The reader familiar with separability will note that disproving reachability of m_f from m_0 amounts to finding a separator between $post^*(m_0)$ and $pre^*(m_f)$. A separator is a set $S \subseteq \mathbb{N}^n$ so that $post^*(m_0) \subseteq S$ and $S \cap pre^*(m_f) = \emptyset$. The difference between separability and linear safety verification is that separators are neither required to be half spaces nor required to be inductive.

The choice for half spaces and inductivity is motivated by the constraint-based approach to safety verification that we pursue. Half spaces can be given in terms of (k, c), a format that is computable by a solver. Inductivity yields a local check for separation. Indeed, if (k, c) is inductive and $m_0 \in \text{Sol}(k, c)$, we already have $post^*(m_0) \subseteq \text{Sol}(k, c)$. Similarly, if (k, c) is inductive and $m_f \notin \text{Sol}(k, c)$, then $pre^*(m_f) \cap \text{Sol}(k, c) = \emptyset$. This means $\text{Sol}(k, c)$ is indeed a separator. But there are separators that are neither half spaces nor inductive. To see the latter, consider a transition that is not enabled in $post^*(m_0)$ but in a separator S. Firing the transition may lead to a marking outside of S and violate inductivity.

While reachability and coverability are decidable for Petri nets, decidability of LSV(R) and LSV(C) is unknown. Our approach semi-decides both problems.

3   Half Spaces

In order to synthesize inductive half spaces, we consider the structure of the space of IHS in more detail. Our goal is to derive a linear constraint system that closely approximates the structure of the space. The system can then be passed to an SMT-solver to synthesize candidates for IHS.

Since IHS require inductivity for all transitions, their structure can be convoluted. Therefore, we do not immediately consider complete inductivity. Instead, we first focus on half spaces that are inductive for a single transition. We derive linear constraints describing these half spaces. They are combined in Sect. 4 in order to obtain the desired SMT-formula for the space of all IHS.

The set of half spaces that are inductive for a given transition splits into two parts: the *trivial* half spaces and the *non-trivial* ones. We first focus on the former. Trivial half spaces were already described in [49,52]. They satisfy one of three conditions that immediately imply inductivity and can be easily synthesized. We provide a formal definition below.

The first condition for triviality describes the fact that the vector k and the transition t point into the same direction. The half space (k, c) is *oriented towards transition* t if $k \cdot t^\triangle \geq 0$. Since the scalar product provides information about the angle between k and $t^\triangle$, the condition means that firing transition t moves a marking in the half space further away from the border. To give an example, consider the half space (k, c) with $k = (3, 2)$ from Fig. 3. It is oriented towards transitions u and v. We have $u^{\Delta} = (-1, 2)$ and $v^{\Delta} = (1, 1)$, so $k \cdot u^{\Delta}$ and $k \cdot v^{\Delta}$ are both non-negative. The half space is not oriented towards t since $t^{\Delta} = (-1, 1)$. Firing t means moving closer to the border of the half space.

It easy to see that a half space which is oriented towards a transition t is actually t-inductive. This observation is a first step in the synthesis of IHS. In fact, note that generating a half space (k, c) that separates two markings m_0 and m_f and that is oriented towards t amounts to finding a solution (k, c) of the linear constraint system $k \cdot m_0 \geq c \land k \cdot m_f < c \land k \cdot t^{\Delta} \geq 0$.

The second condition for triviality uses the fact that for $k \geq 0$, the function $k \cdot m$ is monotone on markings. We call a half space (k, c) *monotone for transition* t if $k \geq 0$ and $k \cdot (t^{-} + t^{\Delta}) \geq c$. Note that with larger markings, $k \cdot m$ grows. This means if the smallest marking in the half space enabling t, namely t^{-}, stays within the half space after firing t, the same holds for all larger markings. The requirement is captured in the inequality $k \cdot (t^{-} + t^{\Delta}) \geq c$. Hence, monotone half spaces are inductive and can be synthesized as solutions of $k \geq 0 \land k \cdot (t^{-} + t^{\Delta}) \geq c$.

The last condition is dual to monotonicity. A half space (k, c) is *antitone for transition* t if $k \leq 0$ and $k \cdot t^{-} < c$. The latter requirement describes that t^{-} does not lie in the half space. Since $k \leq 0$ this means that $\text{Act}(t) \cap \text{Sol}(k, c) = \emptyset$. Hence, antitone half spaces are inductive. Moreover, they can be generated as solutions to the linear constraints $k \leq 0 \land k \cdot t^{-} < c$. We summarize:

Definition 1. *A half space (k, c) is trivial wrt. t if one of the following holds: (k, c) is oriented towards t, (k, c) is monotone for t, or (k, c) is antitone for t.*

Theorem 2. *([52]) If (k, c) is trivial with respect to t then it is t-inductive.*

Of course, not all non-trivial half spaces are t-inductive. As an example, consider the half space from Fig. 3. Recall that $k = (3, 2)$ and $c = 9$. If we replace c by $c' = 8$, we get that (k, c') is a non-trivial half space that is not t-inductive. We have $k \cdot t^{-} = 8 = c'$ but $k \cdot (t^{-} + t^{\Delta}) = 7 < c'$. Hence, when firing t from t^{-}, we leave (k, c'). This has two implications. First, we need an algorithm to test whether a non-trivial half space is indeed inductive. Second, we cannot hope for a simple synthesis as for trivial half spaces. The former is resolved by the algorithm ICA which we show in Sect. 5. For the latter, we develop an independent structure theory in the subsequent section.

3.1 Non-trivial Half Spaces

We consider half-spaces that are non-trivial but inductive. These are neither oriented towards the transition of interest, nor monotone, nor antitone. Our first insight is a structural theorem, which strongly impacts their synthesis. We show that a non-trivial IHS (k, c) cannot have both positive and negative entries in k. This means we can limit the synthesis to $k \geq 0$ or $k \leq 0$ for non-trivial IHS.

Theorem 3. *Let (k, c) be a half space that is not oriented towards a transition t but t-inductive. Then, we have $k \geq 0$ or $k \leq 0$.*

The proof of the theorem relies on the notion of *syzygies* known from commutative algebra [25]. We adapt it to our setting. A *syzygy of k* is a vector $s \in \mathbb{Z}^n$ with $k \cdot s = 0$. This means that adding a syzygy to a marking m does not change the scalar product with k. We have $k \cdot m = k \cdot (m + s)$. Hence, if $m \in \text{Sol}(k, c)$, we get that $m + s \in \text{Sol}(k, c)$ for all syzygies s of k. We proceed with the proof.

Proof. Assume (k, c) is t-inductive and not oriented towards t but there are $i \neq j$ with $k(i) > 0$ and $k(j) < 0$. We show that (k, c) cannot be t-inductive which contradicts the assumption. The idea is as follows. We set $u(i) = \lceil \frac{c}{k(i)} \rceil$ and $u(\ell) = 0$ for $\ell \neq i$. Note that $u \in \text{Sol}(k, c)$. From u, we construct a vector $v \in \mathbb{Z}^n$ that lies in $\text{Sol}(k, c)$ but $v + t^\Delta \notin \text{Sol}(k, c)$. Note that v might not be a proper marking. By adding non-negative syzygies to v, we obtain a marking $m \in \text{Act}(t) \cap \text{Sol}(k, c)$ with $m + t^\Delta \notin \text{Sol}(k, c)$. Hence, (k, c) is not t-inductive.

The vector v is defined by $v = u + \lfloor \frac{c - k \cdot u}{k \cdot t^\Delta} \rfloor \cdot t^\Delta \in \mathbb{Z}^n$. Since (k, c) is not oriented towards t, we have $k \cdot t^\Delta < 0$. Hence, v is well-defined. By $\lfloor x \rfloor \geq x - 1$, we obtain the following inequality showing that $v \in \text{Sol}(k, c)$:

$$k \cdot v \geq k \cdot u + \left(\frac{c - k \cdot u}{k \cdot t^\Delta} - 1 \right) \cdot k \cdot t^\Delta = c - k \cdot t^\Delta \geq c.$$

Similarly, by $\lfloor x \rfloor \leq x$, we obtain that $v + t^\Delta \notin \text{Sol}(k, c)$:

$$k \cdot (v + t^\Delta) \leq k \cdot u + \frac{c - k \cdot u}{k \cdot t^\Delta} \cdot k \cdot t^\Delta + k \cdot t^\Delta = c + k \cdot t^\Delta < c.$$

Note that v is not yet a counter example for t-inductivity. Indeed, we cannot ensure that v is a marking that enables t. But we can construct such a marking by adding syzygies to v. For a place $p \in P$ let e_p denote the p-th unit vector. This means $e_p(p) = 1$ and $e_p(q) = 0$ for $q \neq p$. For any place p, we construct a syzygy s_p defined as follows. If $k(p) > 0$, we set $s_p = -k(j) \cdot e_p + k(p) \cdot e_j$. If $k(p) < 0$, we set $s_p = -k(p) \cdot e_i + k(i) \cdot e_p$. For the case $k(p) = 0$, we simply set $s_p = e_p$. Note that for all places p, we have $s_p \geq 0$ and $k \cdot s_p = 0$.

The syzygies s_p allow for adding non-negative values to each component of v without changing the scalar product with k. Hence, there exist $\mu_p \in \mathbb{N}$ such that $v + \sum_{p \in P} \mu_p \cdot s_p \geq t^-$. By setting $m = v + \sum_{p \in P} \mu_p \cdot s_p$, we get a marking in $\text{Act}(t)$ that satisfies $k \cdot m = k \cdot v \geq c$ and $k \cdot (m + t^\Delta) = k \cdot (v + t^\Delta) < c$. Hence, m proves that (k, c) is not t-inductive which is the desired contradiction. □

The theorem allows us to assume $k \geq 0$ or $k \leq 0$ when synthesizing non-trivial half spaces. However, we cannot hope for a compact linear constraint system like we have for trivial half spaces. The reason is as follows. Assume we have a constraint system $L(k, c)$ of polynomial size describing the space of t-inductive non-trivial half spaces. Each solution of $L(k, c)$ corresponds to such a half space and vice versa. We can then decide, in polynomial time, whether a given half space (k, c) is t-inductive. Indeed, an algorithm would first decide whether (k, c) is trivial or non-trivial. In the former case, t-inductivity follows. In the latter case, the algorithm checks if (k, c) is a solution to $L(k, c)$. All steps can be carried out in polynomial time. However, the algorithm would contradict the coNP-hardness of checking t-inductivity which we prove in Sect. 5. Hence, $L(k,c)$ cannot exist.

Although a concise constraint system for the space of non-trivial IHS seems out of reach, we can give a close linear approximation. To this end, we derive two necessary conditions for non-trivial IHS that can be formulated in terms of linear constraints. The first one is given in the following lemma. The proof follows from Theorem 3 and from inverting the constraints for trivial half spaces.

Lemma 1. *A t-inductive half space (k,c) that is non-trivial for t either satisfies (a) $k \geq 0$ and $k \cdot t^- < c - k \cdot t^\Delta$ or (b) $k \leq 0$ and $k \cdot t^- \geq c$.*

The lemma provides geometric intuition to separate non-trivial from trivial half spaces. If (k,c) is non-trivial, $\mathrm{Sol}(k,t) \cap \mathrm{Act}(t)$ is a strict non-empty subset of the activation space $\mathrm{Act}(t)$. This stands in contrast to the trivial case. Here, (k,c) is either oriented towards t or the following holds. If (k,c) is monotone, we have $\mathrm{Sol}(k,t) \cap \mathrm{Act}(t) = \mathrm{Act}(t)$ and if (k,c) is antitone, we have $\mathrm{Sol}(k,t) \cap \mathrm{Act}(t) = \emptyset$.

We employ Lemma 1 to derive a further necessary condition for non-trivial half spaces. It provides a lower bound for the absolute values of the vector k.

Lemma 2. *Let (k,c) be a t-inductive half space that is non-trivial for t. For all entries $k(i)$ of k, with $|k(i)|$ denoting their absolute values, we have:*

$$k(i) = 0 \vee |k(i)| \geq -k \cdot t^\Delta \qquad (5)$$

The idea behind the lemma is the following. If the absolute value of an entry of k is too small then we can construct a vector $x \in \mathbb{N}^n$ such that $k \cdot x + k \cdot t^-$ lies between c and $c - k \cdot t^\Delta - 1$. This violates the condition stated in Theorem 1.

4   Generating Invariants

We capture the conditions from Sect. 3 in an SMT-formula ϕ. It consists of linear constraints and describes necessary properties of IHS. A solution to ϕ is a vector k that potentially forms an IHS. To keep the constraints in the formula linear, we cannot generate a corresponding constant c immediately. Instead, we replace c by bounds imposed by $\mathsf{LSV(R)}$ and $\mathsf{LSV(C)}$. We generate candidates for c in a second synthesis step with the algorithm CGA, given in Sect. 5.

Recall that in $\mathsf{LSV(R)}$, we are interested in finding an inductive half space (k,c) that separates an initial marking m_0 from a marking m_f. Phrased differently, we want $m_0 \in \mathrm{Sol}(k,c)$ and $m_f \notin \mathrm{Sol}(k,c)$. The former implies that $k \cdot m_0 \geq c$, the latter implies $k \cdot m_f < c$. The inequalities yield that $k \cdot m_0 > k \cdot m_f$ and impose two bounds on c, namely $c \in [k \cdot m_f + 1, k \cdot m_0]$. We apply the bounds to the constraints obtained for trivial half spaces and derive the following conditions:

$$k \cdot m_0 > k \cdot m_f \qquad (0) \qquad k \leq 0 \wedge k \cdot t^- < k \cdot m_0 \qquad (2)$$

$$k \cdot t^\Delta \geq 0 \qquad (1) \qquad k \geq 0 \wedge k \cdot t^- > k \cdot m_f - k \cdot t^\Delta \qquad (3)$$

Each of the conditions (1), (2), and (3) models a type of trivial half spaces. For instance, (1) describes half spaces that are oriented towards t. Together with (0), we ensure t-inductivity for some c within the bounds. To describe non-trivial half spaces, we employ Theorem 3 and Lemma 2. We derive the following constraints:

$$k \geq 0 \vee k \leq 0 \qquad (4) \qquad \forall_i \; k(i) = 0 \vee |k(i)| \geq -k \cdot t^{\Delta} \qquad (5)$$

We collect all the constraints in the SMT-formula ϕ_t in order to find the desired inductive half space. Note that the half space must be separating (0). Moreover, it is either trivial, so it satisfies one out of (1), (2), and (3), or it is non-trivial and satisfies (4) and (5). We construct the formula accordingly:

$$\phi_t := (0) \wedge ((1) \vee (2) \vee (3) \vee ((4) \wedge (5))).$$

As mentioned above, it is not possible to construct a linear constraint system of polynomial size that captures all t-inductive half spaces and yields k and c. However, ϕ_t is a tight approximation. In fact, its solutions are precisely those vectors k that can form a t-inductive half space which separates m_0 from m_f.

Lemma 3. *There exists a constant $c \in \mathbb{Z}$ such that (k, c) is a t-inductive half space with $k \cdot m_0 \geq c$ and $k \cdot m_f < c$ if and only if k is a solution to ϕ_t.*

Our goal is to synthesize an IHS that separates m_0 from m_f. Since IHS are t-inductive for all transitions t, we set $\phi := \bigwedge_{t \in T} \phi_t$. The SMT-formula ϕ describes the desired approximation of the space of IHS. It is a main ingredient of our CEGAR loop outlined in Fig. 1. According to Lemma 3, solutions to ϕ are those vectors k that admit a constant c_t for each transition t so that (k, c_t) is t-inductive. The problem is that these c_t may be different for each t. Hence, ϕ generates half space candidates and it is left to find a single value c such that (k, c) is t-inductive for each t. We can compute all possible values for c with the algorithm CGA that we present in Sect. 5. Once a common c is found, we have synthesized the desired IHS. Otherwise, the CEGAR loop starts the refinement.

If a solution k of ϕ does not have a suitable constant c to form an IHS, then neither does any multiple of k. This means we can exclude all multiples in future iterations of the CEGAR loop. Let $mul(k)$ be the formula satisfied by a $k' \in \mathbb{Z}^n$ if and only if there exists an $a \in \mathbb{N}$ such that $a \cdot k = k'$. Then, the refinement performs the update $\phi := \phi \wedge \neg mul(k)$. The following lemma states correctness.

Lemma 4. *Let $k' := a \cdot k$ with $a \in \mathbb{N}$. If (k', c) is an IHS, then so is $(k, \lceil \frac{c}{a} \rceil)$.*

The presented CEGAR approach generates inductive half spaces. In order to semi-decide LSV(R), our approach needs to yield an IHS whenever we are given a yes-instance. This means we need to ensure that any candidate vector k is generated by the SMT-solver at some point so that we do not miss possible IHS. This is achieved by adding a constraint imposing a bound on the absolute values of the entries of k. If the formula becomes unsatisfiable, the bound is increased. It remains to show how our semi-decider for LSV(R) can be adapted to LSV(C).

Coverability. Recall that a solution k of ϕ satisfies Condition (0). It ensures the existence of a value c such that $k \cdot m_0 \geq c$ and $k \cdot m_f < c$, meaning $m \in \mathrm{Sol}(k, c)$ and $m_f \notin \mathrm{Sol}(k, c)$. While this is sufficient for disproving reachability, it is not for coverability. When we solve LSV(C), we need to additionally guarantee that $\uparrow m_f \cap \mathrm{Sol}(k, c)$ is empty. It turns out that this requirement can be captured by a simple modification of ϕ. We only need to ensure that k is negative.

Theorem 4. *Let (k, c) be a half space (not necessarily inductive) such that $m_f \notin \mathrm{Sol}(k, c)$. Then we have $\uparrow m_f \cap \mathrm{Sol}(k, c) = \emptyset$ if and only if $k \leq 0$.*

The intuition is as follows. If $k \leq 0$ does not hold, then we can start with $m := m_f$ and put tokens into a place i with $k_i > 0$ until $k \cdot m \geq c$. This means $k \leq 0$ is sufficient and necessary. Each solution k of ϕ satisfies $m_f \notin \mathrm{Sol}(k, c)$ for some c. In order to disprove coverability, we apply Theorem 4 and add constraint $k \leq 0$ to ϕ. This ensures that any synthesized IHS separates m_0 from $\uparrow m_f$.

5 Checking Inductivity

We present the algorithms ICA and CGA. The former decides t-inductivity for a given half space (k, c) and transition t. The latter is an instrumentation of ICA capable of synthesizing all constants c such that (k, c) is t-inductive, if only the vector k is given. CGA constitutes the remaining bit of our CEGAR loop. Finally, we show that deciding t-inductivity is an coNP-complete problem. The proof once again employs a connection to discrete mathematics.

5.1 Algorithms

We start with the *inductivity checker* (ICA). Given a half space (k, c) and a transition t, we need to decide whether (k, c) is t-inductive. If (k, c) is trivial with respect to t, then inductivity follows from Theorem 2. Hence, we assume that (k, c) is non-trivial. The idea of ICA is to algorithmically check the constraint formulated in Theorem 1 via dynamic programming. Roughly, we search for a value $k \cdot m$, where $m \in ACT(t)$, that lies in the *target interval* $[c, c - k \cdot t^{\Delta} - 1]$. If such a value can be found, (k, c) is not t-inductive. Otherwise, it is t-inductive.

To describe ICA, we adapt Theorem 1. Let $K := \{k(i) \mid i \in [1, n]\}$ contain all entries of a given vector k. We consider sequences $k_1 \ldots k_\ell \in K^*$ over values of k. Note that k_i does not denote the i-th entry of k but the i-th element in the sequence. Then, (k, c) is t-inductive precisely if there is no sequence $k_1 \ldots k_\ell$ with

$$c \leq k \cdot t^- + \sum_{i=1}^{\ell} k_i < c - k \cdot t^{\Delta}. \tag{6}$$

ICA is stated as Algorithm 1. It searches for a sequence in K^* satisfying (6). Recall that we assumed (k, c) to be non-trivial. According to Theorem 3, k does not contain both positive and negative entries. ICA starts at $k \cdot t^-$ and iteratively

---

Algorithm 1: *Inductivity Checker* (ICA)

```
1  queue.add(k · t⁻);
2  reached[k · t⁻]:= True;
3  repeat
4  │   current:=queue.remove();
5  │   if c ≤ current < c − k · t^Δ then
6  │   │   return Not inductive;
7  │   for k ∈ K do
8  │   │   if (current + k < c − k · t^Δ ∧ k ≥ 0)
9  │   │   ∨(current + k ≥ c ∧ k ≤ 0) then
10 │   │   │   if ¬reached[current + k] then
11 │   │   │   │   queue.add(current + k);
12 │   │   │   │   reached[current + k]:=True
13 until queue.isEmpty;
14 return Inductive
```

---

adds values of K until it either reaches the target interval $[c, c − k \cdot t^\Delta − 1]$ or finds that no value in it is reachable. To this end, ICA employs dynamic programming. This avoids recomputing the same value and speeds up the running time. An example of a run of ICA is illustrated in Fig. 4. If the currently reached value lies below the target interval, then at least one value of K has yet to be added. Once we overshoot the target interval, we can exclude the current value and go to the next one in the queue. When we hit the interval, we can report non-inductivity.

In the extended paper, we show that ICA is correct. Moreover, we prove that it runs in pseudopolynomial time. This means polynomial in the values k, c, and t, or exponential in their bit size. Note that this does not contradict the coNP-hardness of checking t-inductivity which we prove in Sect. 5.2.

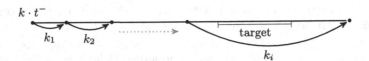

Fig. 4. Example run of Algorithm 1 (ICA) in the case $k \geq 0$. It starts at value $k \cdot t^-$. The algorithm adds values of K until it either overshoots the target interval or hits it.

Constant Generation. Given a vector k and a transition t, ICA can be instrumented to compute all values c such that (k, c) is t-inductive. We refer to the instrumentation as *constant generation algorithm* (CGA). CGA computes all necessary sums $k \cdot t^- + \sum_{i=1}^{\ell} k_i$ with $k_1 \ldots k_i \in K^*$ and returns all c such that $[c, c − k \cdot t^\Delta − 1]$ does not contain any of the sums. Intuitively, we fit the interval between them. Note that each of the returned values c satisfies the characterization of t-inductivity as stated in (6). We show that CGA is correct and terminates.

For termination, we need the so-called *Frobenius number* [5]. Let $a \in \mathbb{N}^n$ be a vector such that $\gcd(a) = \gcd(a(1), \dots, a(n)) = 1$. Here, gcd denotes the *greatest common divisor*. The *Frobenius number of a* is the largest integer that cannot be represented as a positive linear combination of $a(1), \dots, a(n)$. The number exists and is bounded by $a_{\max} \cdot a_{\min}$, where $a_{\max}$ is the largest and $a_{\min}$ the smallest entry of a [5]. Note that this means that each value $x \geq a_{\max} \cdot a_{\min}$ can be represented as a positive linear combination $x = a \cdot m$ with $m \in \mathbb{N}^n$.

This has implications for CGA. Assume we are given $k \geq 0$ with $\gcd(k) = 1$. Possible values of c such that (k, c) is t-inductive cannot exceed $k \cdot t^- + k_{\max} \cdot k_{\min}$. Otherwise, $[c, c - k \cdot t^\Delta - 1]$ contains a linear combination of the form $k \cdot t^- + \sum_{i=1}^{\ell} k_i$ which breaks Requirement (6). The argument can be generalized to $\gcd(k) \geq 1$:

Theorem 5. *Let (k, c) be a non-trivial t-inductive half space and let $k_{\max}, k_{\min}$ denote the entries of k with maximal and minimal absolute value.*

1. *If $k \geq 0$, we have $c < k_{\max} \cdot k_{\min} + k \cdot t^-$.*
2. *If $k \leq 0$, we have $c \geq -k_{\max} \cdot k_{\min} + k \cdot t^-$.*

The theorem enforces termination and correctness of CGA. In fact, we only need to compute sums $k \cdot t^- + \sum_{i=1}^{\ell} k_i$ with $k_1 \dots k_i \in K^*$ up to the limit given in the theorem and still find all values c such that (k, c) is t-inductive. Since the limit is polynomial in the values of k and t, CGA runs in pseudopolynomial time.

We employ CGA within our CEGAR loop. Assume we have a solution k to our SMT-formula ϕ. It is left to decide whether there exists a $c \in \mathbb{N}$ such that (k, c) is an IHS. We apply CGA to k and each transition t. This yields a set C_t containing all c_t such that (k, c_t) is t-inductive and separates m_0 from m_f. Hence, $\bigcap_{t \in T} C_t$ contains all c such that (k, c) is an IHS that separates m_0 from m_f. Algorithmically, we only need to test the intersection for non-emptiness.

5.2   Complexity

We prove that deciding t-inductivity for a transition t is coNP-complete. Membership follows from a non-deterministic variant of ICA. Further analysis shows that the problem is also fixed-parameter tractable and in coCSL, where CSL is the class of languages accepted by context-sensitive grammars. For unary input, it is in coNL and—if the dimension of k is fixed—in L. We provide details in the extended paper. The interesting part is coNP-hardness for which we establish a reduction from the *unbounded subset sum problem* [22].

Theorem 6. *Checking t-inductivity of a half space (k, c) is coNP-complete.*

Before we elaborate on the reduction, we introduce the *unbounded subset sum problem* (USSP). An instance consists of a vector $w \in \mathbb{N}^n$ and an integer $d \in \mathbb{N}$. The task is to decide whether there exists a vector $x \in \mathbb{N}^n$ such that $w \cdot x = d$. The problem is NP-complete [22]. To prove Theorem 6, we reduce from USSP to the complement of checking t-inductivity. This yields the desired coNP-hardness.

Proof. Let (w, d) be an instance of USSP. We construct a half space (k, c) and a Petri net with a transition t such that (k, c) is not t-inductive if and only if there is an $x \in \mathbb{N}^n$ such that $w \cdot x = d$. We rely on the inductivity criterion from Theorem 1. The main difference between this criterion and USSP is that the latter requires reaching a precise value d, while the former requires reaching an interval. The idea is to define an appropriate half space (k, c) and a transition t such that in the corresponding interval only one value might be reachable.

We set $k = w$. Note that we can assume that d is a multiple of $\gcd(k)$. Otherwise, (w, d) is a no-instance of USSP since each linear combination $w \cdot x$ is a multiple of $\gcd(k)$. By using the Euclidean algorithm, we can compute an $a \in \mathbb{Z}^n$ such that $k \cdot a = \gcd(k)$ in polynomial time [2]. We construct a Petri net with n places and one transition t with $t^-(i) := a(i)$ if $a(i) > 0$, $t^+ := -a$ if $a(i) < 0$, and 0 otherwise. It holds $t^\Delta = -a$. Set $c = d + k \cdot t^-$. It is left to show that (k, c) is not t-inductive if and only if (w, d) is a yes-instance of USSP.

Assume that (k, c) is not t-inductive. Then there exists a vector $x \in \mathbb{N}^n$ such that $c \leq k \cdot x + k \cdot t^- < c - k \cdot t^\Delta$. By plugging in the above definitions, we obtain that $d \leq k \cdot x < d + \gcd(k)$. Since d, $d + \gcd(k)$, and $k \cdot x$ are all multiples of $\gcd(k)$, we obtain that $d = k \cdot x = w \cdot x$. Hence, (w, d) is a yes-instance of USSP.

For the other direction, let $w \cdot x = d$. We obtain that $d \leq k \cdot x < d + \gcd(k)$. As above, we can employ the definitions and derive that $c \leq k \cdot x + k \cdot t^- < c - k \cdot t^\Delta$. This shows non-inductivity of (k, c) and proves correctness of the reduction. □

6   Experiments

We implemented the CEGAR loop in our Java prototype tool INEQUALIZER [17]. It employs Z3 [42] as a back-end SMT-solver. The tool makes use of *incremental solving* as well as *minimization*, a feature of Z3 that guides the CEGAR loop towards more likely candidates of IHS. Incremental solving reuses information learned from previous queries to Z3 and minimization prioritizes solutions with minimal values. Before INEQUALIZER starts the CEGAR loop, it uses an SMT-query to check whether there is a separating IHS (k, c) that is trivial for all transitions. We use minimization to get half spaces that are non-trivial with respect to fewer transitions. The reason is that non-trivial half spaces are harder to find and typically only few values for c ensure inductivity in this case. Before we show the applicability of INEQUALIZER on larger benchmarks, let us consider the Petri net in Fig. 2. When executing INEQUALIZER, we find that there are no trivial separating IHS. Using incremental solving, INEQUALIZER performs three iterations of the CEGAR loop and returns the non-trivial separating IHS with $k = (53, 52)$ and $c = 209$. When enabling minimization, we only require two iterations and obtain $k = (8, 5)$, $c = 22$. The difference in iterations is due to that we expect minimization to choose vectors k that are trivial for many transitions. This increases the chance of finding a suitable c. On the other hand, incremental solving improves the running time in executions with more iterations.

We evaluated INEQUALIZER for LSV(C) on a benchmark suite and compared it to various methods for coverability implemented in MIST [13,19,20,23,44].

Table 1. INEQUALIZER vs. MIST.

Benchmark	$|P|$	$|T|$	INEQUALIZER	MIST backward	ic4pn	tsi	eec	eec-cegar
BASICME	5	4	0.6	0.1	0.1	0.1	0.1	0.1
KANBAN	16	14	0.7	0.1	0.2	0.8	0.1	0.2
LAMPORT	11	9	T/O	0.1	0.1	0.1	0.1	0.1
MANUFACTURING	13	6	0.6	1.9	0.1	0.1	0.1	0.1
PETERSSON	14	12	T/O	0.2	0.1	0.1	0.1	0.1
READ-WRITE	13	9	0.5	0.1	1	0.1	0.9	0.5
MESH2x2	32	32	1.2	0.3	0.1	48.6	0.8	0.2
MESH3x2	52	54	2.1	2.2	0.2	T/O	T/O	2.2
MULTIPOOL	18	21	0.8	0.3	2	2.2	1	2.3

Results are given in Table 1. The experiments were performed on a 1,7 GHz Intel Core i7 with 8 GB memory. The running times are given in seconds. For entries marked as T/O, the timeout was reached. The running times of INEQUAL-IZER are similar to MIST although the former has a small overhead from generating the SMT-query. In each of the listed Petri nets, the unsafe marking is not coverable. Except for the mutual exclusion nets PETERSSON and LAMPORT, INEQUALIZER reliably finds separating IHS. Surprisingly, each found IHS is trivial. We suspect that the cases where INEQUALIZER timed out are actually negative instances of LSV(C).

The experiments show that many practical instances admit trivial IHS, which we synthesize using only one SMT-query. To test the generation of non-trivial IHS, we ran INEQUALIZER on a list of nets that do not admit trivial ones. The results are presented in the extended version of the paper and show that INEQUALIZER finds non-trivial IHS within few iterations of the CEGAR loop.

7   Conclusion and Outlook

We considered an invariant-based approach to disprove reachability and coverability in Petri nets. The idea was to synthesize an inductive half space that over-approximates the reachable markings of the net and separates them from unsafe markings. For the synthesis, we established a structure theory of IHS and derived an SMT-formula which approximates the space of IHS using linear constraints. We provided two algorithms, ICA and CGA. The former decides whether a half space is inductive, the latter generates suitable constants that guarantee inductivity. The SMT-formula and the algorithm CGA were then combined in a CEGAR loop which attempts to synthesize IHS. We implemented this into our tool INEQUALIZER. It combines SMT-queries with efficient heuristics and was capable of solving practical instances in our experiments.

While our experiments required only trivial IHS, theoretical studies in the extended paper suggest that we can expect many instances to require non-trivial

IHS. We plan a more detailed study of which practical instances require non-trivial IHS in future work. We expect that further structural studies of IHS will improve the efficiency of the CEGAR loop. This may lead to a tighter approximation of the space of IHS or to an improved refinement step eliminating more than multiples. It is also an intriguing question whether the problems LSV(R) and LSV(C) are decidable. To tackle this, we are currently examining equivalence classes and normal forms of half spaces and their connection to well-quasi orderings.

Acknowledgments. We thank Roland Meyer for his ideas and contributions that greatly influenced the work on inductive half spaces at hand. Moreover, we thank Christian Eder for sharing with us his experience in commutative algebra, and Marvin Triebel for his ideas and the questions that he raised during his visit.

References

1. Abdulla, P.A., Iyer, S.P., Nylén, A.: Sat-solving the coverability problem for Petri nets. Formal Methods Syst. Des. **24**(1), 25–43 (2004)
2. Bach, E., Shallit, J.: Algorithmic Number Theory, Volume I: Efficient Algorithms. MIT Press (1996)
3. Beyer, D., Henzinger, T.A., Majumdar, R., Rybalchenko, A.: Invariant synthesis for combined theories. In: Cook, B., Podelski, A. (eds.) VMCAI 2007. LNCS, vol. 4349, pp. 378–394. Springer, Heidelberg (2007). https://doi.org/10.1007/978-3-540-69738-1_27
4. Blanchet, B., et al.: A static analyzer for large safety-critical software. In: PLDI, pp. 196–207. ACM (2003)
5. Brauer, A.: On a problem of partitions. Am. J. Math. **64**(1), 299–312 (1942)
6. Cardoza, E., Lipton, R., Meyer, A.R.: Exponential space complete problems for Petri nets and commutative semigroups (preliminary report). In: STOC, pp. 50–54. ACM (1976)
7. Chini, P., Furbach, F.: Petri net invariant synthesis. CoRR 2105.03096 (2021)
8. Clarke, E.M., Henzinger, T.A., Veith, H., Bloem, R.: Handbook of Model Checking. Springer, Heidelberg (2018)
9. Clarke, E., Grumberg, O., Jha, S., Lu, Y., Veith, H.: Counterexample-guided abstraction refinement. In: Emerson, E.A., Sistla, A.P. (eds.) CAV 2000. LNCS, vol. 1855, pp. 154–169. Springer, Heidelberg (2000). https://doi.org/10.1007/10722167_15
10. Cousot, P., Halbwachs, N.: Automatic discovery of linear restraints among variables of a program. In: POPL, pp. 84–96. ACM (1978)
11. Czerwinski, W., Lasota, S., Lazic, R., Leroux, J., Mazowiecki, F.: The reachability problem for Petri nets is not elementary. In: STOC, pp. 24–33. ACM (2019)
12. Czerwiński, W., Orlikowski, L.: Reachability in vector addition systems is Ackermann-complete. CoRR, 2104.13866 (2021)
13. Delzanno, G., Raskin, J.-F., Van Begin, L.: Towards the automated verification of multithreaded Java programs. In: Katoen, J.-P., Stevens, P. (eds.) TACAS 2002. LNCS, vol. 2280, pp. 173–187. Springer, Heidelberg (2002). https://doi.org/10.1007/3-540-46002-0_13

14. Esparza, J., Ledesma-Garza, R., Majumdar, R., Meyer, P., Niksic, F.: An SMT-based approach to coverability analysis. In: Biere, A., Bloem, R. (eds.) CAV 2014. LNCS, vol. 8559, pp. 603–619. Springer, Cham (2014). https://doi.org/10.1007/978-3-319-08867-9_40

15. Esparza, J., Römer, S.: An unfolding algorithm for synchronous products of transition systems. In: Baeten, J.C.M., Mauw, S. (eds.) CONCUR 1999. LNCS, vol. 1664, pp. 2–20. Springer, Heidelberg (1999). https://doi.org/10.1007/3-540-48320-9_2

16. Esparza, J., Römer, S., Vogler, W.: An improvement of McMillan's unfolding algorithm. In: Margaria, T., Steffen, B. (eds.) TACAS 1996. LNCS, vol. 1055, pp. 87–106. Springer, Heidelberg (1996). https://doi.org/10.1007/3-540-61042-1_40

17. Furbach, F.: Inequalizer - a prototype tool for linear safety verification of Petri nets. https://github.com/florianfurbach/Inequalizer

18. Floyd, R.W.: Assigning meanings to programs. In: Proceedings of a Symposium on Applied Mathematics, vol. 19, pp. 19–32 (1967)

19. Ganty, P., Meuter, C., Van Begin, L., Kalyon, G., Raskin, J., Delzanno, G.: Symbolic data structure for sets of k-tuples of integers. Technical report (2007)

20. Ganty, P., Raskin, J., Van Begin, L.: From many places to few: automatic abstraction refinement for Petri nets. Fundam. Inform. **88**, 124–143 (2007)

21. Ganty, P., Raskin, J.-F., Van Begin, L.: From many places to few: automatic abstraction refinement for Petri nets. Fundam. Inform. **88**(3), 275–305 (2008)

22. Garey, M.R., Johnson, D.S.: Computers and Intractability; A Guide to the Theory of NP-Completeness. W. H. Freeman & Co. (1990)

23. Geeraerts, G., Raskin, J.-F., Van Begin, L.: Expand, enlarge and check: new algorithms for the coverability problem of WSTS. J. Comput. Syst. Sci. **72**, 180–203 (2006)

24. Geeraerts, G., Raskin, J.-F., Van Begin, L.: On the efficient computation of the minimal coverability set for Petri nets. In: Namjoshi, K.S., Yoneda, T., Higashino, T., Okamura, Y. (eds.) ATVA 2007. LNCS, vol. 4762, pp. 98–113. Springer, Heidelberg (2007). https://doi.org/10.1007/978-3-540-75596-8_9

25. Greuel, G.-M., Pfister, G.: A Singular Introduction to Commutative Algebra. Springer, Heidelberg (2002)

26. Gulwani, S., Srivastava, S., Venkatesan, R.: Program analysis as constraint solving. In: PLDI, pp. 281–292. ACM (2008)

27. Gupta, A., Majumdar, R., Rybalchenko, A.: From tests to proofs. In: Kowalewski, S., Philippou, A. (eds.) TACAS 2009. LNCS, vol. 5505, pp. 262–276. Springer, Heidelberg (2009). https://doi.org/10.1007/978-3-642-00768-2_24

28. Hartmanis, J.: Context-free languages and turing machine computations. In: Symposia in Applied Mathematics, vol. 19, pp. 42–51 (1967)

29. Hoare, C.A.R.: An axiomatic basis for computer programming. Commun. ACM **12**(10), 576–580 (1969)

30. Kaiser, A., Kroening, D., Wahl, T.: Efficient coverability analysis by proof minimization. In: Koutny, M., Ulidowski, I. (eds.) CONCUR 2012. LNCS, vol. 7454, pp. 500–515. Springer, Heidelberg (2012). https://doi.org/10.1007/978-3-642-32940-1_35

31. Karp, R.M., Miller, R.E.: Parallel program schemata. J. Comput. Syst. Sci. **3**(2), 147–195 (1969)

32. Kloos, J., Majumdar, R., Niksic, F., Piskac, R.: Incremental, inductive coverability. In: Sharygina, N., Veith, H. (eds.) CAV 2013. LNCS, vol. 8044, pp. 158–173. Springer, Heidelberg (2013). https://doi.org/10.1007/978-3-642-39799-8_10

33. Rao Kosaraju, S.: Decidability of reachability in vector addition systems. In: STOC, pp. 267–281. ACM (1982)
34. Lambert, J.: A structure to decide reachability in Petri nets. Theor. Comput. Sci. **99**(1), 79–104 (1992)
35. Langerak, R., Brinksma, E.: A complete finite prefix for process algebra. In: Halbwachs, N., Peled, D. (eds.) CAV 1999. LNCS, vol. 1633, pp. 184–195. Springer, Heidelberg (1999). https://doi.org/10.1007/3-540-48683-6_18
36. Leroux, J.: The reachability problem for Petri nets is not primitive recursive. CoRR, 2104.12695 (2021)
37. Leroux, J., Schmitz, S.: Reachability in vector addition systems is primitive-recursive in fixed dimension. In: LICS, pp. 1–13. IEEE (2019)
38. Lipton, R.J.: The reachability problem requires exponential space. Research report (Yale University. Department of Computer Science). Department of Computer Science, Yale University (1976)
39. Mayr, E.: An algorithm for the general Petri net reachability problem. In: STOC, pp. 238–246. ACM (1981)
40. Mayr, E.: An algorithm for the general Petri net reachability problem. SIAM J. Comput. **13**(3), 441–460 (1984)
41. McMillan, K.L.: A technique of state space search based on unfolding. Form. Methods Syst. Des. **6**(1), 45–65 (1995)
42. de Moura, L., Bjørner, N.: Z3: an efficient SMT solver. In: Ramakrishnan, C.R., Rehof, J. (eds.) TACAS 2008. LNCS, vol. 4963, pp. 337–340. Springer, Heidelberg (2008). https://doi.org/10.1007/978-3-540-78800-3_24
43. Murata, T.: Petri nets: properties, analysis and applications. Proc. IEEE **77**(4), 541–580 (1989)
44. Ganty, P.: MIST - a safety checker for Petri nets and extensions. https://github.com/pierreganty/mist
45. Peterson, J.L.: Petri Net Theory and the Modeling of Systems. Prentice Hall, Hoboken (1981)
46. Podelski, A., Rybalchenko, A.: A complete method for the synthesis of linear ranking functions. In: Steffen, B., Levi, G. (eds.) VMCAI 2004. LNCS, vol. 2937, pp. 239–251. Springer, Heidelberg (2004). https://doi.org/10.1007/978-3-540-24622-0_20
47. Rackoff, C.: The covering and boundedness problems for vector addition systems. Theor. Comput. Sci. **6**(2), 223–231 (1978)
48. Reynier, P.-A., Servais, F.: Minimal coverability set for Petri nets: Karp and miller algorithm with pruning. In: Kristensen, L.M., Petrucci, L. (eds.) PETRI NETS 2011. LNCS, vol. 6709, pp. 69–88. Springer, Heidelberg (2011). https://doi.org/10.1007/978-3-642-21834-7_5
49. Sankaranarayanan, S., Sipma, H., Manna, Z.: Petri net analysis using invariant generation. In: Dershowitz, N. (ed.) Verification: Theory and Practice. LNCS, vol. 2772, pp. 682–701. Springer, Heidelberg (2003). https://doi.org/10.1007/978-3-540-39910-0_29
50. Sankaranarayanan, S., Sipma, H.B., Manna, Z.: Scalable analysis of linear systems using mathematical programming. In: Cousot, R. (ed.) VMCAI 2005. LNCS, vol. 3385, pp. 25–41. Springer, Heidelberg (2005). https://doi.org/10.1007/978-3-540-30579-8_2
51. Sipser, M.: Introduction to the Theory of Computation. PWS Publishing Company (1997)

52. Triebel, M., Sürmeli, J.: Characterizing stable inequalities of Petri nets. In: Devillers, R., Valmari, A. (eds.) PETRI NETS 2015. LNCS, vol. 9115, pp. 266–286. Springer, Cham (2015). https://doi.org/10.1007/978-3-319-19488-2_14
53. Turing, A.M.: On computable numbers, with an application to the Entscheidungsproblem. Proc. Lond. Math. Soc. s2–42(1), 230–265 (1937)
54. Valmari, A., Hansen, H.: Old and new algorithms for minimal coverability sets. In: Haddad, S., Pomello, L. (eds.) PETRI NETS 2012. LNCS, vol. 7347, pp. 208–227. Springer, Heidelberg (2012). https://doi.org/10.1007/978-3-642-31131-4_12

Towards Efficient Shape Analysis
with Tree Automata

Martin Hruška$^{(\boxtimes)}$ and Lukáš Holík

Brno University of Technology, Brno, Czech Republic
{ihruska,holik}@fit.vutbr.cz

Abstract. We discuss our proposal of a formalism for representing classes of graphs based on tree automata. We aim at a formalism and an entailment algorithm that could be used in verification of pointer programs, that would be efficient, have well defined completeness guarantees, and be general. We believe that building the formalism on top of tree automata will make it possible to use existing advanced tree automata implementation techniques. We sketch the basic ideas behind the formalism and an entailment decision procedure, and outline some related research challenges.

1 Introduction

The recent 20 years have seen a rise of many approaches to verification of pointer programs, aka shape analysis, up to their industrial deployment (e.g. the technique of [5] in Facebook's Infer). The existing approaches are mainly distinguished by the formalism used to describe sets of memory configurations (shape graphs), which are essentially graphs with nodes being memory locations and edges being pointers. The dominant position, previously held by frameworks such as [28,32], is currently occupied by more automated and scalable approaches based on separation logic (SL) [4,6,31] such as symbolic memory graphs [8], on forest automata [13], and on graph grammars [14]. These approaches clearly identified the importance of local reasoning and modularity in reasoning about memory configurations as the key to scalability.

One of the major bottlenecks in the field is extending the techniques to more complex data structures: with anything beyond relatively simple variants of lists and trees, the existing approaches struggle with scalability and precision or require a non-trivial users assistance. None of the existing formalisms for describing shape graphs have all the following desirable properties. 1. *Expressiveness*: the ability to talk about variants variants lists, trees, structures such as skip-lists, threaded trees, their combinations and overlayer variants. 2. *Local reasoning*: running a program statement on the abstract domain should have only a local effect, it should be possible to reason locally about the affected parts.

This work was supported by the Czech Science Foundation (project No. 19-24397S) and the FIT BUT internal project FIT-S-20-6427.

K. Echihabi and R. Meyer (Eds.): NETYS 2021, LNCS 12754, pp. 206–214, 2021.
https://doi.org/10.1007/978-3-030-91014-3_14

3. *Effectiveness*: satisfiability and entailment should be efficiently decidable, as well as additional graph operations needed, e.g., in (higher-order) bi-abduction [5,25]. 4. *Abstraction and generalization learning mechanisms*: the ability to learn abstractions of inductive invariants with controlled precision.

Separation logic approaches provides the first two qualities, expressiveness and local reasoning, but lacks in the other two. Earlier verification methods are not sufficiently general their decision procedures handle mostly just lists. The approaches such as [11,21,30] are rather restricted and incomplete, while the general approaches [23,27] are theoretical and far from being efficiently implementable. The recent works [9,10,24,29] finally came with an entailment for a large fragment of separation logic. These algorithms have not yet been tried within actual verification of pointer programs but are promising not only in the context of separation logic.

SL approaches so far lack ways to automatically learning shape invariants without a help of user-predefined patterns. The higher-order bi-abduction [25] is a notable exception: it is capable of learning extremely complex shape invariants such as B+ trees, skip-lists, or threaded trees. It is, however, very sensitive to how the code is written (since it is, in a sense, transforming the recursive code to inductive shape predicates) and hence quite fragile, easily failing on seemingly easy examples such as natural implementations of a doubly-linked list reversal.

Outside separation logic, especially the approach based on Forest automata [12,13,18,19] has been shown viable [15–17]. It allows for some degree of local reasoning, it is efficient as it allows to utilise advanced algorithms for tree automata (such as simulation reduction or antichain language inclusion). The main distinguishing advantage is its compatibility with abstraction schemes from abstract regular model checking [15] with counterexample guided refinement loop. The formalism however suffers from some deficiencies, such as that expressible classes shape graphs are limited and that it is not closed under union.

We discuss here our ongoing work on developing a new graph formalism in the spirit of Forest automata that would remedy their weaknesses. We present main ideas on which such formalism can be built. First, we explain how graphs can be encoded into trees and tree automata (as a variation on tree decomposition of graphs [7] and also the formalism used in [20,22]). We then discuss basic ideas for an entailment procedure for the formalism. We believe that this new formalism can eventually combine local reasoning of separation logic and forest automata, strong entailment procedures of [9,10,24,29], efficiency of tree automata [1–3,26], and powerful abstraction schemes of regular model checking.

2   Representing Graphs with Trees and Tree Automata

We will first discuss encoding of graphs as variations on tree decompositions, similar to that used in [20] and also [22].

A Σ-*labeled graph* is a pair $g = (V, E)$ where V is a finite set of nodes, $E \subseteq V \times \Sigma \times V$ is a set of Σ-labeled edges. A graph $g = (V, E)$ is *deterministic* if for every node $n \in V$ and every label $a \in \Sigma$, there is at most one node $n' \in V$

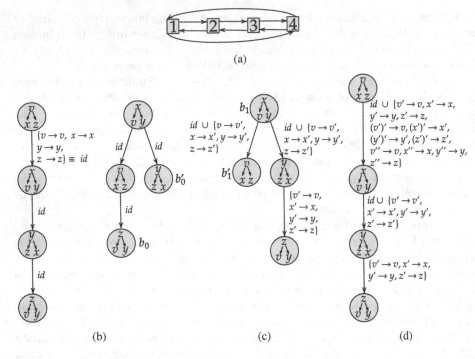

(a)

(b)   (c)   (d)

Fig. 1. Figure shows circular doubly-linked list (a) and its tree decompositions. In the decompositions (b), the variables v, x, y, z represent the nodes $1, 2, 3, 4$ (in this order) of the list. The figure shows that it is possible to transform the right decomposition in (b) to the left one using transformations shown in (c) and (d). Figure (c) illustrates the reconnection operation. It shows the decomposition obtained by reconnection applied to the right-handed side decomposition of (b) where the bag b_0 is reconnected below the bag b'_0. The figure also shows the primed variables introduced by the reconnection to prevent inference of reconnected pipes with pipes along the reconnection path. Analogously, (d) show result of rotation applied to (c) where the bags b_1 and b'_1 were rotated. The operation changes orientation of edge between them and also introduces the new primed variables for each existing variable, e.g., $(x')'$ for x' and since x' already exists, x'' is created for x.

such that $(n, a, n') \in E$. Unless stated otherwise, we will assume all graphs deterministic.

A *tree decomposition* of a labeled graph g over a finite set of variables *Vars* and alphabet Σ is a tree $t = (B, E)$. Nodes B of t are Σ-labeled graphs called *bags*. Nodes of a bag are variables from *Vars*. Edges of t are labelled by partial mappings $\rho : Vars \to Vars$ called *parameter assignments*. The *tree-width* of a decomposition, $tw(t)$, is the maximum cardinality of a parameter assignment in it. A *node occurrence* in t is a pair $(x, b) \in Vars \times B$ where either x is a node of a bag $b \in B$ (i.e., $x \in b$) (so called an *active* occurrence) or x belongs to the image of the parameter assignment on the edge targeting b (then it is a *passive* occurrence), i.e., $x \in img(\rho)$ such that (b, ρ, b') is an edge of t and ρ is a label of the edge. The *alias relation* $\sim$ is the smallest equivalence of occurrences such that if $(b, \rho, b') \in E$ and $x' = \rho(x)$ then $(x, b) \sim (x', b')$. The Σ-labeled *graph represented by t* is the graph $g^t = (V^t, E^t)$ where the nodes V^t are the equivalence classes of $\sim$, called *pipes*, and $E^t = \{([(x, b)]_\sim, a, [(x', b)]_\sim) \mid (x, a, x')$ is an edge of the graph $b \in B\}$ (that is, every edge (x, a, x') of every graph $b \in B$ gives rise to an edge $([(x, b)]_\sim, a, [(x', b)]_\sim)$ of the represented graph g^t).

An example of a graph (representing circular doubly-linked list data structure) and its tree decomposition is shown in Fig. 1(a) and (b).

We will work with the following restrictions of graphs and tree decompositions. If a bag b has an edge originating at x, then the pipe $[(x, b)]_\sim$ is *allocated at b*. A *backbone decomposition* corresponds to an (unoriented) tree backbone of the graph. It has three defining properties: 1. Every bag b allocates *exactly one graph node*. 2. Every graph node is *allocated only once*. 3. The tree is *connected* in the sense that every tree edge corresponds to a graph edge (regardless the edge orientation). That is, for every two adjacent bags, one of them, say b, has an edge adjacent with x, and the other, b', has an active occurrence (b', z) with $(b, x) \sim (b', z)$.

Last, assuming a backbone decomposition, we define so called *pipe child* relation $\triangleright$. Two pipes p and p' of a decomposition t are in the relation $p \triangleright p'$ iff p is allocated in node b and p' in node b' such that $(b, a, b') \in E$ for some a.

A set of tree decompositions can be represented by a tree automaton. Intuitively, a node of a tree represents a node in the tree decomposition, and the label of each tree node records the bag and the labels on the decomposition edges leading from the node to its children.

3   Towards Entailment

The idea for deciding entailment is to use tree automata language inclusion algorithms (such as [1–3, 26]) over tree automata encodings of tree decompositions. The difficulty here is that a single graph has multiple decompositions and the tree automata may accept only some of them, hence simple language inclusion check may underapproximate the inclusion of sets of represented graphs (entailment). We therefore propose means of saturating the tree automata languages with all possible tree decompositions of the represented graphs. Conceptually, we will define a small set of operations which is *complete* for tree decompositions, that is, allows to transform a decomposition into any other decomposition of the same graph. The two tree automata under the entitlement accept decompositions with certain maximum tree width t, that can be easily determined. The entailment procedure will apply the decomposition operations symbolically over the tree automata until they are saturated with all tree decompositions of the represented graphs with the tree width up-to t. Computing the language inclusion of thus saturated automata is then a sound algorithm for entailment.

We will now outline the operations. These operations are essentially meant to complete the *rotation* operation of [22].

Reconnection. The operation of reconnection is parameterised by two *peer* bags b and b' of the decomposition that allocate pipes p and p', respectively (they are peer bags, i.e. not on the same branch of the tree). Its purpose is to create an equivalent decomposition with the child relation on graph nodes being the same up to that p becomes a child of p' in the new decomposition. The operation is implemented as a function $reconnect(b, b', t)$ that transforms a tree decomposition $t = (B, E)$ into t' as follows. First, the pipes that are reaching b in t must be in t' sent to b' through the path between b and b', called the *reconnection path*. Let $\rho_1, \ldots, \rho_k$ be the sequence of edge labels appearing on the reconnection path. Take every label ρ_i of an edge on the reconnection path with $2 < i \leq k$, and replace it by $\rho_i' = \rho_i \cup \{x' \mapsto x' \mid x \in img(\rho_1)\}$. The primed variables must be such that they do not appear anywhere on the reconnection path, neither in the graph bags nor in the variable renamings (they may be fresh variables). This is to stretch the pipes reaching b through the reconnection path towards b'. Replace also ρ_2 by $\rho_2' = \rho_2 \cup \{x' \mapsto y \mid \rho_1(y) = x\}$. This binds the new primed part of the pipes with the original pipes reaching n. Last, replace the edge leading to b by (b', ρ_1', b) where $\rho_1' = \{x' \mapsto x \mid x \in img(\rho_1)\}$. This makes b a child of b' and connects the new primed pipes to the corresponding original pipes of b.

An example of the operation is shown in Fig. 1(c) where the righ-handed side decomposition t_0 from Fig. 1(b) is rotated by $reconnect(b_0, b_0', t_0)$.

Rotation. The rotation operation is parameterised by two bags b and b' of the decomposition. The operation inverts the edges in the path π between b and b'. Then it redirects the incoming edge of b to b'. Intuitively, it takes a subtree with the root b, changes the root to b' and inverts the edges in the subtree. It yields an equivalent tree decomposition with respect to child relation which is inverted between pipes allocated in the nodes of the path π.

The operation is implemented as a function $rotation(b, b', t)$. At the tree level, it works as follows. Consider the path $\pi : b = b_1, \ldots, b_n = b'$. Remove each edge (b_i, ρ, b_{i+1}), where $1 \leq i \leq n$, between nodes in π from E and add (b_{i+1}, ρ, b_i) to E. Moreover, we replace the edge $(m, \rho, b) \in E$ by the edge (m, ρ, b') to E. The labels along the rotated path are changed in the following way. Replace the label ρ_1 of the edge (m, ρ_1, b') by $\rho_1 = \{x \mapsto x' \mid x \in dom(\rho_1)\}$. Replace each label ρ_i in the path π by $\rho_i = \rho_i \cup \{x' \mapsto x' \mid x' \in img(\rho_1')\}$ for $2 \leq i \leq b$. Finally, replace the label ρ_b over the edge leading to b by $\rho_b' = \rho_b \cup \{x' \mapsto x \mid x' \in img(\rho_1')\}$.

An example of the operation is shown in Fig. 1(d) where the decomposition t_1 from Fig. 1(c) is rotated by $rotation(b_1, b_1', t_1)$.

Phase. The operations from one tree decomposition to an equivalent one may take unboundedly many operations. We will devide them into *phases*. One phase can still perform unboundedly many operations, but the set is restricted: the reconnection paths of all reconnections must be node disjoint. Formally, a phase is characterised by a set of operation parameters, pairs of nodes $\{(b_1, b_1'), \ldots, (b_k, b_k')\}$ of a decomposition t such the for any two $1 \leq i, j \leq k, i \neq j$, the operation paths between b_i and b_i' and between b_j and b_j' are node disjoint. A result of the phase is any decomposition which arises by performing the appropriate operation on (b_i, b'_i) for each i (in any order). An example of two phases are show in Figs. 1(c) and (d) where the right-handed side decomposition from Fig. 1(b) is transformed in the left-handed side decomposition of the same figure in these two phases.

An important consequence of the disjointness of the reconnection paths is that all the operations can be implemented while only doubling the number of variables and the tree-width of the original decomposition. Particularly, all the operations can use the same set of fresh primed versions of the variables *Vars* on the operation path without fearing a conflict with the names of the existing pipes.

Lemma 1. *One phase at most doubles the number of variables.*

We conjecture that the number of phases needed depends on only on the tree-width:

Conjecture 1. An equivalent decomposition t' can be obtained from t in a number of phases that depends only on $\max(tw(t), tw(t'))$.

Next, we discuss implementation of a phase over tree automata (TA) representation. We design phase over TA in such way that its results is an automaton that encodes all possible results of phase applied at any possible decomposition represented by the original automaton. We will briefly sketch the basic idea of the operation.

Namely, saturation with reconnections can be implemented by a tree transducer. Seen as a top-down machine, it oscillates between two routines, *idle*, and *reconnecting*. In the idle state, it is just traversing the tree. At any node r, it may non-deterministically chose to start reconnecting. When reconnecting, it non-deterministically selects two peer descendants of r, nodes b and b', and performs the reconnection on them. The reconnection, roughly, involves adding of primed versions of existing pipes on the path from b to b' and reconnecting the subtree of b below b'. After that, the reconnecting phase stops and the transducer continues traversing the tree in the idle phase. The requirement in the definition of phase on the disjointness of the reconnection maths makes this doable—when reconnecting, the transducer needs to worry only about one reconnection path at a time. Saturation with rotations can be then implemented similarly as in [21].

We conjecture, based partially on Lemma 1, that such implementation of a phase over tree automata representation is cheap:

Conjecture 2. The implementation of a tree automata phase at most doubles the number of variables and leads to an automaton that is of a polynomial size assuming a fixed tree-width of the original automaton.

Based on Conjecture 1 and 2, the saturation of a tree automata representation with all decompositions until a fixed tree-width t can be done in a time that is polynomial (when the t is fixed). Recall that deciding entailment between two

TA representations, we saturate both of them up to the maximum tree-width t (in fact, it is enough to use the maximum tree with of the automaton that is supposed to entail), and then we compute language inclusion between the saturated automata. Since language inclusion of tree automata is EXPTIME-complete, Conjectures 1 and 2 give an entailment algorithm singly exponential, assuming a fixed maximum tree-width t.

4   Conclusions and Future Work

We have presented basic outline of a formalism for representing shape graphs based on tree automata. The ideas should lead to an entailment algorithm, and conjectures that, if true, would imply that the algorithm is relatively fast assuming fixed maximum tree-width of the graph representations.

We plan to perfect these ideas and to prove Conjectures 1 and 2. The conjectures are somewhat optimistic, but not in a direct contradiction with the recent 2-EXPTIME-hardness result of [9]. If the conjectures turn out to be false, we wish to search for (1) restrictions under which they are true, and/or (2) to prove termination of the described entailment algorithm regardless its complexity. Our long term plan is to develop a shape analysis framework based on this formalism and entailment check in the spirit of [19] and also [25].

References

1. Abdulla, P.A., Bouajjani, A., Holík, L., Kaati, L., Vojnar, T.: Computing simulations over tree automata. In: Ramakrishnan, C.R., Rehof, J. (eds.) TACAS 2008. LNCS, vol. 4963, pp. 93–108. Springer, Heidelberg (2008). https://doi.org/10.1007/978-3-540-78800-3_8
2. Abdulla, P.A., Chen, Y.-F., Holík, L., Mayr, R., Vojnar, T.: When simulation meets antichains. In: Esparza, J., Majumdar, R. (eds.) TACAS 2010. LNCS, vol. 6015, pp. 158–174. Springer, Heidelberg (2010). https://doi.org/10.1007/978-3-642-12002-2_14
3. Almeida, R., Holík, L., Mayr, R.: Reduction of nondeterministic tree automata. In: Chechik, M., Raskin, J.-F. (eds.) TACAS 2016. LNCS, vol. 9636, pp. 717–735. Springer, Heidelberg (2016). https://doi.org/10.1007/978-3-662-49674-9_46
4. Berdine, J., et al.: Shape analysis for composite data structures. In: Damm, W., Hermanns, H. (eds.) CAV 2007. LNCS, vol. 4590, pp. 178–192. Springer, Heidelberg (2007). https://doi.org/10.1007/978-3-540-73368-3_22
5. Calcagno, C., Distefano, D., O'Hearn, P.W., Yang, H.: Compositional shape analysis by means of bi-abduction. ACM Trans. Comput. Log. **58**, 26:1–26:66 (2011)
6. Chang, B.-Y.E., Rival, X., Necula, G.C.: Shape analysis with structural invariant checkers. In: Nielson, H.R., Filé, G. (eds.) SAS 2007. LNCS, vol. 4634, pp. 384–401. Springer, Heidelberg (2007). https://doi.org/10.1007/978-3-540-74061-2_24
7. Courcelle, B.: The monadic second-order logic of graphs. I. Recognizable sets of finite graphs. Inf. Comput. **85**, 12–75 (1990)
8. Dudka, K., Peringer, P., Vojnar, T.: Byte-precise verification of low-level list manipulation. In: Logozzo, F., Fähndrich, M. (eds.) SAS 2013. LNCS, vol. 7935, pp. 215–237. Springer, Heidelberg (2013). https://doi.org/10.1007/978-3-642-38856-9_13

9. Echenim, M., Iosif, R., Peltier, N.: Entailment checking in separation logic with inductive definitions is 2-exptime hard. In: LPAR 2020. EPiC Series in Computing, vol. 73, pp. 191–211. EasyChair (2020)

10. Echenim, M., Iosif, R., Peltier, N.: Decidable entailments in separation logic with inductive definitions: beyond establishment. In: CSL 2021. LIPIcs, vol. 183, pp. 20:1–20:18. Schloss Dagstuhl - Leibniz-Zentrum für Informatik (2021)

11. Habermehl, P., Holík, L., Rogalewicz, A., Šimáček, J., Vojnar, T.: Forest automata for verification of heap manipulation. In: Gopalakrishnan, G., Qadeer, S. (eds.) CAV 2011. LNCS, vol. 6806, pp. 424–440. Springer, Heidelberg (2011). https://doi.org/10.1007/978-3-642-22110-1_34

12. Habermehl, P., Holík, L., Rogalewicz, A., Šimáček, J., Vojnar, T.: Forest automata for verification of heap manipulation. In: Gopalakrishnan, G., Qadeer, S. (eds.) CAV 2011. LNCS, vol. 6806, pp. 424–440. Springer, Heidelberg (2011). https://doi.org/10.1007/978-3-642-22110-1_34

13. Habermehl, P., Holík, L., Rogalewicz, A., Šimáček, J., Vojnar, T.: Forest automata for verification of heap manipulation. Formal Methods Syst. Design 1, 83–106 (2012)

14. Heinen, J., Jansen, C., Katoen, J.-P., Noll, T.: Juggrnaut: using graph grammars for abstracting unbounded heap structures. Formal Methods Syst. Design 47(2), 159–203 (2015)

15. Holík, L., Hruška, M., Lengál, O., Rogalewicz, A., Šimáček, J., Vojnar, T.: Forester: shape analysis using tree automata. In: Baier, C., Tinelli, C. (eds.) TACAS 2015. LNCS, vol. 9035, pp. 432–435. Springer, Heidelberg (2015). https://doi.org/10.1007/978-3-662-46681-0_37

16. Holík, L., Hruška, M., Lengál, O., Rogalewicz, A., Šimáček, J., Vojnar, T.: Run forester, run backwards! In: Chechik, M., Raskin, J.-F. (eds.) TACAS 2016. LNCS, vol. 9636, pp. 923–926. Springer, Heidelberg (2016). https://doi.org/10.1007/978-3-662-49674-9_61

17. Holík, L., Hruška, M., Lengál, O., Rogalewicz, A., Šimáček, J., Vojnar, T.: Forester: from heap shapes to automata predicates. In: Legay, A., Margaria, T. (eds.) TACAS 2017. LNCS, vol. 10206, pp. 365–369. Springer, Heidelberg (2017). https://doi.org/10.1007/978-3-662-54580-5_24

18. Holík, L., Hruška, M., Lengál, O., Rogalewicz, A., Vojnar, T.: Counterexample validation and interpolation-based refinement for forest automata. In: Bouajjani, A., Monniaux, D. (eds.) VMCAI 2017. LNCS, vol. 10145, pp. 288–309. Springer, Cham (2017). https://doi.org/10.1007/978-3-319-52234-0_16

19. Holík, L., Lengál, O., Rogalewicz, A., Šimáček, J., Vojnar, T.: Fully automated shape analysis based on forest automata. In: Sharygina, N., Veith, H. (eds.) CAV 2013. LNCS, vol. 8044, pp. 740–755. Springer, Heidelberg (2013). https://doi.org/10.1007/978-3-642-39799-8_52

20. Iosif, R., Rogalewicz, A., Simacek, J.: The tree width of separation logic with recursive definitions. In: Bonacina, M.P. (ed.) CADE 2013. LNCS (LNAI), vol. 7898, pp. 21–38. Springer, Heidelberg (2013). https://doi.org/10.1007/978-3-642-38574-2_2

21. Iosif, R., Rogalewicz, A., Vojnar, T.: Deciding entailments in inductive separation logic with tree automata. In: Cassez, F., Raskin, J.-F. (eds.) ATVA 2014. LNCS, vol. 8837, pp. 201–218. Springer, Cham (2014). https://doi.org/10.1007/978-3-319-11936-6_15

22. Iosif, R., Rogalewicz, A., Vojnar, T.: Deciding entailments in inductive separation logic with tree automata. In: Cassez, F., Raskin, J.-F. (eds.) ATVA 2014. LNCS, vol. 8837, pp. 201–218. Springer, Cham (2014). https://doi.org/10.1007/978-3-319-11936-6_15

23. Iosif, R., Rogalewicz, A., Simacek, J.: The tree width of separation logic with recursive definitions. In: Bonacina, M.P. (ed.) CADE 2013. LNCS (LNAI), vol. 7898, pp. 21–38. Springer, Heidelberg (2013). https://doi.org/10.1007/978-3-642-38574-2_2

24. Katelaan, J., Matheja, C., Zuleger, F.: Effective entailment checking for separation logic with inductive definitions. In: Vojnar, T., Zhang, L. (eds.) TACAS 2019. LNCS, vol. 11428, pp. 319–336. Springer, Cham (2019). https://doi.org/10.1007/978-3-030-17465-1_18

25. Le, Q.L., Gherghina, C., Qin, S., Chin, W.-N.: Shape analysis via second-order bi-abduction. In: Biere, A., Bloem, R. (eds.) CAV 2014. LNCS, vol. 8559, pp. 52–68. Springer, Cham (2014). https://doi.org/10.1007/978-3-319-08867-9_4

26. Lengál, O., Šimáček, J., Vojnar, T.: VATA: a library for efficient manipulation of non-deterministic tree automata. In: Flanagan, C., König, B. (eds.) TACAS 2012. LNCS, vol. 7214, pp. 79–94. Springer, Heidelberg (2012). https://doi.org/10.1007/978-3-642-28756-5_7

27. Matheja, C., Jansen, C., Noll, T.: Tree-like grammars and separation logic. In: Feng, X., Park, S. (eds.) APLAS 2015. LNCS, vol. 9458, pp. 90–108. Springer, Cham (2015). https://doi.org/10.1007/978-3-319-26529-2_6

28. Møller, A., Schwartzbach, M.I.: The pointer assertion logic engine, vol. 36, pp. 221–231. ACM, New York (2001)

29. Pagel, J., Matheja, C., Zuleger, F.: Complete entailment checking for separation logic with inductive definitions (2020)

30. Piskac, R., Wies, T., Zufferey, D.: Automating separation logic with trees and data. In: Biere, A., Bloem, R. (eds.) CAV 2014. LNCS, vol. 8559, pp. 711–728. Springer, Cham (2014). https://doi.org/10.1007/978-3-319-08867-9_47

31. Reynolds, J.C.: Separation logic: a logic for shared mutable data structures. In: LICS 2002, pp. 55–74. IEEE Computer Society (2002)

32. Sagiv, S., Reps, T.W., Wilhelm, R.: Parametric shape analysis via 3-valued logic. ACM Trans. Program. Lang. Syst. **24**(3), 217–298 (2002)

Deciding S1S: Down the Rabbit Hole and Through the Looking Glass

Vojtěch Havlena, Ondřej Lengál$^{(\boxtimes)}$, and Barbora Šmahlíková

Faculty of Information Technology, Brno University of Technology,
Brno, Czech Republic
lengal@fit.vutbr.cz

Abstract. Monadic second-order logic of one successor (S1S) is a logic for specifying ω-regular languages in a concise way. In this paper, we revisit the classical decision procedure based on translating S1S formulae into Büchi automata and employ state-of-the-art algorithms for their manipulation, in particular complementation and size reduction. We compare our implementation to the one based on loop-deterministic finite automata and observe cases where the classical approach scales better.

1 Introduction

The study of formalisms allowing reasoning about ω-regular languages still attracts a lot of attention. For instance, ω-regular languages are often used for specifying properties of reactive systems via the formalisms of linear-time temporal logics such as LTL [1] or QPTL [2]. In addition to that, ω-regular languages have also been used for formal verification of programs [3] and, recently, in the context of automated theorem proving, for reasoning about properties of Sturmian words [4,5]. A prominent logic allowing to describe the whole class of ω-regular properties is *monadic second-order logic of one successor* (S1S). The decidability of S1S was proven by Büchi in 1962 by introducing a connection of the logic with automata over infinite words called Büchi automata (BAs) [6]. S1S offers immense succinctness for the price of nonelementary worst-case complexity.

The many applications of ω-regular languages, often represented using BAs, together with BAs' nice theoretical properties have attracted a lot of attention towards developing efficient algorithms for their manipulation. Unlike the ones for automata over finite words, algorithms for BAs are often much more involved. In particular, the problem of efficiently complementing BAs has been approached from several sides [2,7–29] and so has been the problem of BA reduction [30–33].

In this paper we revisit the original automata-based decision procedure for S1S and exploit state-of-the-art approaches for handling BAs, in particular approaches for their reduction and techniques of complementation, to obtain an efficient decision procedure. We summarize our observations with the implementation, identify the bottlenecks, and provide an experimental comparison with an approach deciding S1S based on deterministic-loop automata [34].

© Springer Nature Switzerland AG 2021
K. Echihabi and R. Meyer (Eds.): NETYS 2021, LNCS 12754, pp. 215–222, 2021.
https://doi.org/10.1007/978-3-030-91014-3_15

2   Preliminaries

Functions, Words, and Alphabets. We use ω to denote the first infinite ordinal $\omega = \{0, 1, \ldots\}$. An (infinite) word α over alphabet Σ is represented as a function $\alpha \colon \omega \to \Sigma$ where the i-th symbol is denoted as α_i. We abuse notation and sometimes also represent α as an infinite sequence $\alpha = \alpha_0 \alpha_1 \ldots$ We use Σ^ω to denote the set of all infinite words over Σ.

Büchi Automata. A (nondeterministic) *Büchi automaton* (BA) over Σ is a quadruple $\mathcal{A} = (Q, \delta, I, F)$ where Q is a finite set of *states*, δ is a *transition function* $\delta \colon Q \times \Sigma \to 2^Q$, and $I, F \subseteq Q$ are the sets of *initial* and *accepting* states respectively. We sometimes treat δ as a set of transitions of the form $p \xrightarrow{a} q$, for instance, we use $p \xrightarrow{a} q \in \delta$ to denote that $q \in \delta(p, a)$. A *run* of $\mathcal{A}$ from $q \in Q$ on an input word α is an infinite sequence $\rho \colon \omega \to Q$ that starts in q and respects δ, i.e., $\rho(0) = q$ and $\forall i \geq 0 \colon \rho(i) \xrightarrow{\alpha_i} \rho(i+1) \in \delta$. Let $\inf(\rho)$ denote the states occurring in ρ infinitely often. We say that ρ is *accepting* iff $\inf(\rho) \cap F \neq \emptyset$. A word α is accepted by $\mathcal{A}$ if there is an accepting run ρ of $\mathcal{A}$ from some initial state, i.e., $\rho(0) \in I$. The set $\mathcal{L}(\mathcal{A}) = \{\alpha \in \Sigma^\omega \mid \mathcal{A} \text{ accepts } \alpha\}$ is called the *language* of $\mathcal{A}$.

Simulation. The *(maximum) direct simulation* on $\mathcal{A}$ is the relation $\preceq_{di} \subseteq Q \times Q$ defined as the largest relation s.t. $p \preceq_{di} q$ implies (i) $p \in F \Rightarrow q \in F$ and (ii) $p \xrightarrow{a} p' \in \delta \Rightarrow \exists q' \in Q \colon q \xrightarrow{a} q' \in \delta \wedge p' \preceq_{di} q'$ for each $a \in \Sigma$.

3   Monadic Second-Order Logic of One Successor (S1S)

In this section we briefly introduce monadic second-order logic of one successor, denoted as S1S, used for expressing ω-regular properties of linear structures.

3.1   Syntax and Semantics

In this paper we build S1S formulae from *atomic formulae* of the form (i) $0 \in X$, (ii) $X \subseteq Y$, (iii) $X = Succ(Y)$, and (iv) $Sing(X)$ where X and Y are *second-order* variables. Formulae are then obtained as a Boolean combination of atomic formulae and existential quantification. Other connectives and universal quantification can be obtained as a syntactic sugar, e.g., we can define $\varphi \to \psi$ to denote $\neg \varphi \vee \psi$ and $\forall X. \varphi$ to denote $\neg \exists X. \neg \varphi$.

S1S formulae are interpreted over the set of natural numbers. In particular, second-order variables range over (possibly infinite) subsets of ω. For an S1S formula $\varphi(\mathbb{X})$ with free variables $\mathbb{X}$ an *assignment* is a mapping $\sigma \colon \mathbb{X} \to 2^\omega$. The *satisfaction* of an atomic formula φ by an assignment σ, denoted as $\sigma \vDash \varphi$, is inductively defined as follows: (i) $\sigma \vDash 0 \in X$ iff 0 is in $\sigma(X)$, (ii) $\sigma \vDash X \subseteq Y$ iff $\sigma(X)$ is a subset of $\sigma(Y)$, (iii) $\sigma \vDash X = Succ(Y)$ iff $\sigma(X) = \{y + 1 \mid y \in \sigma(Y)\}$, and (iv) $\sigma \vDash Sing(X)$ iff $|X| = 1$. Satisfaction of an S1S formula by σ is then defined inductively as usual. Formula φ is called *satisfiable* if there is an assignment σ such that $\sigma \vDash \varphi$.

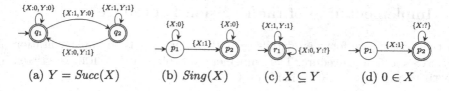

(a) $Y = Succ(X)$ (b) $Sing(X)$ (c) $X \subseteq Y$ (d) $0 \in X$

Fig. 1. BAs for atomic formulae.

3.2 Encoding Models as Words

The first step towards automata-based decision procedure is an encoding of assignments as words. In the following, we fix a formula φ with free variables $\mathbb{X}$. A symbol ξ over $\mathbb{X}$ is a mapping $\xi \colon \mathbb{X} \to \{0,1\}$, e.g., $\xi = \{X{:}0, Y{:}1\}$. We use $\Sigma_{\mathbb{X}}$ to denote the set of all symbols over $\mathbb{X}$. Furthermore, for a set of variables $\mathbb{Y}$, we define the *projection* of ξ wrt. $\mathbb{Y}$ as $\pi_{\mathbb{Y}}(\xi) = \xi_{|\mathbb{X}\setminus\mathbb{Y}}$. An assignment σ of φ is then encoded as the word α^{σ} over $\Sigma_{\mathbb{X}}$ s.t. for each variable $X \in \mathbb{X}$ and for each $i \in \omega$ the following two conditions hold (i) if $i \in \sigma(X)$ then α_i^{σ} contains $X{:}1$ and (ii) if $i \notin \sigma(X)$ then α_i^{σ} contains $X{:}0$. The language of the formula φ is then defined as $\mathcal{L}(\varphi) = \{\alpha^{\sigma} \mid \sigma$ is a model of $\varphi\}$.

3.3 Automata-Based Decision Procedure

The automata-based decision procedure for S1S takes an input formula φ and inductively builds the BA $\mathcal{A}_{\varphi}$ accepting the same language as φ. Checking satisfiability of φ is then equivalent to testing whether $\mathcal{L}(\mathcal{A}_{\varphi}) \neq \emptyset$. The automaton $\mathcal{A}_{\varphi}$ is defined as follows: (i) If φ is an atomic formula, then $\mathcal{A}_{\varphi}$ is a predefined BA (see Fig. 1). (ii) If $\varphi = \psi_1 \wedge \psi_2$, then, in the first step, both $\mathcal{A}_{\psi_1}$ and $\mathcal{A}_{\psi_1}$ are adjusted to accept the original models extended to symbols over $\Sigma_{\mathbb{X}_1 \cup \mathbb{X}_2}$ ($\mathbb{X}_1$ and $\mathbb{X}_2$ are the free variables of ψ_1 and ψ_2 respectively). This step is called *cylindrification* and can be implemented by modifying the transition functions of $\mathcal{A}_{\psi_1}$ and $\mathcal{A}_{\psi_1}$. In particular, for $\mathcal{A}_{\psi_1}$, each transition over a symbol ξ is replaced by multiple transitions over all symbols $\xi' \in \Sigma_{\mathbb{X}_1 \cup \mathbb{X}_2}$ s.t. $\pi_{\mathbb{X}_1}(\xi') = \xi$. The BA $\mathcal{A}_{\varphi}$ is then obtained as $\mathcal{A}_{\varphi} = \mathcal{A}'_{\psi_1} \cap \mathcal{A}'_{\psi_2}$ where $\mathcal{A}'_{\psi_1}$ and $\mathcal{A}'_{\psi_2}$ are cylindrified BAs and $\cap$ is the standard operation of intersection of two BAs. (iii) If $\varphi = \psi_1 \vee \psi_2$ then $\mathcal{A}_{\varphi} = \mathcal{A}'_{\psi_1} \cup \mathcal{A}'_{\psi_2}$ where $\mathcal{A}'_{\psi_1}$ and $\mathcal{A}'_{\psi_2}$ are cylindrified BAs and $\cup$ is the standard operation of union over BAs. (iv) If $\varphi = \neg\psi$, then $\mathcal{A}_{\varphi} = \mathcal{A}_{\psi}^{\complement}$ where $\mathcal{A}^{\complement}$ denotes the BA accepting $\Sigma^{\omega}\setminus\mathcal{L}(\mathcal{A})$. (v) If $\varphi = \exists X.\,\psi$, then $\mathcal{A}_{\varphi} = \pi_X(\mathcal{A}_{\psi})$ where $\pi_X(\mathcal{A})$ is the BA obtained from $\mathcal{A}$ by modifying its transition function, applying $\pi_{\{X\}}$ on the symbol of each transition.

Handling of First-Order Variables. Although the definition of S1S presented in Sect. 3.1 uses second-order variables only, first-order variables (denoted by lowercase letters) can be handled using the support of the *Sing* predicate as follows: $\exists x.\,\varphi$ is transformed into $\exists X.\,\varphi \wedge Sing(X)$ and $\forall x.\,\varphi$ is transformed into $\forall X.\,Sing(X) \to \varphi$.

4   Implementation of the Decision Procedure

In this section, we focus on details related to our prototype implementation of the S1S decision procedure. The implemented tool, called ALICE, is written in PYTHON and is publicly available on GITHUB[1].

Automata-Based Decision Procedure. ALICE implements the classical decision procedure as described in Sect. 3.3. In particular, it uses the two-copy product construction for intersection and performs union by simply uniting the input automata (making sure they have disjoint sets of states). BA complementation (corresponding to using negation in the input formula) is performed either by Schewe's optimal construction [10, Sect. 3] improving the original rank-based construction [11,23] or by determinization-based complementation implemented within SPOT [35]. Although the used complementation algorithms meet the lower bound of BA complementation $2^{\mathcal{O}(n \log n)}$, the complexity is still a bottleneck of the decision procedure. Note that development of efficient complementation algorithms for BAs is still a hot topic of current research [8,13]. In order to avoid the state explosion during complementation, we keep the automata as small as possible using (i) *lightweight reductions*, such as quotienting wrt. the direct simulation equivalence, i.e., two states p, q are merged if $p \preceq_{di} q$ and $q \preceq_{di} p$, or disconnecting little brother states [33], i.e., if there are transitions $p \xrightarrow{a} q$ and $p \xrightarrow{a} r$ with $q \preceq_{di} r$, we can remove the transition $p \xrightarrow{a} q$ from the automaton, and (ii) *heavyweight reductions*, based on a 10-step lookahead simulation relation combined with advanced transition pruning, implemented in the tool RABIT [30].

Alphabet Handling. When working with BAs, the number of states is not the only issue. Recall from Sect. 3.2 that if we consider a formula with n free variables, there are 2^n symbols that can occur in the corresponding automaton. For this reason we implement *symbolic* handling of symbols using a "don't care" flag (denoted by "?"). For instance two transitions $p \xrightarrow{\xi_1} q$ and $p \xrightarrow{\xi_2} q$ where $\xi_1 = \{X{:}1, Y{:}0\}$ and $\xi_2 = \{X{:}1, Y{:}1\}$ are represented by a single transition $p \xrightarrow{\kappa} q$ where $\kappa = \{X{:}1, Y{:}?\}$. In future, we might consider handling alphabets via *binary decision diagrams* in the similar way as MONA [36].

5   Experimental Evaluation

In this section, we compare our tool with, to the best of our knowledge, the only other existing implementation of a decision procedure for S1S, which is based on loop-deterministic finite automata (denoted as L-DFA) [34]. The evaluation uses a benchmark that consists of 26 hand-crafted S1S formulae obtained from [34]. We compared the approaches with respect to the number

---

[1] https://github.com/barbora4/projektova-praxe.

Table 1. Comparison of Alice and L-DFA on S1S formulae. In addition to the atomic formulae from Sect. 3.1, Alice also considers $x < y$ to be atomic.

	Formula	Alice-Rank	Alice-Spot-Light	Alice-Spot	L-DFA
1.	$(x \in Y \wedge x \notin Z) \vee (x \in Z \wedge x \notin Y)$	2	5	2	9
2.	$\neg \exists x.((x \in Y \wedge x \notin Z) \vee (x \in Z \wedge x \notin Y))$	1	1	1	9
3.	$after(X, Y) := \forall x.(x \in X \rightarrow \exists y.(y > x \wedge y \in Y))$	5	3	3	9
4.	$fair(X, Y) := after(X, Y) \wedge after(Y, X)$	24	5	5	9
5.	$\forall X.(fair(X, Y) \rightarrow fair(Y, Z))$	OOM	29	21	14
6.	$suc(x, y) := x < y \wedge \forall z.(\neg x < z \vee \neg z < y))$	3	4	3	10
18.	$offset(X, Y) := \forall i \forall j.(suc(i, j) \wedge i \in X \rightarrow j \in Y)$	2	2	2	11
19.	$offset(X, Y) \wedge offset(Y, Z) \wedge offset(Z, X)$	8	8	8	107
20.	$offset(V, W) \wedge offset(W, X) \wedge offset(X, Y) \wedge$ $offset(Y, Z) \wedge offset(Z, V)$	32	32	32	2331
22.	$insm(i, j, U, V, W) := (j \in U \rightarrow i \in V \vee i \in W)$	8	13	8	15
23.	$\forall i \forall j (suc(i, j) \rightarrow insm(i, j, U, V, Z) \wedge$ $insm(i, j, V, X, Y) \wedge insm(i, j, X, Y, V) \wedge$ $insm(i, j, Y, Z, X) \wedge insm(i, j, Z, U, Y))$	OOM	TO	TO	198
26.	$\forall x \forall y.(x < y \wedge y \in X \wedge y \in Y) \wedge \forall x \forall y.(x <$ $y \wedge y \in X \wedge y \notin Y) \wedge \forall x \forall y.(x < y \wedge y \notin X \wedge y \in$ $Y) \wedge \forall x \forall y.(x < y \wedge y \notin X \wedge y \notin Y)$	21	11	11	18

of states of the automaton $\mathcal{A}_\varphi$ (either BA or L-DFA) corresponding to the formula φ.[2]

In the comparison, we use the following three settings of our tool: Alice-Rank denotes the setting with Schewe's complementation and reduction by Rabit, Alice-Spot denotes Spot's complementation and reduction by Rabit, and, lastly, Alice-Spot-Light denotes Spot's complementation and lightweight reduction. The timeout (TO) was set to 1 h. Selected results are in Table 1.

Discussion. Our tool usually gives better results than L-DFA in terms of state count, as shown in Table 1. In particular, for the case of Alice-Spot, the state counts of the resulting automata were in the vast majority of cases lower than for L-DFA. There were only two worse cases, one of them being formula 23 that did not finish in a day (complementing an automaton having 33 states). Furthermore, parametric formulae 18–20 are worth noticing; the number of states of L-DFA grows much faster than in our case. If we compare the variants Alice-Rank and Alice-Spot, the setting Alice-Spot gives overall better results—e.g., for formula 5, Alice-Rank ran out of memory (OOM), yet Alice-Spot yields an automaton having 21 states. On the other hand, lightweight reduction behaves surprisingly well: Alice-Spot outperforms Alice-Spot-Light just in 7 cases (most significantly on formulae 7, 9, and 22).

By analyzing the results, we found that the bottleneck of our approach is indeed BA complementation—it caused all the TOs and OOMs in the bench-

---

[2] We do not compare other measurements such as the execution time or the sum of sizes of all automata obtained during the construction of $\mathcal{A}_\varphi$, because we were not able to obtain the L-DFA tool and [34] does not provide these values.

mark. For instance the TO in result of ALICE-SPOT for formula 23 is caused by complementing a BA with 33 states. To keep the sizes of automata small, their reduction is a crucial operation. Therefore, as a future work, we would like to investigate state-of-the-art techniques for BA complementation and identify the most suitable approach in connection with advanced minimization techniques. Although ALICE often produces smaller automata than L-DFA, the number of states is not the only possible measure: with a missing run time the comparison is incomplete, since dealing with BAs is usually harder than dealing with DFAs.

As far as we know, ALICE is the only off-the-shelf publicly available S1S solver. We intend to use it in the following settings: (i) educational (students input S1S formulae and observe the corresponding BAs) and (ii) research (we wish to study the structure of the created BAs and search for potential heuristics).

Acknowledgment. This work has been supported by the Czech Science Foundation project 19-24397S and the FIT BUT internal project FIT-S-20-6427.

References

1. Vardi, M.Y., Wolper, P.: An automata-theoretic approach to automatic program verification. In: Proceedings of the First Symposium on Logic in Computer Science, pp. 322–331. IEEE (1986)
2. Sistla, A.P., Vardi, M.Y., Wolper, P.: The complementation problem for Büchi automata with applications to temporal logic. In: Brauer, W. (ed.) ICALP 1985. LNCS, vol. 194, pp. 465–474. Springer, Heidelberg (1985). https://doi.org/10.1007/BFb0015772
3. Heizmann, M., Hoenicke, J., Podelski, A.: Termination analysis by learning terminating programs. In: Biere, A., Bloem, R. (eds.) CAV 2014. LNCS, vol. 8559, pp. 797–813. Springer, Cham (2014). https://doi.org/10.1007/978-3-319-08867-9_53
4. Oei, R., Ma, D., Schulz, C., Hieronymi, P.: Pecan: an automated theorem prover for automatic sequences using büchi automata. CoRR abs/2102.01727 (2021)
5. Hieronymi, P., Ma, D., Oei, R., Schaeffer, L., Schulz, C., Shallit, J.O.: Decidability for Sturmian words. CoRR abs/2102.08207 (2021)
6. Büchi, J.R.: On a decision method in restricted second order arithmetic. In: Proceedings of International Congress on Logic, Method, and Philosophy of Science 1960, Stanford University Press, Stanford (1962)
7. Gurumurthy, S., Kupferman, O., Somenzi, F., Vardi, M.Y.: On complementing nondeterministic Büchi automata. In: Geist, D., Tronci, E. (eds.) CHARME 2003. LNCS, vol. 2860, pp. 96–110. Springer, Heidelberg (2003). https://doi.org/10.1007/978-3-540-39724-3_10
8. Chen, Y.-F., Havlena, V., Lengál, O.: Simulations in rank-based Büchi automata complementation. In: Lin, A.W. (ed.) APLAS 2019. LNCS, vol. 11893, pp. 447–467. Springer, Cham (2019). https://doi.org/10.1007/978-3-030-34175-6_23
9. Chen, Y., et al.: Advanced automata-based algorithms for program termination checking. In: Foster, J.S., Grossman, D. (eds.) Proceedings of the 39th ACM SIGPLAN Conference on Programming Language Design and Implementation, PLDI 2018, Philadelphia, PA, USA, 18–22 June 2018, pp. 135–150. ACM (2018)

10. Schewe, S.: Büchi complementation made tight. In: Albers, S., Marion, J. (eds.) 26th International Symposium on Theoretical Aspects of Computer Science, STACS 2009, 26–28 February 2009, Freiburg, Germany, Proceedings. LIPIcs, vol. 3, pp. 661–672. Schloss Dagstuhl - Leibniz-Zentrum fuer Informatik, Germany (2009)
11. Kupferman, O., Vardi, M.Y.: Weak alternating automata are not that weak. ACM Trans. Comput. Log. **2**(3), 408–429 (2001)
12. Allred, J.D., Ultes-Nitsche, U.: A simple and optimal complementation algorithm for Büchi automata. In: Proceedings of the Thirty third Annual IEEE Symposium on Logic in Computer Science (LICS 2018), pp. 46–55. IEEE Computer Society Press (July 2018)
13. Blahoudek, F., Duret-Lutz, A., Strejček, J.: Seminator 2 can complement generalized Büchi automata via improved semi-determinization. In: Lahiri, S.K., Wang, C. (eds.) CAV 2020. LNCS, vol. 12225, pp. 15–27. Springer, Cham (2020). https://doi.org/10.1007/978-3-030-53291-8_2
14. Tsai, M.-H., Fogarty, S., Vardi, M.Y., Tsay, Y.-K.: State of Büchi complementation. In: Domaratzki, M., Salomaa, K. (eds.) CIAA 2010. LNCS, vol. 6482, pp. 261–271. Springer, Heidelberg (2011). https://doi.org/10.1007/978-3-642-18098-9_28
15. Li, Y., Turrini, A., Zhang, L., Schewe, S.: Learning to complement Büchi automata. In: Dillig, I., Palsberg, J. (eds.) VMCAI 2018. LNCS, vol. 10747, pp. 313–335. Springer, Cham (2018). https://doi.org/10.1007/978-3-319-73721-8_15
16. Vardi, M.Y., Wilke, T., Kupferman, O., Fogarty, S.J.: Unifying Büchi complementation constructions. Log. Methods Comput. Sci. **9** (2013)
17. Blahoudek, F., Heizmann, M., Schewe, S., Strejček, J., Tsai, M.-H.: Complementing semi-deterministic Büchi automata. In: Chechik, M., Raskin, J.-F. (eds.) TACAS 2016. LNCS, vol. 9636, pp. 770–787. Springer, Heidelberg (2016). https://doi.org/10.1007/978-3-662-49674-9_49
18. Sistla, A.P., Vardi, M.Y., Wolper, P.: The complementation problem for Büchi automata with applications to temporal logic. Theoret. Comput. Sci. **49**(2–3), 217–237 (1987)
19. Fogarty, S., Vardi, M.Y.: Büchi complementation and size-change termination. In: Kowalewski, S., Philippou, A. (eds.) TACAS 2009. LNCS, vol. 5505, pp. 16–30. Springer, Heidelberg (2009). https://doi.org/10.1007/978-3-642-00768-2_2
20. Fogarty, S., Vardi, M.Y.: Efficient Büchi universality checking. In: Esparza, J., Majumdar, R. (eds.) TACAS 2010. LNCS, vol. 6015, pp. 205–220. Springer, Heidelberg (2010). https://doi.org/10.1007/978-3-642-12002-2_17
21. Piterman, N.: From nondeterministic Büchi and Streett automata to deterministic parity automata. In: Proceedings of LICS 2006, pp. 255–264. IEEE (2006)
22. Kähler, D., Wilke, T.: Complementation, disambiguation, and determinization of Büchi automata unified. In: Aceto, L., Damgård, I., Goldberg, L.A., Halldórsson, M.M., Ingólfsdóttir, A., Walukiewicz, I. (eds.) ICALP 2008. LNCS, vol. 5125, pp. 724–735. Springer, Heidelberg (2008). https://doi.org/10.1007/978-3-540-70575-8_59
23. Friedgut, E., Kupferman, O., Vardi, M.: Büchi complementation made tighter. Int. J. Found. Comput. Sci. **17**, 851–868 (2006)
24. Vardi, M.Y.: The Büchi complementation saga. In: Thomas, W., Weil, P. (eds.) STACS 2007. LNCS, vol. 4393, pp. 12–22. Springer, Heidelberg (2007). https://doi.org/10.1007/978-3-540-70918-3_2
25. Breuers, S., Löding, C., Olschewski, J.: Improved Ramsey-based Büchi complementation. In: Birkedal, L. (ed.) FoSSaCS 2012. LNCS, vol. 7213, pp. 150–164. Springer, Heidelberg (2012). https://doi.org/10.1007/978-3-642-28729-9_10

26. Li, Y., Vardi, M.Y., Zhang, L.: On the power of unambiguity in Büchi complementation. In Raskin, J.F., Bresolin, D. (eds.) Proceedings 11th International Symposium on Games, Automata, Logics, and Formal Verification, Brussels, Belgium, September 21–22, 2020. Electronic Proceedings in Theoretical Computer Science, vol. 326, pp. 182–198. Open Publishing Association (2020)
27. Kurshan, R.P.: Complementing deterministic Büchi automata in polynomial time. J. Comput. Syst. Sci. **35**(1), 59–71 (1987)
28. Havlena, V., Lengál, O.: Reducing (To) the ranks: efficient rank-based Büchi automata complementation. In: Haddad, S., Varacca, D. (eds.) 32nd International Conference on Concurrency Theory (CONCUR 2021). (LIPIcs). vol. 203, pp. 2:1–2:19. Schloss Dagstuhl – Leibniz-Zentrum für Informatik, Germany (2021). https://drops.dagstuhl.de/opus/volltexte/2021/14379. https://doi.org/10.4230/LIPIcs.CONCUR.2021.2
29. Karmarkar, H., Chakraborty, S.: On minimal odd rankings for Büchi complementation. In: Liu, Z., Ravn, A.P. (eds.) ATVA 2009. LNCS, vol. 5799, pp. 228–243. Springer, Heidelberg (2009). https://doi.org/10.1007/978-3-642-04761-9_18
30. Abdulla, P.A., et al.: Advanced Ramsey-based Büchi automata inclusion testing. In: Katoen, J.-P., König, B. (eds.) CONCUR 2011. LNCS, vol. 6901, pp. 187–202. Springer, Heidelberg (2011). https://doi.org/10.1007/978-3-642-23217-6_13
31. Etessami, K.: A hierarchy of polynomial-time computable simulations for automata. In: Brim, L., Křetínský, M., Kučera, A., Jančar, P. (eds.) CONCUR 2002. LNCS, vol. 2421, pp. 131–144. Springer, Heidelberg (2002). https://doi.org/10.1007/3-540-45694-5_10
32. Gurumurthy, S., Bloem, R., Somenzi, F.: Fair simulation minimization. In: Brinksma, E., Larsen, K.G. (eds.) CAV 2002. LNCS, vol. 2404, pp. 610–623. Springer, Heidelberg (2002). https://doi.org/10.1007/3-540-45657-0_51
33. Bustan, D., Grumberg, O.: Simulation based minimization. ACM Trans. Comput. Log. **4**, 181–206 (2000)
34. Barth, S.: Deciding monadic second order logic over ω-words by specialized finite automata. In: Ábrahám, E., Huisman, M. (eds.) IFM 2016. LNCS, vol. 9681, pp. 245–259. Springer, Cham (2016). https://doi.org/10.1007/978-3-319-33693-0_16
35. Duret-Lutz, A., Lewkowicz, A., Fauchille, A., Michaud, T., Renault, É., Xu, L.: Spot 2.0—a framework for LTL and ω-automata manipulation. In: Artho, C., Legay, A., Peled, D. (eds.) ATVA 2016. LNCS, vol. 9938, pp. 122–129. Springer, Cham (2016). https://doi.org/10.1007/978-3-319-46520-3_8
36. Klarlund, N., Møller, A., Schwartzbach, M.I.: MONA implementation secrets. Int. J. Found. Comput. Sci. **13**(4), 571–586 (2002)

BAM: Efficient Model Checking for Barriers

Michalis Kokologiannakis$^{(\boxtimes)}$ and Viktor Vafeiadis

MPI-SWS, Kaiserslautern, Germany
{michalis,viktor}@mpi-sws.org

Abstract. Stateless Model Checking (SMC) and Dynamic Partial Order Reduction (DPOR) are prominent techniques that are often used together to verify safety properties of concurrent programs under a variety of different memory models. Although existing SMC/DPOR implementations excel at verifying parallel algorithms, they scale extremely poorly once *barriers* are used to synchronize the participating threads.

In response, we develop BAM (Barrier-Aware Model-checker), a DPOR extension that explores exponentially fewer executions for programs that employ synchronization schemes involving barriers. We have implemented BAM in a verification tool for C programs, and show that it greatly outperforms the state-of-the-art for programs with barriers.

1 Introduction

Barriers (as in e.g., pthread_barrier [24]) are synchronization primitives used to ensure that the execution of a program will continue only after all threads have reached a certain point ("a barrier"). Their usage is best understood with an example:

$$
\begin{array}{c}
\texttt{barrier\_init}(b, N); \\
\begin{array}{c|c|c}
\begin{array}{l} m[1] := ...; \\ \texttt{barrier\_wait}(b); \\ n[1] := ...; \end{array} & ... & \begin{array}{l} m[N] := ...; \\ \texttt{barrier\_wait}(b); \\ n[N] := ...; \end{array}
\end{array}
\end{array}
\qquad (\textsc{Barrier-}N\textsc{-Sync})
$$

In this program, the main thread first initializes a barrier object to N, indicating that N threads will meet together ("rendezvous") at the barrier. Each thread calculates a part of the array m, and waits for all the other threads using a barrier_wait call: no thread gets past barrier_wait until all threads have executed their respective barrier_wait call. After all threads have met at the barrier, each thread continues and calculates a part of the array n, which (potentially) uses the array m that was calculated in the previous step. Such iterative parallel computations are common in scientific applications, e.g., simulations.

More generally, barriers are useful when we want to wait for the threads to perform some calculations before continuing. Upon continuation, all calculations performed by one thread will be visible to all other threads. In contrast to joining

© Springer Nature Switzerland AG 2021
K. Echihabi and R. Meyer (Eds.): NETYS 2021, LNCS 12754, pp. 223–239, 2021.
https://doi.org/10.1007/978-3-030-91014-3_16

the threads, using barriers does not cause the threads to be terminated, but rather blocked; this can be crucial for performance reasons.

But while the usage of barriers is straightforward, verifying programs with barriers is not always so. Suppose that we want to verify the BARRIER-N-SYNC program from above automatically, and that we want to use *Stateless Model Checking* (SMC) [12,21] coupled with *Dynamic Partial Order Reduction* (DPOR) [1,11] to do so. This combination has been proven to scale very well for parallel programs [13,17,22], and also takes into account the effects of the underlying weak memory model [2–4,14,16].

Alas, all existing SMC/DPOR techniques explore an exponential number of executions for this program, as they examine all possible orderings in which different threads arrive at the barrier (see Sect. 2). Even worse, they do so even though the order in which the threads rendezvous is irrelevant. In fact, the order in which threads reach the barrier is not even observable by the user program; the only thing that *is* observable according to the pthread_barrier documentation [24], is whether a thread was the last one to reach the barrier. However, for the programs we are aware of, even that condition is never used.

Leveraging this insight, we develop BAM (Barrier-Aware Model-checker), a memory-model-agnostic DPOR extension that reconciles SMC/DPOR with barriers. By avoiding the exploration of executions that only differ in the order in which threads execute barrier_wait, BAM explores exponentially fewer executions than state-of-the-art SMC/DPOR tools. Concretely, we make the following contributions:

In Sect. 3, we introduce BAM, an SMC/DPOR extension that does not order calls to barrier_wait, and yet models barrier semantics correctly: all instructions executed after a rendezvous at a barrier will see the effects of all instructions executed before the rendezvous.

In Sect. 4, we implement BAM as an extension of the state-of-the-art GENMC model checker [16], and show that BAM is exponentially faster than vanilla GENMC in programs with barriers.

We start with an overview of how barriers are handled by the state-of-the-art stateless model checkers. To simplify the presentation, we assume a model of sequential consistency (SC) [20]. Our results carry over to all other axiomatic memory models.

2   State-of-the-Art

Why is it that SMC/DPOR experiences an exponential slowdown in programs with barriers? To answer this question, we first have to review the fundamentals of SMC/DPOR.

2.1   SMC and DPOR

SMC verifies a program by checking all of its thread interleavings. For example, for the W+R+W program below, an SMC algorithm would enumerate all 6 interleavings of the program, and validate that all of them satisfy the desired properties.

$$x := 1 \ \| \ r := x \ \| \ y := 1 \qquad\qquad (\text{W+R+W})$$

Of course, enumerating interleavings does not scale as programs become larger. Hence, SMC is usually coupled with *Dynamic Partial Order Reduction* (DPOR) [1,11,16], which avoids exploring an interleaving if an equivalent one has already been explored. DPOR considers two interleavings equivalent if one can be obtained from the other by swapping adjacent, non-conflicting instructions. While many notions of conflict have been proposed in the literature [1,8–10,16], the simplest one considers two instructions as conflicting if they access the same memory location, and at least one of them is a write. For W+R+W, the only conflicting instructions are $x := 1$ and $r := x$. Thus, a DPOR algorithm would verify the program by exploring only 2 interleavings: one where $x := 1$ is executed before $r := x$, and one where the order is reversed.

SMC/DPOR provides an excellent solution for verifying concurrent programs as it does not explicitly store the states of the program that have already been visited, and its notion of conflict has been extended to weak memory models [2–4,14,16,25]. In particular, SMC/DPOR scales very well for programs with few conflicts, such as parallel algorithms. We will not go into details of how SMC/D-POR works, as SMC/DPOR has been thoroughly studied in the literature, and the exact details are not important for this paper. Instead, we only provide a high-level overview of DPOR later on (see Sect. 3.4), and refer interested readers to Kokologiannakis et al. [16].

2.2 Barriers in SMC/DPOR

The reason why barriers and SMC/DPOR do not work well together is that barriers *inhibit* DPOR. Existing DPOR algorithms consider barrier_wait calls conflicting, and thus explore an exponential number of interleavings, even for a barrier program doing the bare minimum:

$$\begin{array}{c}\texttt{barrier\_init}(b, N); \\ \texttt{barrier\_wait}(b); \ \| \ \dots \ \| \ \texttt{barrier\_wait}(b);\end{array} \qquad (\text{Barrier-}N)$$

For Barrier-N, an SMC/DPOR algorithm would explore $N!$ executions, effectively rendering DPOR a useless addition to SMC.

To understand why barriers are considered conflicting operations by DPOR, however, we have to examine how barriers are implemented. Typically, barriers are implemented using condition variables or futexes: a thread executing barrier_wait acquires a lock, manipulates a variable indicating the number of threads that have reached the barrier, and then waits on a futex/condition variable. Such implementations, however, while standard for barrier libraries, are suboptimal for model checking: each barrier_wait call would boil down to many different instructions, thus unnecessarily increasing the number of different events a model checker would have to generate.

```
barrier_init(b, N) :
  b := N;

barrier_wait(b) :
  atomic { if (b = 1) b := N else b := b−1; }; assume(b = N);
```

Fig. 1. Implementation of barrier_init and barrier_wait.

Since we are only interested in verifying programs that *use* barriers, we can get away with a much more abstract barrier implementation, such as the one in Fig. 1. We model each barrier_init(b, N) as a plain write that initializes a shared variable b to N, and each barrier_wait(b) as an atomic read-modify-write (RMW) instruction followed by an assume instruction. For the barrier_wait call, the RMW instruction decrements b each time it is called, apart from when the value read is 1, at which point it resets is back to N (so that the barrier can be subsequently reused). For the same call, the assume reads b and blocks the calling thread if the value read was different than N.

Given this implementation, it becomes clear that programs like BARRIER-N lead to an exponential blowup in the state space. Since the RMW instructions all write to the same location (b), they are considered conflicting, and so the model checker will examine all their $N!$ possible orderings. In addition to these $N!$ executions, some state-of-the-art DPOR implementations, such as GENMC [16], may also consider an exponential number of blocked executions (see Sect. 4).

3   BAM: Barrier Model Checking

We now present BAM and explain how it improves over baseline DPOR for programs with barriers. After presenting the key idea behind BAM (Sect. 3.1), we provide a formal framework in which the executions of a program can be modeled, and show how BAM's modeling of barriers leads to exponential savings when verifying programs with barriers, while at the same time maintaining the guarantees that barriers provide (Sect. 3.3).

3.1   Key Idea

We note that, although the barrier implementation effectively records the order in which different thread call barrier_wait by counting the number of threads that have joined the barrier, programs that use barriers do not care about this order. In fact, even though barrier implementations typically provide a distinct value returned by the barrier_wait call that resets the barrier to its initial value, the user programs we are aware of do not make use of that.

We further observe that programs using barriers typically initialize the barrier to the number of threads in the system, and so there is never a case with more parallel calls to barrier_wait than the barrier's initial value. Intuitively, this is because the standard scenario for barrier synchronization is to arrange a rendezvous between all threads participating in a parallel computation. With that in mind, it does not really make sense to initialize a barrier with a value smaller than the number of threads calling `barrier_wait`, as that would imply that only some threads will be unblocked after reaching the barrier, while the others will remain blocked.

The key insight behind BAM is that, for programs satisfying the two conditions described above, tracking the order between `barrier_wait` calls is unnecessary. BAM models `barrier_wait` calls as *dummy events* that are not considered conflicting, thus enabling the underlying DPOR algorithm to consider fewer executions. More specifically, when a thread executes `barrier_wait` it simply checks how many threads have reached the barrier: if not all threads have arrived, the thread blocks; otherwise all program threads unblock and continue their execution. Notice that, when all threads unblock, all the instructions before the respective `barrier_wait` statements will have been executed, thereby satisfying the fundamental guarantee provided by barriers i.e., instructions executed after the threads have rendezvoused will see the effects of the instructions executed before the rendezvous.

Let us now make the above idea formal in the framework of axiomatic memory models.

3.2   Execution Graphs

Although the executions of a concurrent program under SC are usually thought of as interleavings, we model them using *execution graphs* [7]. Execution graphs allow for a flexible formalization that can easily be extended to weak memory models, but also, as we will shortly see, abstract away the notion of a "conflict" used by the DPOR algorithm.

Execution graphs have two basic components:

(i) a set of events (nodes), modeling the memory accesses performed by the program, and
(ii) some relations on these events (edges).

Standard relations included in all memory models are the *program order* (po) and *reads-from* (rf) relations: po relates events in the same thread according to their serial execution order, while rf relates reads to writes they are reading from. In this paper, we also assume the existence of a *happens-before* (hb) relation, a strict partial order that includes po, and which models ordering due to synchronization between events.

Let us now formally describe events and execution graphs.

Definition 1. *An event, $e \in$ Event, is either an initialization event $\langle \text{init } l \rangle \in$ $\text{Event}_0 \subseteq$ Event for a location $l \in$ Loc or a thread event $\langle t, i, lab \rangle$ where $t \in$ Tid is a thread identifier, $i \in \text{Idx} \triangleq \mathbb{N}$ is a serial number inside each thread, and $lab \in$ Lab is a label that takes one of the following forms:*

- *Read label: $R(l, v)$ where $l \in$ Loc is the location accessed, and $v \in \text{Val} \triangleq \mathbb{Z}$ is the value read.*

- *Write label:* $W(l, v)$ *where* $l \in$ Loc *is the location accessed, and* $v \in$ Val *is the value written.*
- *Read-modify-write label:* $RMW(l, v_1, v_2)$ *where* $l \in$ Loc *is the location accessed,* v_1 *is the value read, and* $v_2 \in$ Val *is the value written. This label models a single atomic RMW operation.*
- *Error label:* error, *denoting a safety violation.*

The functions tid, idx, loc, valr, valw *return (when applicable) the thread identifier, serial number, location, read-value and written-value of an event, respectively.*

Given the above representation of events, we induce the *program order*, which is a strict partial order on events given by:

$$po \triangleq \text{Event}_0 \times (\text{Event} \setminus \text{Event}_0) \cup \left\{ \langle\langle t_1, i_1\rangle, \langle t_2, i_2\rangle\rangle \mid t_1 = t_2 \land i_1 < i_2 \right\}$$

Intuitively, initialization events precede all non-initialization events, while events in the same thread are ordered according to their serial numbers.

Definition 2. *An* execution graph G *consists of:*

1. *a set* $G.E$ *of events that includes initialization events for all locations accessed by the program, and*
2. *a relation* $G.\text{rf} \subseteq G.E \times G.E$, *called the* reads-from *relation, that relates each write event to the same-location reads that read from it.*

We write $G.R, G.W$ *to denote the set of events of the respective type (RMW events belong both to* $G.R$ *and* $G.W$*), and use subscripts to further restrict these sets (e.g.,* $G.W_x = \{w \in G.W \mid \text{loc}(w) = x\}$*).*

Definition 3 (Well-formedness). *An execution graph* G *is* well-formed *if the following hold for* $G.\text{rf}$*:*

1. rf *only relates writes and reads with matching locations and values, i.e., for every* $\langle w, r\rangle \in G.\text{rf}$ *it is* $w \in G.W$, $r \in G.R$, $\text{loc}(w) = \text{loc}(r)$ *and* $\text{valw}(w) = \text{valr}(r)$,
2. rf *is functional on its range, i.e., if* $\langle w_1, r\rangle, \langle w_2, r\rangle \in G.\text{rf}$ *it is* $w_1 = w_2$, *and*
3. *each read reads a value, i.e.,* $\forall r \in G.R. \exists w. \langle w, r\rangle \in G.\text{rf}$.

The semantics of a program P is simply given by the set of well-formed execution graphs that satisfy a consistency predicate dictated by the memory-model. For instance, *sequential consistency* (SC) [20] can be defined using a coherence order as follows.

Definition 4 (Coherence order). *A relation* co *is a* coherence order *for an execution* G *iff* co *is a strict partial order,* $co \subseteq \bigcup_{l \in \text{Loc}} G.W_l \times G.W_l$, *and for every location* $l \in$ Loc, co *is total on* $G.W_l$.

Definition 5 (SC). G *is* sequentially consistent, *written* $\text{cons}_{SC}(G)$, *iff there is a coherence order* co *for* G *such that* $hb \triangleq po \cup \text{rf} \cup co \cup \text{fr}$ *is acyclic, where* $\text{fr} \triangleq \{(a, b) \mid a \neq b \land \exists c. (c, a) \in \text{rf} \land (c, b) \in co\}$ *is the* from-reads *relation.*

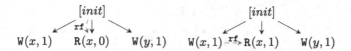

Fig. 2. Execution graphs of W+R+W under SC.

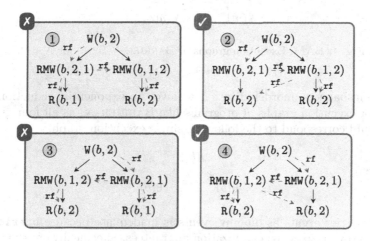

Fig. 3. Execution graphs of BARRIER-N for $N = 2$.

As an example, Fig. 2 shows the two sequentially consistent execution graphs of W+R+W. Notice that each of these graphs corresponds to multiple interleavings. In effect, the graphs subsume the notion of a conflict used by DPOR algorithms; each linearization of hb in these graphs yields a possible interleaving. Thus, an SMC/DPOR algorithm can alternatively be seen as a procedure that verifies a program by enumerating its execution graphs.

As a further example, Fig. 3 shows the sequentially consistent executions of BARRIER-N for $N = 2$ with the conventional modeling of barriers shown in Fig. 1. The two execution graphs on the left are blocked because one `assume` condition is violated. By contrast, the two graphs on the right satisfy the `assume` conditions and are thus non-blocked. SMC/DPOR algorithms will thus have to generate at least the two non-blocked executions, though actual implementations typically generate all four (blocked and non-blocked) executions.

3.3 BAM: Keeping Barriers Unordered

To model barriers, we extend the definition of events (Definition 1) to allow for a new kind of label modeling calls to the `barrier_wait` operation:

- Barrier-wait label: $B(l)$ where $l \in \mathsf{Loc}$ is the barrier location accessed.

We write $G.B$ for all the barrier events of an execution graph G. Barrier events do not participate in the rf relation of execution graphs.

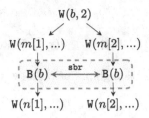

Fig. 4. BAM: Execution graphs of BARRIER-N-SYNC for $N = 2$.

Keeping barriers unordered by rf achieves an exponential reduction in the number of execution graphs of programs like BARRIER-N, as all four graphs of Fig. 3 would correspond to the following single execution graph.

$$W(b, 2)$$
$$B(b) \qquad B(b)$$

Treating barrier events as dummy events is inadequate because barrier_wait calls also provide some synchronization guarantees. Specifically, every event po-before a barrier call is guaranteed to happen before every event po-after a barrier call in the same rendezvous. Recall the BARRIER-N-SYNC program from Sect. 1:

$$
\begin{array}{l}
m[1] := \dots; \\
\texttt{barrier\_wait}(b); \\
n[1] := \dots;
\end{array}
\;\Big\|\; \dots \;\Big\|\;
\begin{array}{l}
m[N] := \dots; \\
\texttt{barrier\_wait}(b); \\
n[N] := \dots;
\end{array}
\qquad (\text{BARRIER-}N\text{-SYNC})
$$

Here, merely treating B events as dummy events is unsound. As B events do not contribute to hb between different threads, each thread will only see its own calculation of a single part of m. By contrast, had we used the conventional barrier representation, the rf edges across threads would ensure that the calculation of m is visible when n is calculated.

To solve this problem, we extend the definition of execution graphs (Definition 2) with a new component:

– a partial equivalence relation $G.\texttt{sbr}$, called *same-barrier-round*, that relates barrier events that synchronize with each other in a rendezvous. Events related by $G.\texttt{sbr}$ act on the same (barrier) location.

We will use the sbr relation to enforce synchronization between the events executed before the threads meet at the barrier, and the events executed after the rendezvous at the barrier. But before presenting how barrier synchronization works, we assume two basic conditions about the sbr relation.

Given a graph G and a barrier location b initialized with value N (i.e., there is a unique write $w \in G.\text{E}$ such that $\texttt{lab}(w) = W(b, N)$, and that $\langle w, n \rangle \in G.\text{hb}$, for all $n \in G.B_b$), we further require that $G.\text{sbr}$ satisfy the following conditions:

$$|G.B_b \setminus dom(G.\text{sbr})| < N \qquad\qquad \text{(SBR-MUST-MEET)}$$

$$\forall e \in G.B_b.\, |\text{succ}_{G.\text{sbr}}(e)| = N \vee \text{succ}_{G.\text{sbr}}(e) = \text{succ}_{G.\text{po}}(e) = \emptyset \quad \text{(SBR-BLOCK)}$$

where $\text{succ}_r(e)$ denotes the set $\{e' \mid \langle e, e'\rangle \in r\}$, i.e., set of successors of e in r.

The SBR-MUST-MEET condition captures the basic guarantee provided by the barrier implementation that once N `barrier_wait` calls are issued, then they will meet in a rendezvous round. A consistent graph can therefore contain at most $N - 1$ barrier calls that do not belong to any barrier round.

The purpose of the SBR-BLOCK condition is twofold. First, it dictates that exactly N calls to `barrier_wait` participate in the same barrier round. That is, each event e either belongs in the same round with N events or does not have any events in the same round. Second, it dictates that no thread is allowed past a `barrier_wait` call before all threads rendezvous at the barrier. In other words, if an event does not participate in a (full) barrier round, it is blocked and has no po-successors in the graph. This condition renders graphs like the one below for BARRIER-N-SYNC and $N = 2$ invalid:

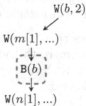

As soon as all threads reach the barrier, all corresponding barrier events become part of `sbr`, and events past the barrier may be added.

We next discuss how barrier synchronization contributes to the happens-before (hb) relation. We extend the (model-specific) definition of hb with sbr; po and po; sbr. That is, a barrier happens before the po-successors of any barriers it synchronizes with and after their po-predecessors. Since hb is transitive, this means that all events that are po-before a given barrier round happen before all events that are po-after the same barrier round. For example, for the BARRIER-N-SYNC program (cf. Fig. 4), all events po-after the highlighted barrier round will also be hb-after the events that are po-before the highlighted barrier round.

Synchronization ensures that the `barrier_wait` events related by sbr belong to the same barrier round. To see how this is achieved, consider the program below where two threads rendezvous at a barrier twice:

$$\text{barrier\_init}(b, N);$$
$$\begin{array}{l|l} \text{barrier\_wait}(b); & \text{barrier\_wait}(b); \\ \text{barrier\_wait}(b); & \text{barrier\_wait}(b); \end{array} \qquad \text{(BARRIER2-}N)$$

For this example, graphs like the one below, where sbr includes `barrier_wait` events from different rounds of the same barrier acquisition, are invalid:

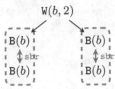

The reason why this graph is invalid, is that $G.\mathtt{sbr}; G.\mathtt{po}$ is included in $G.\mathtt{hb}$. This condition implies that, e.g., the second barrier event of the first thread is $\mathtt{hb}$-before itself (since we can take an $\mathtt{sbr};\mathtt{po}$ step), which contradicts the fact that $\mathtt{hb}$ is a strict partial order.

Finally, let us end this section by formalizing the conditions under which BAM can be used (see Sect. 3.1). These are expressed by the notion of barrier well-formedness, as described below.

Definition 6 (Barrier Well-formedness). *An execution graph G is barrier-well-formed on a barrier location b if $G.\mathtt{B}_b = \emptyset$ or if the following hold.*

1. *There is a unique write event $w_0 \in G.\mathtt{E} \setminus \mathtt{Event}_0$ with $\mathtt{loc}(w_0) = b$.*
2. *w_0 is a plain write event: $\mathtt{lab}(w_0) = \mathtt{W}(b, N)$ for some $N \in \mathbb{N}$.*
3. *w_0 is $\mathtt{hb}$-before all $\mathtt{B}_b$ events: $\langle w_0, e \rangle \in G.\mathtt{hb}$ for all $e \in G.\mathtt{B}_b$.*
4. *For all $S \subseteq G.\mathtt{B}_b$ with $|S| > \mathtt{valw}(w_0)$, there exist $e, e' \in S$ s.t. $\langle e, e' \rangle \in G.\mathtt{hb}$.*

Barrier well-formedness ensures that there is a unique initializing write for each barrier location, and that no more threads than the barrier's initializing value call $\mathtt{barrier\_wait}$ concurrently. Note that the latter precludes the usage of BAM in programs like the following:

$$\mathtt{barrier\_init}(b, 2);$$
$$\mathtt{barrier\_wait}(b); \; \| \; \mathtt{barrier\_wait}(b); \; \| \; \mathtt{barrier\_wait}(b);$$

That said, as already mentioned, we do not expect such programs to show up often in practice, as they are built on the (not very useful) premise that some subset of the threads meeting at the barrier will continue past the barriers, while the rest will remain blocked.

3.4 BAM: Extending DPOR for Barriers

We now explain how DPOR can be extended to accommodate for BAM.

Algorithm 1 shows the general structure of a DPOR algorithm with BAM's extensions highlighted. VERIFY verifies a program P by enumerating its execution graphs, and ensuring that none of them contains an **error** label. VERIFY achieves this by repeatedly calling VISITONE (Line 4): the latter will explore one full execution of P, and at the same time populate an *environment* Γ (initially empty; cf. Line 2) with alternative exploration options. These exploration options will be subsequently explored by VERIFY (Line 5).

VISITONE is the workhorse of the DPOR algorithm. At each step, as long as G remains consistent according to the memory model (Line 7), VISITONE uses $\mathtt{next}_P(G)$ to extend the current graph G by an event a from a non-blocked

Algorithm 1. Dynamic Partial Order Reduction

```
1: procedure VERIFY(P)
2:     ⟨G, Γ⟩ ← ⟨G₀, ∅, Γ₀⟩
3:     do
4:         VISITONE(P, G, Γ)
5:     while ⟨G, Γ⟩ ← pop(Γ)

6: procedure VISITONE(P, G, Γ)
7:     while consₘ(G) ∧ a ← nextₚ(G) do
8:         G.E ← G.E ∪ {a}
9:         if a ∈ error then exit("error")
10:        if a ∈ G.R then CALCRFS(G, Γ, a)
11:        if a ∈ G.W then CALCREVISITS(G, Γ, a)
12:        if a ∈ G.B then
13:            N ← valw(w) where w ∈ G.Wₗₒ꜀₍ₐ₎
14:            S ← G.Bₗₒ꜀₍ₐ₎ \ dom(G.sbr)
15:            if |S| = N then G.sbr ← G.sbr ∪ {⟨e, e'⟩ | e ∈ S, e' ∈ S}
```

thread. A thread is considered blocked if it contains a barrier event that is not in the domain of $G.\mathtt{sbr}$. (By construction, such events are po-maximal.) When there are no more events to add, then G is complete, and VISITONE returns.

Depending on the type of an added event a, VISITONE takes appropriate action. Specifically, if a denotes an error (e.g., an assertion violation), it is reported to the user and the verification terminates (Line 9). If a is a read, then we need to find an appropriate $\mathtt{rf}$ edge for it from G. To that end, VISITONE calls CALCRFS (Line 10), which will calculate possible $\mathtt{rf}$ options for a, set one, and push the rest to Γ. If a is a write, it needs to revisit existing reads of the same location in G, because a was not present in the graph when VISITONE was considering possible reads-from options for these reads. To that end, VISITONE calls CALCREVISITS (Line 11), which extends Γ with such alternative explorations.

If a is a barrier-wait event, BAM-specific code takes over. First, BAM finds this barrier's initializing value N (Line 13). Well-formed programs contain a unique initialization of barrier, and so their execution graphs have a unique write event w to each barrier location. Then, BAM collects in the set S all barrier events to the same location as a that are not related by $G.\mathtt{sbr}$ (Line 14. This set contains a as well as all blocked events to the same location. If the number of such events is N, then they form a rendezvous and are thus added to $G.\mathtt{sbr}$, which has the effect of unblocking the waiting threads (Line 15).

As can be seen, BAM can be seamlessly integrated into existing DPOR algorithms. The additional work performed—a linear scan over the graph—does not incur any overhead as it is dominated by the DPOR's consistency checks.

4   Evaluation

4.1   Implementation

We have implemented BAM as an extension of the state-of-the-art stateless model checker GENMC [16]. GENMC operates at the level of LLVM-IR, and can verify C/C++ programs under different (weak) memory models such as RC11 [19] and IMM [23]. We have made our implementation publicly available at https://github.com/MPI-SWS/genmc.

4.2   Experiments

In what follows we compare BAM against the baseline GENMC implementation. We do not directly compare BAM against other tools as 1) most other tools do not offer built-in support for barriers and would thus yield similar results to the baseline GENMC encoding, and 2) GENMC has been extensively compared with other model checking tools in the past (e.g., [16,18]).

Instead, we set out to show that BAM yields exponential benefits compared to the baseline GENMC implementation for programs with barriers, while at the same time imposes zero overhead.

Experimental Setup. We conducted all experiments on a Dell PowerEdge M620 blade system, with two Intel Xeon E5-2667 v2 CPUs (8 cores @ 3.3 GHz) and 256 GB of RAM, running a custom Debian-based distribution. We used LLVM 7 for GENMC (v0.5.3). All reported times are in seconds, unless explicitly noted otherwise. We set the timeout limit to 30 min.

Benchmarks. We evaluate the effectiveness of BAM using a variety of synthetic benchmarks, ranging from simple benchmarks containing a single rendezvous round with no additional computation to benchmarks that involve multiple rendezvous rounds. The results are reported in Tables 1 and 2. As expected, BAM achieves exponential gains over GENMC for all these benchmarks, and scales very well to larger programs. By contrast, the baseline GENMC implementation frequently times out, especially on benchmarks with multiple rendezvous rounds.

Let us first focus on Table 1. Starting with `barrier`, we see that GENMC explores exponentially more executions than BAM, most of which correspond to blocked executions. Indeed, as explained in Sect. 2.2, since the `barrier_wait` operations are considered conflicting, GENMC explores an exponential number of executions for this benchmark. In fact, GENMC explores $(N!)^2$ executions for `barrier(N)`, of which $(N!)^2 - N!$ are blocked.

These numbers might come off as a surprise at first, since it would suffice for GENMC to explore precisely $(N!)$ executions, and no blocked executions. The discrepancy is due to the modeling of `barrier_wait` calls. As described in Sect. 2.2 and Fig. 1, each `barrier_wait` comprises an RMW operation, but also an `assume(b == N)` statement that re-reads the value of the barrier, and ensures that the value read is N. This second read, however, has another $N!$ consistent

Table 1. Synthetic benchmarks containing only barrier operations

	Executions		Blocked		Time	
	GENMC	BAM	GENMC	BAM	GENMC	BAM
barrier(4)	24	1	552	0	0.02	0.01
barrier(5)	120	1	14280	0	0.21	0.01
barrier(6)	720	1	517680	0	7.03	0.01
barrier2(4)	576	1	36816	0	0.74	0.01
barrier2(5)	14400	1	5156880	0	114.63	0.01
barrier2(6)	⏲	1	⏲	0	⏲	0.01
barrier3(4)	13824	1	907152	0	26.07	0.01
barrier3(5)	⏲	1	⏲	0	⏲	0.01
barrier3(6)	⏲	1	⏲	0	⏲	0.01

barrier(N): N threads rendezvous at a barrier.
barrier2(N): N threads rendezvous twice at a barrier.
barrier3(N): N threads rendezvous thrice at a barrier

rf options, which GENMC subsequently has to explore. And at this point, one may wonder: isn't it possible to pack the assume statement into the atomic block, and use the value already read for b for the assume? Unfortunately, the answer is no. Although we will not go into further details here, we mention in passing that the second read statement is necessary under weak memory models to ensure synchronization between the events before and after the barrier rendezvous.

The differences between GENMC and BAM are magnified once we consider benchmarks with multiple rendezvous rounds. Starting with 4 threads, GENMC explores 5 orders of magnitude more executions than BAM for barrier2, and 6 orders of magnitude more for barrier3. As the number of threads increases, the performance gap between GENMC and BAM increases even more, despite the fact that most of the executions that GENMC explores are blocked; as it turns out, the cost of enumerating blocked executions quickly becomes exorbitant.

We move on to Table 2, which contains some typical use cases of barriers. The observations here are similar to the ones made for Table 1. The simplest case is that of barrier-det that includes a single rendezvous round and only local computations. GENMC scales similarly to the barrier benchmark, but takes much more time because of the higher cost per execution. By contrast, the number of threads has a negligible effect to BAM's execution time.

The other three benchmarks use multiple rendezvous rounds to synchronize some computations, while still maintaining a high cost per execution. As expected, this makes GENMC quickly time out. In addition, observe that in the case of barrier-lock and barrier-count barriers are used to synchronize computations that have additional sources for an exponential number of executions. As the state space of these benchmarks is large to begin with (even disregarding barriers), GENMC quickly exceeds the time limit, while BAM is able to scale to a larger number of threads. We note that the blocked executions that BAM

Table 2. Benchmarks with realistic barrier use cases

	Executions		Blocked		Time	
	GENMC	BAM	GENMC	BAM	GENMC	BAM
barrier-det(3)	6	1	30	0	107.34	17.87
barrier-det(4)	24	1	552	0	424.04	17.87
barrier-det(5)	⊕	1	⊕	0	⊕	17.89
barrier-transc(3)	46656	1	671790	0	18min	0.02
barrier-transc(4)	⊕	1	⊕	0	⊕	0.02
barrier-transc(5)	⊕	1	⊕	0	⊕	0.02
barrier-lock(3)	1296	36	7140	105	0.70	0.03
barrier-lock(4)	331776	576	4340784	3100	417.58	0.42
barrier-lock(5)	⊕	14400	⊕	143385	⊕	18.99
barrier-count(3)	55296	64	715878	0	88.33	0.04
barrier-count(4)	⊕	4992	⊕	0	⊕	2.57
barrier-count(5)	⊕	2276352	⊕	0	⊕	28

`barrier-det(N)`: Given a matrix M, calculates the determinant of M^4. The calculation of M^4 is split among N threads, which rendezvous after calculating M^2.

`barrier-transc(N)`: N threads calculate the transitive closure of a matrix via a fixpoint. They rendezvous twice per fixpoint iteration.

`barrier-lock(N)`: N threads test a simple lock implementation: after they rendezvous at a barrier, the threads concurrently attempt to enter their critical section, and mutual exclusion is checked.

`barrier-count(N)`: Contains N threads, with each thread i waiting at barriers b_k, where $i \leq k \leq N$. Counts the number of threads getting through at each round

explores in `barrier-lock` are not due to barriers, but rather due to spinloops that can block in the lock implementation under test.

We end this section with a remark on scalability. While it can be argued that scaling up to a large number of threads is unimportant (since e.g., these benchmarks are symmetric), this is not always the case. Often, concurrent implementations tune their behavior depending on the number of threads spawned, and concurrency bugs cannot be manifested with a few threads. Being able to verify programs that employ a large number of threads can therefore be crucial.

5   Summary and Related Work

We presented BAM, a DPOR extension that explores exponentially fewer executions than state-of-the-art stateless model checkers for programs that use synchronization barriers. BAM is based on the key insight that, for most programs, the order in which different threads rendezvous at the barrier is irrelevant, and thus `barrier_wait` statements can be seen as non-conflicting operations by the underlying DPOR algorithms.

After the inception of SMC with tools like VERISOFT [12] and CHESS [21], a growing number of different DPOR techniques has been proposed [1–6,8–11,14–16,18]. Some of these extend DPOR to weak memory models (e.g., [2,3]), others achieve a coarser equivalence partitioning (e.g., [6,8,15]), while others do both (e.g., [4,16]).

While we are not aware of any other technique that extends DPOR for programs that use barriers, the two works that are closer to ours are CDPOR [6] and LAPOR [15], as they both extend DPOR to scale for particular classes of programs. CDPOR exploits conditional independence between atomic blocks: if the execution of two concurrent atomic blocks leads to the same state under some conditions $\bar{C}$, then the two blocks are deemed independent whenever $\bar{C}$ holds. Thus, if each `barrier_wait` is modeled as an atomic block, CDPOR would be able to explore only 1 execution in programs like BARRIER-N, assuming that the atomic blocks are proven (unconditionally) independent. Proving independence for CDPOR, however, is done using an SMT solver, which might not always be able to prove independence. Alternatively, such conditions would have to be provided manually by the user. LAPOR exploits a similar key idea to BAM and avoids exploring executions that only differ in the order that two critical sections were executed, assuming that these critical sections do not have any conflicting events.

Acknowledgements. We thank the anonymous reviewers for their feedback. We are grateful to Xiaowei Ren for his initial implementation of BAM. This work was supported by a European Research Council (ERC) Consolidator Grant for the project "PERSIST" under the European Union's Horizon 2020 research and innovation programme (grant agreement No. 101003349).

References

1. Abdulla, P., Aronis, S., Jonsson, B., Sagonas, K.: Optimal dynamic partial order reduction. In: POPL 2014, pp. 373–384. ACM, New York (2014). https://doi.org/10.1145/2535838.2535845. http://doi.acm.org/10.1145/2535838.2535845
2. Abdulla, P.A., Aronis, S., Atig, M.F., Jonsson, B., Leonardsson, C., Sagonas, K.: Stateless model checking for TSO and PSO. In: Baier, C., Tinelli, C. (eds.) TACAS 2015. LNCS, vol. 9035, pp. 353–367. Springer, Heidelberg (2015). https://doi.org/10.1007/978-3-662-46681-0_28
3. Abdulla, P.A., Atig, M.F., Jonsson, B., Leonardsson, C.: Stateless model checking for POWER. In: Chaudhuri, S., Farzan, A. (eds.) CAV 2016. LNCS, vol. 9780, pp. 134–156. Springer, Cham (2016). https://doi.org/10.1007/978-3-319-41540-6_8
4. Abdulla, P.A., Atig, M.F., Jonsson, B., Ngo, T.P.: Optimal stateless model checking under the release-acquire semantics. Proc. ACM Program. Lang. 2(OOPSLA), 135:1–135:29 (2018). https://doi.org/10.1145/3276505. http://doi.acm.org/10.1145/3276505. ISSN 2475-1421
5. Albert, E., Arenas, P., de la Banda, M.G., Gómez-Zamalloa, M., Stuckey, P.J.: Context-sensitive dynamic partial order reduction. In: Majumdar, R., Kunčak, V.

(eds.) CAV 2017. LNCS, vol. 10426, pp. 526–543. Springer, Cham (2017). https://doi.org/10.1007/978-3-319-63387-9_26. ISBN 978-3-319-63387-9

6. Albert, E., Gómez-Zamalloa, M., Isabel, M., Rubio, A.: Constrained dynamic partial order reduction. In: Chockler, H., Weissenbacher, G. (eds.) CAV 2018. LNCS, vol. 10982, pp. 392–410. Springer, Cham (2018). https://doi.org/10.1007/978-3-319-96142-2_24. ISBN 978-3-319-96142-2

7. Alglave, J., Maranget, L., Tautschnig, M.: Herding cats: modelling, simulation, testing, and data mining for weak memory. ACM Trans. Program. Lang. Syst. **36**(2), 7:1–7:74 (2014). https://doi.org/10.1145/2627752. http://doi.acm.org/10.1145/2627752. ISSN 0164-0925

8. Aronis, S., Jonsson, B., Lång, M., Sagonas, K.: Optimal dynamic partial order reduction with observers. In: Beyer, D., Huisman, M. (eds.) TACAS 2018. LNCS, vol. 10806, pp. 229–248. Springer, Cham (2018). https://doi.org/10.1007/978-3-319-89963-3_14

9. Chalupa, M., Chatterjee, K., Pavlogiannis, A., Sinha, N., Vaidya, K.: Data-centric dynamic partial order reduction. Proc. ACM Program. Lang. **2**(POPL), 31:1–31:30 (2017). https://doi.org/10.1145/3158119. http://doi.acm.org/10.1145/3158119. ISSN 2475-1421

10. Chatterjee, K., Pavlogiannis, A., Toman, V.: Value-centric dynamic partial order reduction. Proc. ACM Program. Lang. **3**(OOPSLA) (2019). https://doi.org/10.1145/3360550

11. Flanagan, C., Godefroid, P.: Dynamic partial-order reduction for model checking software. In: POPL 2005, pp. 110–121. ACM, New York (2005). https://doi.org/10.1145/1040305.1040315. http://doi.acm.org/10.1145/1040305.1040315

12. Godefroid, P.: Software model checking: the VeriSoft approach. Formal Methods Syst. Des. **26**(2), 77–101 (2005). https://doi.org/10.1007/s10703-005-1489-x

13. Godefroid, P., Hanmer, R.S., Jagadeesan, L.J.: Model checking without a model: an analysis of the heart-beat monitor of a telephone switch using VeriSoft. In: ISSTA 1998, pp. 124–133. ACM, Clearwater Beach (1998). https://doi.org/10.1145/271771.271800. http://doi.acm.org/10.1145/271771.271800. ISBN 0-89791-971-8

14. Kokologiannakis, M., Lahav, O., Sagonas, K., Vafeiadis, V.: Effective stateless model checking for C/C++ concurrency. Proc. ACM Program. Lang. **2**(POPL), 17:1–17:32 (2017). https://doi.org/10.1145/3158105. http://doi.acm.org/10.1145/3158105. ISSN 2475-1421

15. Kokologiannakis, M., Raad, A., Vafeiadis, V.: Effective lock handling in stateless model checking. Proc. ACM Program. Lang. **3**(OOPSLA) (2019). https://doi.org/10.1145/3360599

16. Kokologiannakis, M., Raad, A., Vafeiadis, V.: Model checking for weakly consistent libraries. In: PLDI 2019. ACM, New York (2019). https://doi.org/10.1145/3314221.3314609

17. Kokologiannakis, M., Sagonas, K.: Stateless model checking of the Linux kernel's read-copy update (RCU). Int. J. Soft. Tool. Tech. Transf. (2019). https://doi.org/10.1007/s10009-019-00514-6. ISSN 1433-2787

18. Kokologiannakis, M., Vafeiadis, V.: HMC: model checking for hardware memory models. In: ASPLOS 2020. pp. 1157–1171. ACM, Lausanne (2020). https://doi.org/10.1145/3373376.3378480. ISBN 9781450371025

19. Lahav, O., Vafeiadis, V., Kang, J., Hur, C.-K., Dreyer, D.: Repairing sequential consistency in C/C++11. In: PLDI 2017, pp. 618–632. ACM, Barcelona (2017). https://doi.org/10.1145/3062341.3062352. http://doi.acm.org/10.1145/3062341.3062352. ISBN 978-1-4503-4988-8

20. Lamport, L.: How to make a multiprocessor computer that correctly executes multiprocess programs. IEEE Trans. Comput. **28**(9), 690–691 (1979). https://doi.org/10.1109/TC.1979.1675439
21. Musuvathi, M., Qadeer, S., Ball, T., Basler, G., Nainar, P.A., Neamtiu, I.: Finding and reproducing Heisenbugs in concurrent programs. In: OSDI 2008. USENIX Association, pp. 267–280 (2008). https://www.usenix.org/legacy/events/osdi08/tech/full_papers/musuvathi/musuvathi.pdf. Accessed 16 Nov 2020
22. Oberhauser, J., et al.: VSync: push-button verification and optimization for synchronization primitives on weak memory models. In: ASPLOS 2021 (2021)
23. Podkopaev, A., Lahav, O., Vafeiadis, V.: Bridging the gap between programming languages and hardware weak memory models. Proc. ACM Program. Lang. **3**(POPL), 69:1–69:31 (2019). https://doi.org/10.1145/3290382. http://doi.acm.org/10.1145/3290382. ISSN 2475-1421
24. pthread.h man page (2017). https://man7.org/linux/man-pages/man0/pthread.h.0p.html. Accessed 19 Mar 2021
25. Zhang, N., Kusano, M., Wang, C.: Dynamic partial order reduction for relaxed memory models. In: PLDI 2015, pp. 250–259. ACM, New York (2015). https://doi.org/10.1145/2737924.2737956. http://doi.acm.org/10.1145/2737924.2737956

Verifying and Optimizing the HMCS Lock for Arm Servers

Jonas Oberhauser[1,2]([✉]), Lilith Oberhauser[1,2], Antonio Paolillo[1,2], Diogo Behrens[1,2], Ming Fu[1,2], and Viktor Vafeiadis[3]

[1] Huawei Dresden Research Center, 01067 Dresden, Germany
{jonas.oberhauser,lilith.oberhauser,antonio.paolillo,
diogo.behrens,ming.fu}@huawei.com
[2] Huawei OS Kernel Lab, Shenzhen, China
[3] Max Planck Institute for Software Systems, 67663 Kaiserslautern, Germany
viktor@mpi-sws.org

Abstract. To optimize the performance of some of our systems running on non-uniform memory architecture (NUMA) servers with Arm processors, we have implemented multiple versions of the HMCS lock, an advanced NUMA-aware lock that has been identified in the literature as particularly scalable.

This is a highly non-trivial task because of the many implementation choices for interlocked operations, alignment, and memory barrier placement, affecting not only the lock's performance but also its correctness. The published HMCS lock does not discuss choices that affect performance, but it does present a choice of barriers. We observe that this choice is wrong, leading to hangs on Kunpeng Arm servers. We repair the barriers and implement the first formally-verified HMCS lock with VSync, an automated formal verification and optimization tool for weak consistency. We explain the barrier bugs in detail and report our experience of barrier optimizations for Arm servers.

Keywords: Consistency models · Verification · Optimization · NUMA-aware locks

1 Introduction

Arm is making inroads on many-core servers [4,11]. To achieve a high level of parallelism, these many-core servers are implemented as non-uniform memory architectures (NUMA) in which CPUs are clustered on NUMA nodes. In these architectures, communication between CPUs within a single node is much faster than across nodes. Software therefore needs to ensure *locality* to scale well, i.e., avoid communication across NUMA nodes.

One strategy to achieve locality is through so-called NUMA-aware locks, which favor CPUs within the same NUMA node when passing the lock. Among these, we have chosen the NUMA-aware HMCS lock [6], which has been shown to be very scalable [5,9]. We have implemented the NUMA-aware HMCS lock on

© Springer Nature Switzerland AG 2021
K. Echihabi and R. Meyer (Eds.): NETYS 2021, LNCS 12754, pp. 240–260, 2021.
https://doi.org/10.1007/978-3-030-91014-3_17

Arm with the goal of improving the performance of Huawei products running on Kunpeng Arm servers. Implementing the HMCS lock for use in industry involved two main challenges.

The first challenge is the *weak consistency*. To improve single-core performance, Arm CPUs commit and propagate memory operations out-of-order: for example, memory operations issued after a cache miss can be performed while the missing cache line is being fetched. Such optimizations can be fatal to the HMCS lock, which relies on the order of a few crucial memory operations. To avoid bugs, one needs to selectively turn these optimizations off through so-called memory barriers; these include stand-alone explicit fences (e.g., DMB) as well as implicit barriers attached to the memory operations (e.g., LDAR and STLR). Turning off the optimizations everywhere is relatively easy, e.g., by using sequentially consistent C11 atomics to insert barriers for every memory access. The excessive use of barriers, however, does degrade performance. Therefore, experts attempt to identify precisely the operations that need to be executed in-order, and insert only barriers needed to enforce those orders. Indeed, the original HMCS lock paper "shows the fences necessary for the HMCS lock on systems with processors that use weak ordering" [6, p. 218], as identified by its authors. Our investigation reveals that these fences are wrong, potentially leading to hangs on Arm, Power, and RISC-V. We have reproduced the hang on a Kunpeng Arm server.

The second challenge is *performance-tuning*. We investigate two main factors that influence the performance of the HMCS lock: (1) the implementation of atomic SWAP and CAS operations and (2) the placement of barriers. These atomic operations can be implemented on Arm either through built-in interlocked SWP and CAS instructions (introduced in Arm's LSE extension [8]), which perform the operation in memory, or with load/store-exclusive LDXR/STXR instruction pairs, which perform the operation inside the CPU. For barrier placement there are similarly various implementation choices, e.g., between fences and implicit barriers. As the performance implications of these choices are not well-understood, the best choice needs to be identified by trial-and-measure.

In this paper, we show how to solve both challenges with the help of VSync [15], a formal verification and optimization tool for weak consistency. We generate formally verified barrier placements with VSync (Fig. 2). Since precise Arm support is not yet implemented in VSync, our barriers are verified against the slightly weaker IMM (intermediate memory model [16]) model, which forms the least common denominator of several weak consistency models including Arm, Power, and RISC-V. Thus the verified barriers are correct but not optimal for Arm. In fact, VSync detects a second hang and a mutual exclusion violation on IMM, but we manually verify that these bugs cannot occur on Arm weak consistency.

In the following, we present the HMCS algorithm (Sect. 2) and discuss the set of barriers necessary for its correctness (Sect. 3), showing what goes wrong if some barriers are omitted. We then briefly describe our verification and optimization setup (Sect. 4). Finally, we measure the performance impact of the implementation choices mentioned above as well as the conservative barriers introduced by VSync (Sect. 5) on a microbenchmark and on LevelDB [7].

In summary, we make the following contributions:

- We have discovered a bug in the fences proposed in the HMCS lock from the literature, and present a formally verified fix.
- We propose various barrier optimizations for the HMCS lock and investigate their impact on performance.
- We present the following insights:
  - Barriers optimizations make little difference for scalability; sequentially consistent C11 atomics are good enough for Arm.
  - If barrier optimizations are desired, they should be left to an automatic tool like VSync.
  - Arm's interlocked instructions (LSE) degrade performance.

2   Background

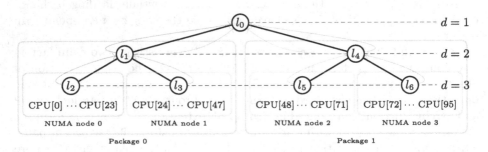

Fig. 1. NUMA topology and lock trees for 96-core Kunpeng Arm server

2.1   HMCS Lock

The HMCS lock is a tree of MCS locks, configured to model the NUMA topology tree of the target machine; in our case, we consider a Kunpeng 920 Arm server with four NUMA nodes (24 CPUs each), organized in two packages. As illustrated in Fig. 1, the lock tree for this topology is a binary tree of depth DEPTH=3. We now explain the MCS lock, which is the main component of the HMCS lock, and the acquire and release protocols of the HMCS lock. The code of the HMCS lock is shown in Fig. 2.

MCS Lock. The MCS lock [14] forms a queue so that threads enter the critical section in a FIFO manner. Acquiring and releasing the MCS lock are performed by the AcqReal<1> and RelReal<1> functions in Fig. 2. A thread enqueues its QNode, which contains a status field that is used as means of communication with its predecessor. Before enqueueing, a thread sets its status to 🔒 (Line 14),

```
1   enum LockStatus {                    43   Release(HNode *L , QNode *I){
2     🔒=UINT64_MAX-1,                    44     ---- RELEASE FENCE ----
3     🔓=0x1,                             45     RelReal<DEPTH>(L, I);
4     🔒=0x0,                             46   }
5     n ∈ [2 : THRESHOLD]                 47
6   };                                    48   ReleaseHelper(HNode *L, QNode *I,
7                                         49     LockStatus st) {
8   Acquire(HNode *L, QNode *I){          50     QNode *succ = I->next_acq;
9     AcqReal<DEPTH>(L, I);               51     ---- ACQUIRE FENCE IMM ----
10    ---- ACQUIRE FENCE ----             52     if (succ) {
11  }                                     53       succ->status =_rel st;
12                                        54     } else {
13  AcqReal<1>(HNode *L, QNode *I){       55       if (CAS_sc(& L->tail, I, ⊥))
14    I->status = 🔒; I->next = ⊥;        56         return;
15    ---- RELEASE FENCE ARMIMM ----      57       while ((succ = I->next) == ⊥);
16    QNode *pred;                        58       succ->status =_rel st;
17    pred = SWAP_sc(& L->tail, I);       59     }
18    if (!pred) {                        60   }
19      I->status = 🔓;                   61
20    } else {                            62   RelReal<1>(HNode *L, QNode *I){
21      pred->next =_rel I ;              63     ReleaseHelper(L, I, 🔓);
22      while (I->status_acq == 🔒);      64   }
23    }                                   65
24  }                                     66   RelReal<d>(HNode *L , QNode *I){
25                                        67     uint64_t curCount = I->status;
26  AcqReal<d>(HNode *L, QNode *I) {      68     if (curCount == THRESHOLD[d]) {
27    I->status = 🔒; I->next = ⊥;        69       RelReal<d-1>(L->parent, & L->N);
28    ---- RELEASE FENCE ----             70       ---- RELEASE FENCE ----
29    QNode *pred;                        71       ReleaseHelper(L, I, 🔒);
30    pred = SWAP_sc(& L->tail, I);       72       return;
31    if (pred) {                         73     }
32      pred->next =_rel I;               74     QNode *succ = I->next_acq;
33      LockStatus curStatus;             75     ---- ACQUIRE FENCE IMM ----
34      do curStatus = I->status_acq      76     if (succ) {
35      while (curStatus == 🔒);          77       curCount += 1;
36      ---- ACQUIRE FENCE IMM ----       78       succ->status =_rel curCount;
37      if (curStatus < 🔒) return;       79       return;
38    }                                   80     }
39    I->status = 1;                      81     RelReal<d-1>(L->parent, & L->N);
40    AcqReal<d-1>(L->parent, & L->N);    82     ---- RELEASE FENCE ----
41  }                                     83     ReleaseHelper(L, I, 🔒);
42                                        84   }
```

Fig. 2. Pseudo-code of the HMCS Lock from [6] except for barrier placement and cosmetic changes

then it advances the tail pointer (Line 17). If it finds a predecessor p it waits in Line 22 for p to give the signal status = 🔓; otherwise it unlocks itself (Line 19) and enters the critical section. Once it is done, it releases the lock. If it is the tail (i.e., it has no successor), it does so by setting the tail pointer to ⊥ (Line 55). Otherwise, if it has a successor s, it signals s by setting the status of s to 🔓 (Line 53 or Line 58).

HMCS Lock Acquisition. The critical section is protected by the root lock l_0 at depth $d = 1$. To initiate the lock acquisition protocol, the HMCS lock client calls Acquire on the leaf lock that belongs to the NUMA on which the thread is running; e.g., in Fig. 1 a thread bound to CPU[0] calls Acquire on l_2. This calls AcqReal<3> on l_2, which recursively calls AcqReal<2> on l_1 (l_2's parent)

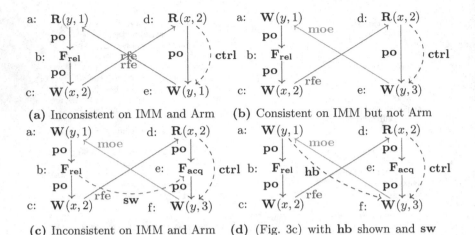

Fig. 3. Execution graphs

and `AcqReal<1>` on l_0 (l_1's parent). Note that this means a) all threads in the queue of a leaf lock are on the same NUMA node, b) all threads in the queue of a lock at depth $d = 2$ are in the same package but different NUMA nodes, and c) all threads in the queue of lock l_0 are in different packages. Enqueueing at any MCS lock in the tree requires a `QNode`. Each thread T_i has its own `QNode` N_i with which it enqueues at its leaf lock. Each lock l protects a `QNode` $l.N$ which is used to enqueue at the parent of l. For example, if T_1 is running on NUMA node 0, it uses N_1 to enqueue at l_2. Once it owns that lock, it can use $l_2.N$ to enqueue at l_1, and so on.

HMCS Lock Release. Invoking `Release` initiates the release protocol. This recursively calls `RelReal<3>`, `RelReal<2>` and `RelReal<1>`. The lock can be passed at any depth $d \in \{1, 2, 3\}$, if a successor is found at depth d. To maximize throughput, the lock should be passed within the NUMA node, i.e., at depth $d = 3$. However, this would lead to starvation in the other NUMA nodes. `THRESHOLD[d]` defines the maximum number of times a lock is passed at depth $d \in \{2, 3\}$. If `THRESHOLD[d]` has been reached, the lock owner sets the **status** of the successor at depth d to ↥. This signals to the successor that the lock is passed at a depth $d' < d$. In contrast, when a successor is found and `THRESHOLD[d]` is not reached, the lock is passed directly to the successor by setting its **status** to $n \in [2 : \text{THRESHOLD}[d]]$, counting the number of times the lock was acquired at depth d.

2.2   Weak Consistency and Execution Graphs

A standard way to define weak consistency models is through execution graphs such as those in Fig. 3. **Nodes** in these graphs represent events such as reads

($\mathbf{R}$, $\mathbf{R_{sc}}$, $\mathbf{R_{acq}}$), fences ($\mathbf{F_{acq}}$ and $\mathbf{F_{rel}}$), and writes ($\mathbf{W}$, $\mathbf{W_{sc}}$, $\mathbf{W_{rel}}$), and **edges** specify various relations between these events, such as moe (**modification order external**) and rfe (**reads-from external**) edges which indicate the order in which reads and writes to the same location are committed, and **po** (**program order**) edges which indicate the order in which instructions are issued (but not necessarily committed). In this paper, we only give high-level explanations for differences between Arm and IMM; motivated readers will find more detailed explanations in Appendix A. A weak consistency model is defined by the execution graphs it permits. For IMM and Arm, this is done by forbidding graphs in which any event "happens before" itself, where "happens before" is defined by model-specific relations that indicate the order in which events happen. The difference between the models can be explained in terms of when one event "happens before" another according to the model.

Arm has one "happens before" relation (called **ob**, ordered-before), which respects among other things: a) the order in which writes are committed (moe), b) the order between a write and a read that observes the write (rfe), c) fences such as DMB.ISH (implied by a release fence ($\mathbf{F_{rel}}$) in the code), and d) control dependencies from a read influencing the position of control, e.g., through an if-condition, to a write occurring after the condition (written **ctrl** in graphs). In Figs. 3(a) and 3(b), this means that Event a "happens before" itself on Arm.

On IMM, there are two "happens before" relations, both weaker than that of Arm. A graph is forbidden if an event "happens before" itself according to either relation. The first is the **acyclic** relation (**ar**) which critically does not respect moe. According to this relation, Event a "happens before" itself only in Fig. 3(a), not in Fig. 3(b). The second relation (which is nameless in IMM, but which we will call $\mathbf{hb_{IMM}}$) respects moe, but critically ignores control dependencies; thus according to this relation, Event a "happens before" itself neither in Figs. 3(a) and 3(b). Indeed, none of the other events "happen before" themselves in Fig. 3(b) with either definition, and Fig. 3(b) is consistent on IMM.

Unfortunately, behaviors like that in Fig. 3(b) lead to various bugs in the HMCS lock. To forbid this behavior on IMM, one has to add an acquire fence $\mathbf{F_{acq}}$ along the **ctrl** edge (Fig. 3(c)). The existing $\mathbf{F_{rel}}$ fence synchronizes-with (**sw**) this $\mathbf{F_{acq}}$, creating a happens-before (**hb**) edge from Event a to Event f (Fig. 3(d)); together with the moe edge in the opposite direction, Event a "happens before" itself according to $\mathbf{hb_{IMM}}$. Thus Fig. 3(c) is inconsistent on IMM.

3   Barriers on Arm and IMM

Figure 2 shows two formally verified barrier placements: one uses the highlighted implicit barriers, and the other uses the highlighted fences. Both use sequentially consistent (**sc**) SWAP and CAS operations. Further barrier optimizations on these operations are possible but bring no performance benefit (see hmcs-amo in Fig. 10, or a detailed discussion in Appendix B) and are thus not shown.

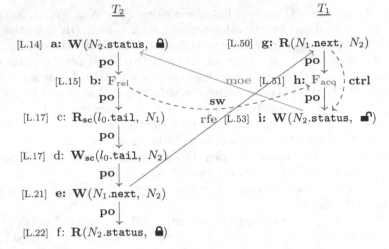

Fig. 4. Bug on Arm and IMM: Non-terminating execution due to missing fences at Events b and h

In addition to the fences already presented in [6], VSync introduces fences at Lines 36, 75 and 51 (for IMM) and Line 15 (for IMM and Arm) to solve three bugs. In the following sections we discuss these in more detail.

3.1 Termination Violation

To simplify the discussion of the hang we consider only an HMCS lock L with maximum depth one (DEPTH = 1), with two threads T_1 and T_2. We first discuss the desired behavior in which the lock is passed correctly. Initially thread T_1 owns the lock and is about to release it, while thread T_2 is attempting to acquire the lock. T_2 first prepares its node (Line 14), writing 🔒 to its status to indicate that it does not yet have permission to enter the critical section. It proceeds to append itself to the queue by moving the tail pointer (Line 17) and updating T_1's next pointer (Line 21). T_1 sees its successor in Line 50 or after failing to set the tail pointer to ⊥ in Lines 55 and 57. Subsequently T_1 will set T_2's status to 🔓, indicating that T_2 can enter the critical section (Line 58).

On Arm, T_2's initialization to its own node can happen *after* it informs T_1 that it has a successor. In this case T_1 can unlock T_2 before T_2 initializes its node (locking itself again). An execution graph for this case is shown in Fig. 4. Perhaps surprisingly, the SWAP operation in Line 17 does not prevent this reordering even if it is sequentially consistent (note that the original presentation in [6] does not mention whether the atomic SWAP and CAS operations have any ordering semantics). The reason for this is that sequentially consistent atomic operations are compiled to LDAXR/STLXR instruction pairs which generate the Events c and d. Intuitively speaking, Arm only preserves the order 1) between Event c and subsequent events, 2) between Event d and preceding events, and 3) between all

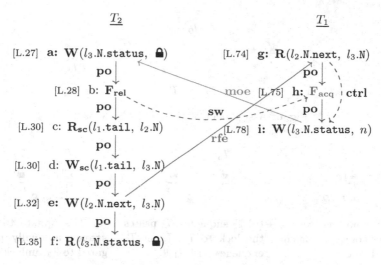

Fig. 5. Bug on IMM: Non-terminating execution

sequentially consistent events on the same processor. But, it does not preserve the order between Event a and Event c, or Event d and Event e. Thus the events can be committed in the order c, e, a, d. In this commit order, the node initialization (Event a) happens after T_2 informs T_1 (Event e).

We repair the bug by adding a $\mathbf{F_{rel}}$ fence in Line 15 in `AcqReal<1>`. With this fence, the Events a, b, e, g and i map directly to the events in Fig. 3(b). Thus (with the fence) the buggy execution becomes inconsistent on Arm, and the bug can not occur anymore. On IMM we additionally need to add an $\mathbf{F_{acq}}$ fence at Event h. With both fences, the events Events a, b, e and g to i map directly to the events in Fig. 3(c), showing that the bug is fixed also on IMM.

Note that for higher depths $d > 1$, the corresponding $\mathbf{F_{rel}}$ fence already exists (in Line 28), but the corresponding $\mathbf{F_{acq}}$ fence is also missing. Indeed VSync reports the analogous termination bug (Fig. 5) at greater depths. Analogously to before, we can see that this bug only exists on IMM and that it can be fixed by inserting the $\mathbf{F_{acq}}$ fence in Line 75.

3.2   Mutual Exclusion Violation

This bug only occurs with three threads T_1, T_2 and T_3 on separate NUMA nodes, as indicated in Fig. 6. In a nutshell, T_2 enqueues behind T_3 at l_0 with the QNode $l_1.N$, which was previously used by T_1 (Fig. 7(b)). When T_1 entered the critical section (Fig. 7(a)), it had no predecessor and therefore set the status of $l_1.N$ to 🔓 (Line 19). Due to a missing fence, this operation is only propagated to T_2 after T_2 enqueued behind T_3, giving T_2 the false signal that it can enter the critical section even though T_3 is still holding the lock (Fig. 7(c)).

A more detailed execution leading to the bug is shown in Fig. 8. Note that Events d, e, g, i and l map to the events in Fig. 3(b), implying that the bug is

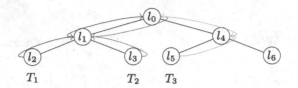

Fig. 6. Assignment of threads to NUMAs

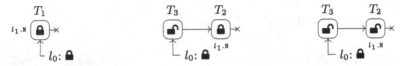

(a) T_1 has no predecessor, but does not set **status** of l_1.N to 🔓 yet.

(b) T_3 enqueues. T_1 passes the lock to T_3. Later T_2 enqueues with l_1.N.

(c) T_1's update to l_1.N's **status** is finally propagated to T_2, unlocking it.

Fig. 7. Mutual exclusion violation on IMM due to a missing acquire fence.

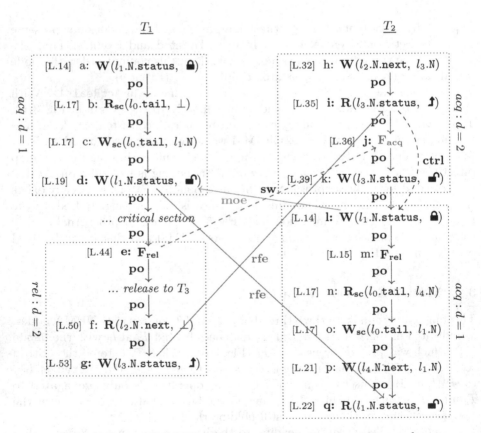

Fig. 8. Bug on IMM: Mutual exclusion violation execution graph.

T_1 $\qquad\qquad$ T_2 $\qquad\qquad$ T_3

```
Acquire(l₂, N₁);      Acquire(l₃, N₂);      Acquire(l₅, N₃);
counter++;            counter++;            counter++;
Release(l₂, N₁);      Release(l₃, N₂);      Release(l₅, N₃);

Acquire(l₂, N₁);
counter++;
Release(l₂, N₁);
```

Fig. 9. The client code for verifying and optimizing HMCS with VSync

not possible on Arm. On IMM it can be fixed like indicated in Figs. 3(c) and 3(d) by adding $\mathbf{F_{acq}}$ at Event j.

4 Verifying and Optimizing HMCS with VSync

Vsync [15] is a fully automated verification and optimization tool that accepts (bounded) concurrent C/C++ programs as input. In its verification mode, it exhaustively enumerates all the possible executions of the input program following the GenMC model checking algorithm [12], and checks that these are terminating, memory-safe, and satisfy all user-supplied assertions. In its optimization mode, it uses an iterative algorithm to find minimal barrier placements that ensure program correctness (i.e., successful verification).

Like all model checkers, VSync does not verify locks abstractly: one must provide client code that uses the lock appropriately. A reasonable client must visit all functions and paths of a lock. For example, if in our client we configured HMCS with maximum depth − 1, the verification would cover only AcqReal<1> and RelReal<1>, and if we created only one thread then we would miss all concurrency bugs.

Verification time is generally super-exponential in the number of threads and acquire/release calls. We thus need to find the minimum number of threads with which we can still generate all bug-prone execution graphs. In general, finding this number is an open problem. We simply choose the maximum number of threads for which verification time is within reason. We experimentally justify this bound by adding an additional thread and observe that no additional bugs are found by VSync.

Our client code is shown in Fig. 9: it uses three threads, maximum DEPTH = 3, and thresholds THRESHOLD[d] = 2. We choose this maximum DEPTH = 3 because it covers the case where a lock (at depth $d = 2$) has both children and a parent. We assign threads to NUMA nodes as in Fig. 6. Each thread in our client acquires the lock, increments a shared variable, and then releases the lock. One thread (T_1) repeats this twice. This way we cover the case of a thread entering the critical section twice. With this setup, verification with VSync takes 10 s. Adding a fourth thread T_4 on NUMA 0 (respectively, NUMA 3) increases verification time to 1300 s (respectively, 2800 s).

To verify mutual exclusion, we assert that our shared counter has the expected value after all threads are done (`assert(counter==4)`). If any execution graph of the client violates this assertion or indicates a non-terminating run of the program (such as the graphs in Figs. 4, 5 and 8), VSync prints that graph in text form. We note that debugging such graphs is non-trivial.

A simpler and more elegant way to use VSync is to implement the lock with only sequentially consistent memory operations (without fences). This ensures that there will be no bugs related to the consistency model. VSync then optimizes these barriers and reports to the user which barriers can be relaxed and/or removed. With our client code with three threads, optimization takes one second. With four threads, optimization takes less than 100 s.

5   Performance Evaluation

We evaluated HMCS on our Arm server and studied implementation choices that we expect are affecting performance. In particular, we tackle the following questions:

- Do the Large System Extensions (LSE) of Armv8.1 bring the promised performance improvements?
- Is there a performance penalty of unnecessary barriers, e.g., those introduced by VSync when optimizing for IMM rather than Arm?
- Do implicit barriers provide better performance than fences?

5.1   Experimental Setup

Environment. We ran the experiments on a Huawei TaiShan 200 (Model 2280) [1] with two HiSilicon Kunpeng 920-4826 processors [2] (2 packages), each of them with 48 Armv8.2 64-bit cores organized in 2 NUMA nodes and running at 2.6 GHz. The experiments reported in this section were conducted on openEuler 20.09 [3]. We reproduced similar results on Ubuntu 18.04 LTS.

Benchmarks. We conducted userspace experiments with LevelDB (`readrandom` benchmark) [7] and with a custom microbenchmark. In the microbenchmark, each thread repeatedly acquires a `pthread_mutex_lock`, increments a shared counter (causing a cache miss), and releases the mutex. In each experiment, we vary the number of threads and the lock implementation. We interpose calls to `pthread` functions with `LD_PRELOAD` in order to replace the lock implementation without modifying the benchmarks—in a similar fashion as [10]. We run each experiment for 3 s, repeat the experiment 10 times, and report the median throughput (number of iterations per second). We pin threads to cores from core 95 downwards, always keeping core 0 free to serve other OS tasks.

Lock Variants. We compare the following variants of the HMCS-lock in regard to barrier/fence placements:

- hmcs-arm with a minimal set of fences required for Arm,
- hmcs-imm with a minimal set of fences required for IMM,
- hmcs-sc in which all racy accesses use sequentially consistent implicit barriers,
- hmcs-vsync with VSync-optimized implicit barriers, and
- hmcs-amo with optimized barriers on CAS and SWAP in hmcs-vsync.

We use the mcs lock with optimized barriers as a baseline. To avoid false sharing and ensure reliable results, we also cache-align and pad the shared data structures (QNode and HNode). All locks are implemented using C11 atomics (stdatomic.h).

5.2 Experimental Results for Low Contention

We start by exploring the performance of the HMCS variants with our microbenchmark running a single thread (see Fig. 10).

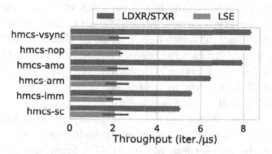

Fig. 10. Low contention scenario: single-threaded microbenchmark with several HMCS variants compiled with and without LSE instructions.

LSE Versus LDXR/STXR. HMCS variants that employ LSE instructions perform poorly in comparison to those that employ the conventional LDXR/STXR pair. The current hardware implementation of LSE degrades the performance of all HMCS variants. For example, in the case of hmcs-vsync, the throughput of the LSE version is 27% of the LDXR/STXR throughput. We also observe that LSE implementations tend to have a higher variance than the LDXR/STXR implementations (see standard deviation reported on Fig. 10).

Due to the importance of the single-thread scenario, we only consider the implementations with LDXR/STXR in the remainder of the evaluation.

Performance Penalty of No Optimization. The performance of hmcs-sc shows that exclusively employing sequentially consistent implicit barriers incurs a considerable cost under low contention. As we will see below, the performance of hmcs-sc is comparable to the other variants under high contention.

Performance Penalty of Targeting IMM. We observe that hmcs-arm has 16% higher throughput than hmcs-imm, implying that the additional fences required by IMM impact the performance negatively. In contrast, the additional implicit barriers in hmcs-vsync do not reduce throughput compared to hmcs-amo, which has been manually optimized for Arm. This suggests that the performance penalty of using IMM as the verification target depends on the type of barriers and not simply on the number of additional barriers required by IMM.

Implicit Barriers Versus Fences. Automatically-selected implicit barriers perform better than fences: hmcs-vsync shows 49% higher throughput than hmcs-imm, and 28% higher throughput than hmcs-arm. Note that replacing fences with implicit barriers reduces the code length, which in turn can shorten single-threaded runs and improve the instruction cache usage. To validate that the shorter code length is not the source of the improved performance of hmcs-vsync over hmcs-arm, we create a variant based on hmcs-vsync, in which we introduce a NOP instruction for every removed fence (NOP and fences have the same length in Arm); we call this variant hmcs-nop. Figure 10 shows a negligible difference between hmcs-vsync and hmcs-nop, corroborating the claim that implicit barriers improve performance for *single* threaded code [13]. Nevertheless, whether implicit barriers or fences perform better for *multiple* threads may depend on the benchmark, as we will see below.

The reason for this discrepancy is not clear; besides micro-architectural implementation details, a possible reason may lie in the weak consistency model of Arm itself. For the correctness of the HMCS lock, the order between specific loads and subsequent memory operations needs to be enforced. On the Kunpeng 920 server, these loads can be implemented either as a load with a trailing DMB LD instruction (acquire fence), or as LDAR/LDAXR load instructions with implicit acquire barriers. Both are unsatisfactory. The DMB LD instruction needlessly orders *all previous* loads with subsequent operations. The LDAR/LDAXR instructions needlessly order all previous stores with implicit release barriers with that load. These non-comparable unnecessary ordering constraints might be the reason both implementation choices are sometimes the better choice. Armv8.3 (not supported on Kunpeng 920) introduces the LDAPR load instruction, which only introduces the necessary order. Perhaps the comparison between implicit barriers and fences would be more clear-cut with this instruction.

5.3   Experimental Results for High Contention

We now explore higher contention scenarios with our microbenchmark and with LevelDB benchmark. In the following experiments, we consider hmcs-arm, hmcs-vsync, hmcs-sc, and mcs.

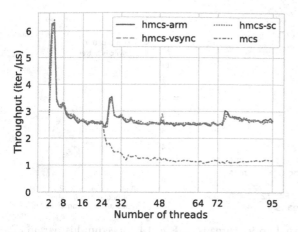
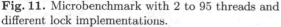

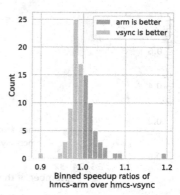

Fig. 11. Microbenchmark with 2 to 95 threads and different lock implementations.

Fig. 12. Speedup histogram of hmcs-arm over hmcs-vsync for the microbenchmark with 1 to 95 threads.

Figure 11 shows the performance of our benchmark running with 2 to 95 threads. (Single-threaded runs were evaluated in Sect. 5.2, and one core is left free to handle interrupts, which are otherwise a source of noise.) We assign threads to cores sequentially.

After filling a complete NUMA node (with 24 threads), the performance of mcs drops considerably; for example at 95 threads, mcs throughput is about 45% of hmcs-vsync throughput. The performance spike with 4 threads is due to the higher cache locality achieved when threads share the same L3 cache region. Kunpeng 920 processors split the L3 cache in regions shared by groups of 4 cores. The spike with 28 threads is caused by the interplay of the HMCS policy to keep the lock in the NUMA node and the fact that 4 cores of the second NUMA node share the same L3 cache region. HMCS enforces that both NUMA nodes have the same share of the lock with a user-configured threshold (see Sect. 2.1). Therefore, the first NUMA node executes half of the benchmark iterations with 24 cores, whereas the second NUMA node executes the other half with 4 threads and few L3 cache misses, improving the overall throughput. The spike repeats at lower intensities when the other NUMA nodes only use 4 cores.

The different HMCS variants perform in most configurations less than 10% apart. Figure 12 shows the histogram of speedups of hmcs-arm over hmcs-vsync for 2 to 95 threads. The single case around 0.89 is with 49 threads, where hmcs-arm is slower than hmcs-vsync. The single case around 1.20 is with 2 threads, where hmcs-arm is faster than hmcs-vsync: this is caused by the slowpath of MCS lock release, which is triggered more often with hmcs-vsync and 2 threads. For the other cases, we observe that hmcs-arm tends to be slightly slower than hmcs-vsync, but the difference is below 8%.

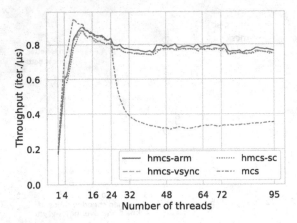

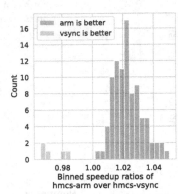

Fig. 13. LevelDB benchmark with 1 to 95 threads and different lock implementations.

Fig. 14. Speedup histogram of hmcs-arm over hmcs-vsync for the LevelDB benchmark with 1 to 95 threads.

Figure 13 shows the performance of the LevelDB benchmark running 1 to 95 threads. The benchmark contains parallel work and can scale up to around 8 threads. Up to 9 threads, mcs performs up to 20% faster than hmcs-vsync, but continuously degrades its throughput when more than 24 threads are running (or more than one NUMA node is used in the application). For example at 95 threads, the mcs throughput is 47% of hmcs-vsync throughput. Between 10 and 24 threads, hmcs-vsync and mcs are at most 6% apart.

Figure 14 shows again the histogram of speedups of hmcs-arm over hmcs-vsync for 1 to 95 threads. In the range from 1 to 5 threads, hmcs-vsync performs up to 4% faster than hmcs-arm. With 6 threads or more, hmcs-arm performs up to 5% faster than hmcs-vsync.

Finally, hmcs-vsync performs up to 8% faster than hmcs-sc in the range from 1 to 10 threads. With 11 threads or more, the hmcs-vsync throughput is between 0.99 to 1.03 times the hmcs-sc throughput.

6   Discussion

Already with sequentially consistent barriers, the NUMA-aware HMCS lock considerably outperforms the MCS lock at high levels of contention. At these levels, the performance impact of barrier optimization is negligible. On the other hand, incorrect optimizations can lead to heisenbugs. For this reason, we recommend simply using sequentially consistent barriers on all racy accesses, and not worrying about weak consistency. In cases of low contention, however, barrier optimizations can show substantial performance improvements. In these cases, the automatic and formally verified optimizations by VSync outperform manual optimizations (both our own and the repaired fences from the literature). This shows that barrier optimization, if desired, should be left to the machine.

A   Arm vs. IMM Consistency Model

$$obs \supseteq \text{rfe} \cup \text{moe} \tag{1}$$

$$dob \supseteq \textbf{ctrl}; [W] \tag{2}$$

$$bob \supseteq \textbf{po}; [\textbf{F}_{\textbf{rel}}]; \textbf{po} \tag{3}$$

$$lob \supseteq dob \cup bob \tag{4}$$

$$\textbf{ob} \supseteq obs \cup lob \cup \textbf{ob}; \textbf{ob} \tag{5}$$

$$\textbf{ob} \text{ is irreflexive} \tag{6}$$

Fig. 15. A Subset of the Arm Consistency Model. The key derived relation is ordered-before (**ob**), which is irreflexive in consistent graphs.

$$F \supseteq \textbf{F}_{\textbf{rel}} \cup \textbf{F}_{\textbf{acq}} \tag{7}$$

$$deps \supseteq \textbf{ctrl} \tag{8}$$

$$ppo \supseteq [\textbf{R}]; deps; [\textbf{W}] \tag{9}$$

$$bob \supseteq [F]; \textbf{po} \cup \textbf{po}; [F] \tag{10}$$

$$\textbf{ar} \supseteq \text{rfe} \cup bob \cup ppo \tag{11}$$

$$\textbf{ar} \text{ is acyclic} \tag{12}$$

$$release \supseteq [\textbf{F}_{\textbf{rel}}]; \textbf{po} \tag{13}$$

$$\textbf{sw} \supseteq release; \text{rfe}; \textbf{po}; [\textbf{F}_{\textbf{acq}}] \tag{14}$$

$$\textbf{hb} \supseteq \textbf{po} \cup \textbf{sw} \cup \textbf{hb}; \textbf{hb} \tag{15}$$

$$\textbf{eco} \supseteq \text{rfe} \cup \text{moe} \tag{16}$$

$$\textbf{hb}; \textbf{eco} \text{ is irreflexive} \tag{17}$$

Fig. 16. A Subset of the IMM Consistency Model. Key relations are the acyclic relation (**ar**) which is acyclic in consistent graphs, as well as synchronizes-with (**sw**), extended coherence order (**eco**), and happens-before (**hb**), where **hb**; **eco** is irreflexive in consistent graphs.

A standard way to define weak consistency models is through execution graphs. **Nodes** in these graphs represent events such as reads and writes, and **edges** specify various relations between these events, e.g., the order in which reads and writes to the same location are committed. Memory models are defined by a) the edges that exist in the graph and b) restrictions on these edges. For brevity, we introduce only the event and edge types of Arm and IMM that are relevant to the bugs we mention in this paper. We consider write events $\textbf{W}_X(loc, val)$, read events $\textbf{R}_X(loc, val)$, and fence events $\textbf{F}_X$, where $X \in \{\textbf{sc}, \textbf{acq}, \textbf{rel}, \textbf{rlx}\}$ is the so called **mode** of the event, loc is the shared memory location on which the event operates, and val is the value written or read in the event. The mode denotes the type of memory barrier (if any) represented by the event: **rlx** indicates that no barrier is present, **acq** represents acquire, **rel** release, and **sc** sequentially consistent barriers. Mode **rlx** is the default mode and omitted.

We consider the following types of fundamental edges:

- **rfe** (**r**ead **f**rom **e**xternal) edges $\mathbf{W}_X(x,a) \xrightarrow{\text{rfe}} \mathbf{R}_Y(x,a)$ connect a write event of a thread to a read event of another thread that reads from it.
- **moe** (**mo**dification order **e**xternal) edges $\mathbf{W}_X(x,a) \xrightarrow{\text{moe}} \mathbf{W}_Y(x,b)$ connect write events (writing to the same location) of different threads indicating the order in which they were committed.
- **po** (**p**rogram **o**rder) edges connect events of the same thread in the order in which they are issued by the program.
- **ctrl** (**c**on**tr**o**l** dependency) edges connect a read $\mathbf{R}_X(x,a)$ that influences a condition (e.g., if- or while-condition) evaluation to every event of the same thread that is issued after the condition.
- event-type self-loops $e \xrightarrow{[E]} e$ for event type $E \in \{\mathbf{R}, \mathbf{W}, \mathbf{F}_{\text{rel}}, \mathbf{F}_{\text{acq}}\}$ connect every event e of type E to itself.

Other edges are derived from these fundamental edges according to the rules of the consistency model (Figs. 15 and 16). For instance, the edge a $\xrightarrow{\text{moe}}$ e in Fig. 3(b) implies an **eco** edge a $\xrightarrow{\text{eco}}$ e on IMM (with Eq. (16)). Such derived rules are often defined with the composition operator ';', which for arbitrary edge types R and S is defined by

$$a \xrightarrow{R;S} c \iff a \xrightarrow{R} b \xrightarrow{S} c$$

The meaning of barriers is defined by the derived edges they imply; for example, the meaning of $\mathbf{F}_{\text{rel}}$ (which maps to the full DMB.ISH fence) on Arm is defined through the **ob** edge it implies between preceding and subsequent operations (with Eqs. (3) to (5)).

In Figs. 15 and 16 we have collected the rules of IMM and Arm consistency that are relevant to our discussion. In [16] it is shown that Arm consistency implies IMM consistency; thus any bug on Arm is also present on IMM, and verification on IMM implies correctness on Arm. The converse is not true, and bugs on IMM are not always bugs on Arm. Indeed, some of the bugs identified by VSync on the HMCS lock on IMM are not bugs on Arm. The key difference relevant to these bugs is that moe edges imply an **ob** edge on Arm, but do not imply an **ar** edge on IMM. Thus they contribute to **ob** cycles but not to **ar** cycles.

We illustrate the implications at hand of the execution graphs in Fig. 3. In Fig. 3(a), we have an rfe edge from Event e to Event a; in Fig. 3(b), we instead have an moe edge from Event e to Event a. Other than those events and the edge between them, the graphs are the same. Thus in both graphs, the following imply **ob** edges:

- a $\xrightarrow{\text{po}}$ b $\xrightarrow{[\mathbf{F}_{\text{rel}}]}$ b $\xrightarrow{\text{po}}$ c (Eqs. (3) to (5))
- c $\xrightarrow{\text{rfe}}$ d (Eqs. (1) and (5))
- d $\xrightarrow{\text{ctrl}}$ e $\xrightarrow{[\mathbf{W}]}$ e (Eqs. (2), (4) and (5))

The only edge missing for an **ob** cycle is e $\xrightarrow{ob}$ a. This edge is implied by the e $\xrightarrow{rfe}$ a edge in Fig. 3(a) and the e $\xrightarrow{moe}$ a edge in Fig. 3(b) (with Eqs. (1) and (5)). Note that due to transitivity (Eq. (5)) the cycle a $\xrightarrow{ob}$ ... $\xrightarrow{ob}$ a implies a reflexive edge a $\xrightarrow{ob}$ a, which contradicts the irreflexivity of **ob** (Eq. (6)). Thus both graphs are inconsistent on Arm.

On IMM, the following imply **ar**-edges:

- a $\xrightarrow{po}$ b $\xrightarrow{[F_{rel}]}$ b and b $\xrightarrow{[F_{rel}]}$ b $\xrightarrow{po}$ c (Eqs. (7), (10) and (11))
- c $\xrightarrow{rfe}$ d (Eq. (11))
- d $\xrightarrow{[R]}$ d $\xrightarrow{ctrl}$ e $\xrightarrow{[W]}$ e (Eqs. (8), (9) and (11))

Analogous to before, only an e $\xrightarrow{ar}$ a is missing for an **ar** cycle. In Fig. 3(a) this edge is implied by the e $\xrightarrow{rfe}$ a edge with Eq. (11), and this graph is inconsistent on IMM. But in Fig. 3(b), the moe edge does not contribute an **ar** edge. Indeed, there is no **ar** cycle in Fig. 3(b), which is consistent on IMM. Unfortunately, two of the bugs detected by VSync on IMM appear only in graphs that look like Fig. 3(b). These bugs therefore only appear on IMM, but can not appear on Arm.

We proceed to discuss how to fix these bugs on IMM. Consider the third graph (see Fig. 3(c)) which is almost identical to the second (see Fig. 3(b)). We only added a $\mathbf{F_{acq}}$ fence between the d and f. Adding this fence does not eliminate the **ob**-cycle we inferred previously, and this graph is also inconsistent with Arm. On IMM we derive the following edges:

- b $\xrightarrow{[F_{rel}]}$ b $\xrightarrow{po}$ c $\xrightarrow{rfe}$ d $\xrightarrow{po}$ e $\xrightarrow{[F_{acq}]}$ e thus b $\xrightarrow{sw}$ e (Eqs. (13) and (14))
- a $\xrightarrow{hb}$ b, b $\xrightarrow{hb}$ e and e $\xrightarrow{hb}$ f thus a $\xrightarrow{hb}$ f (Eq. (15))
- f $\xrightarrow{eco}$ a (Eq. (16))

As shown in (Fig. 3(d)) we end up with a $\xrightarrow{hb}$ f $\xrightarrow{eco}$ a and thus a $\xrightarrow{hb;eco}$ a. But the **hb;eco** relation is irreflexive (Eq. 17). We conclude that this graph is inconsistent with IMM. In other words, due to the $\mathbf{F_{acq}}$ fence the execution with the bug cannot occur on IMM.

B   Optimizing Barriers on Atomic Operations

The implicit sc barriers on CAS and SWAP in Fig. 2 are not optimal. VSync reports that they are already too strong for IMM, and indeed they can be optimized further for Arm. The exact optimization depends on the variant. Manual analysis shows that when using fences, all barriers on the atomic operations can be removed. When using implicit barriers, release barriers on Lines 17 and 30 are needed to avoid non-termination (with similar bugs as those in Sect. 3.1) and

Table 1. Possible optimizations on Arm for atomic operations when using fences or implicit barriers.

	SWAP [Line 17]	SWAP [Line 30]	CAS [Line 55]
Fences	-	-	-
Implicit	sc	rel	rel
Implicit (LSE)	rel; F_{acq}	rel	rel

acquire and release barriers are needed on Line 17 resp. Line 55 to ensure that operations in the critical section can not leak out of the lock (resulting in loss of mutual exclusion). The resulting barriers are shown in Table 1. That table also shows a variant that may be more optimal when using interlocked LSE instructions. Unlike load/store-exclusive pairs, on which sc implicit barriers do not act like a full barrier (see discussion in Sect. 3.1), LSE interlocked operations have been strengthened in a recent change to Arm specifications to provide the same semantics for sc implicit barriers as a DMB.ISH (see Eq. (10)) through the rule

$$bob \supseteq \mathbf{po}; ([\mathbf{A}]; amo; [\mathbf{L}]); \mathbf{po}$$

where *amo* relates a the read event of an atomic memory operation (such as SWAP) to its write event, and [**A**] and [**L**] are event-type self-loops for acquire resp. release events. This contrasts the earlier definition in [17], in which LSE instructions provide the same ordering guarantees as load/store-exclusive pairs.

However, this stronger ordering is not necessary for the HMCS lock, and thus we optimize barriers further by relegating the acquire barrier to a trailing fence. This variant is what is denoted by hmcs-amo in Sect. 5. As demonstrated in Fig. 10, this optimization does not currently improve performance compared to hmcs-vsync (which uses sc barriers on atomic operations). Perhaps if LSE operations become more efficient for low-contention cases in the future, these optimizations will become more interesting.

For the sake of completeness we also implement a variant hmcs-armamo which applies the optimization to hmcs-arm, i.e., in which as described in Table 1 all implicit barriers on atomic operations are removed. Performance results (without LSE) are shown in Figs. 17 and 18. While minor improvements can be measured in the microbenchmark, these improvements also do not translate to the larger benchmark.

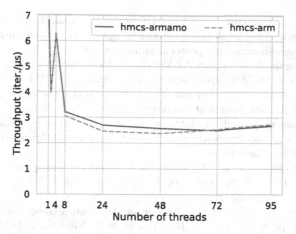

Fig. 17. Performance of AMO-optimizations with fences on microbenchmark

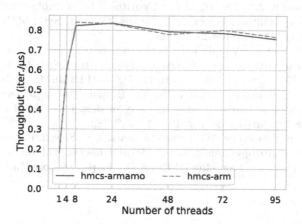

Fig. 18. Performance of AMO-optimizations with fences on LevelDB

References

1. https://e.huawei.com/uk/products/servers/taishan-server/taishan-2280-v2
2. https://en.wikichip.org/wiki/hisilicon/kunpeng/920-6426
3. https://openeuler.org
4. Amazon Web Services: AWS Graviton Processor - Enabling the best price performance in Amazon EC2 (2020). https://aws.amazon.com/ec2/graviton
5. Chabbi, M., Amer, A., Wen, S., Liu, X.: An efficient abortable-locking protocol for multi-level NUMA systems. SIGPLAN Not. **52**(8), 61–74 (2017). https://doi.org/10.1145/3155284.3018768
6. Chabbi, M., Fagan, M.W., Mellor-Crummey, J.M.: High performance locks for multi-level NUMA systems. In: PPoPP 2015, New York, USA, pp. 215–226. ACM (2015). https://doi.org/10.1145/2688500.2688503
7. Dean, J., Ghemawat, S.: Leveldb (2021). https://github.com/google/leveldb

8. Defilippi, J.: Introducing AMBA 5 CHI protocol enhancements (2017). https://community.arm.com/developer/ip-products/system/b/soc-design-blog/posts/introducing-new-amba-5-chi-protocol-enhancements

9. Dice, D., Kogan, A.: Compact NUMA-aware locks. In: EuroSys 2019, New York, USA. ACM (2019). https://doi.org/10.1145/3302424.3303984

10. Guiroux, H., Lachaize, R., Quéma, V.: Multicore locks: the case is not closed yet. In: USENIX Annual Technical Conference, pp. 649–662 (2016)

11. Huawei: Huawei unveils industry's highest-performance ARM-based CPU, January 2019. https://www.huawei.com/en/news/2019/1/huawei-unveils-highest-performance-arm-based-cpu

12. Kokologiannakis, M., Raad, A., Vafeiadis, V.: Model checking for weakly consistent libraries. In: PLDI 2019, New York, USA, pp. 96–110. ACM (2019). https://doi.org/10.1145/3314221.3314609

13. Liu, N., Zang, B., Chen, H.: No barrier in the road: a comprehensive study and optimization of ARM barriers. In: PPoPP 2020, New York, USA, pp. 348–361. ACM (2020). https://doi.org/10.1145/3332466.3374535

14. Mellor-Crummey, J.M., Scott, M.L.: Algorithms for scalable synchronization on shared-memory multiprocessors. ACM Trans. Comput. Syst. **9**(1), 21–65 (1991). https://doi.org/10.1145/103727.103729

15. Oberhauser, J., et al.: VSync: push-button verification and optimization for synchronization primitives on weak memory models. In: ASPLOS 2021, New York, USA. ACM (2021). https://doi.org/10.1145/3445814.3446748

16. Podkopaev, A., Lahav, O., Vafeiadis, V.: Bridging the gap between programming languages and hardware weak memory models. Proc. ACM Program. Lang. **3**(POPL) (2019). https://doi.org/10.1145/3290382

17. Pulte, C., Flur, S., Deacon, W., French, J., Sarkar, S., Sewell, P.: Simplifying ARM concurrency: multicopy-atomic axiomatic and operational models for ARMv8. Proc. ACM Program. Lang. **2**(POPL) (2017). https://doi.org/10.1145/3158107

Author Index

Abegg, Jean-Philippe   129

Bayramzadeh, Zahra   88
Behrens, Diogo   240
Benkaouz, Yahya   144
Bieniusa, Annette   3
Bramas, Quentin   95, 129, 161

Chini, Peter   187

Devismes, Stéphane   95
Dietsch, Daniel   169

El Abid, Imane   144

Fokkink, Wan   71
Frey, Davide   111
Fu, Ming   240
Fuchs, Per   71
Furbach, Florian   187

Georgiou, Chryssis   36

Havlena, Vojtěch   215
Heizmann, Matthias   169
Hoenicke, Jochen   169
Holík, Lukáš   206
Hood, Kendric   19
Hruška, Martin   206

Karlos, Georgios   71
Khoumsi, Ahmed   144
Kokologiannakis, Michalis   223
Kshemkalyani, Ajay D.   88

Lafourcade, Pascal   95
Lamani, Anissa   95
Lengál, Ondřej   215
López, Carlos   54

Marcoullis, Ioannis   36
Molla, Anisur Rahaman   88

Nesterenko, Mikhail   19
Noël, Thomas   129
Nutz, Alexander   169

Oberhauser, Jonas   240
Oberhauser, Lilith   240
Oglio, Joseph   19

Paolillo, Antonio   240
Pilet, Amaury Bouchra   111
Podelski, Andreas   169

Rajsbaum, Sergio   54
Rauch, Arthur   95
Raynal, Michel   36, 54
Rezae, Ahmad Hussein   3

Schiller, Elad M.   36
Sharma, Gokarna   19, 88
Šmahlíková, Barbora   215

Taïani, François   111

Vafeiadis, Viktor   223, 240
Vargas, Karla   54

Yanakieva, Elena   3
Youssef, Michael   3

Printed in the United States
by Baker & Taylor Publisher Services